Encyclopaedia of Mathematical Sciences

Volume 60

Editor-in-Chief: R.V. Gamkrelidze

Books of Related Interest by Serge Lang

Fundamentals of Diophantine Geometry

A systematic account of fundamentals, including the basic theory of heights, Roth and Siegel's theorems, the Néron–Tate quadratic form, the Mordell–Weil theorem, Weil and Néron functions, and the canonical form on a curve as it related to the Jacobian via the theta function.

Introduction to Complex Hyperbolic Spaces

Since its introduction by Kobayashi, the theory of complex hyperbolic spaces has progressed considerably. This book gives an account of some of the most important results, such as Brody's theorem, hyperbolic imbeddings, curvature properties, and some Nevanlinna theory. It also includes Cartan's proof for the Second Main Theorem, which was elegant and short.

Elliptic Curves: Diophantine Analysis

This systematic account of the basic diophantine theory on elliptic curves starts with the classical Weierstrass parametrization, complemented by the basic theory of Néron functions, and goes on to the formal group, heights and the Mordell–Weil theorem, and bounds for integral points. A second part gives an extensive account of Baker's method in diophantine approximation and diophantine inequalities which were applied to get the bounds for the integral points in the first part.

Other Books by Lang Published by Springer-Verlag

Introduction to Arakelov Theory • Riemann-Roch Algebra (with William Fulton) • Complex Multiplication • Introduction to Modular Forms • Modular Units (with Daniel Kubert) • Introduction to Algebraic and Abelian Functions • Cyclotomic Fields I and II • Elliptic Functions • Algebraic Number Theory • $SL_2(R)$ • Abelian Varieties • Differential Manifolds • Complex Analysis • Undergraduate Analysis • Undergraduate Algebra • Linear Algebra • Introduction to Linear Algebra • Calculus of Several Variables • First Course in Calculus • Basic Mathematics • Geometry: A High School Course (with Gene Murrow) • Matn! Encounters with High School Students • The Beauty of Doing Mathematics • THE FILE

Serge Lang (Ed.)

Number Theory III

Diophantine Geometry

Springer-Verlag
Berlin Heidelberg New York
London Paris Tokyo
Hong Kong Barcelona

Title of the Russian edition:
Itogi nauki i tekhniki, Sovremennye problemy matematiki,
Fundamental'nye napravleniya.
Vol. 60, Teoriya Chisel 3
Publisher VINITI, Moscow.

Mathematics Subject Classification (1980): 11Dxx, 11Gxx,
14Dxx, 14Gxx, 14Hxx, 14Kxx

ISBN 3-540-53004-5 Springer-Verlag Berlin Heidelberg New York
ISBN 0-387-53004-5 Springer-Verlag New York Berlin Heidelberg

Library of Congress Cataloging-in-Publication Data
Number theory III : diophantine geometry / Serge Lang (ed.).
p. cm. — (Encyclopaedia of mathematical sciences : v. 60)
Includes bibliographical references and index.
ISBN 0-387-53004-5
1. Diophantine analysis. I. Lang, Serge, 1927– . II. Title:
Number theory 3. III. Title: Diophantine geometry. IV. Series.
QA242.N85 1991
512'.7—dc20 91-17718
CIP

Printed in the United States of America
Typesetting: Asco Trade Typesetting Ltd., Hong Kong
2141/3140-543210—Printed on acid-free paper

List of Editors and Contributors

Editor-in-Chief

R. V. Gamkrelidze, Academy of Sciences of the USSR, Steklov Mathematical Institute, ul. Vavilova 42, 117966 Moscow, Institute for Scientific Information (VINITI), Usievich ul. 20, 125219 Moscow, USSR

Consulting Editor and Author

S. Lang, Department of Mathematics, Yale University, New Haven, CT 06520, USA

Preface

In 1988 Shafarevich asked me to write a volume for the Encyclopaedia of Mathematical Sciences on Diophantine Geometry. I said yes, and here is the volume.

By definition, diophantine problems concern the solutions of equations in integers, or rational numbers, or various generalizations, such as finitely generated rings over **Z** or finitely generated fields over **Q**. The word Geometry is tacked on to suggest geometric methods. This means that the present volume is not elementary. For a survey of some basic problems with a much more elementary approach, see [La 90c].

The field of diophantine geometry is now moving quite rapidly. Outstanding conjectures ranging from decades back are being proved. I have tried to give the book some sort of coherence and permanence by emphasizing structural conjectures as much as results, so that one has a clear picture of the field. On the whole, I omit proofs, according to the boundary conditions of the encyclopedia. On some occasions I do give some ideas for the proofs when these are especially important. In any case, a lengthy bibliography refers to papers and books where proofs may be found. I have also followed Shafarevich's suggestion to give examples, and I have especially chosen these examples which show how some classical problems do or do not get solved by contemporary insights. Fermat's last theorem occupies an intermediate position. Although it is not proved, it is not an isolated problem any more. It fits in two main approaches to certain diophantine questions, which will be found in Chapter II from the point of view of diophantine inequalities, and Chapter V from the point of view of modular curves and the Taniyama–Shimura conjecture. Some people might even see a race between the two approaches: which one will prove Fermat first? It is actually conceivable that diophantine inequalities might prove the Taniyama–Shimura conjecture, which would give a high to everybody. There are also two approaches to Mordell's conjecture that a curve of genus $\geqq 2$ over the rationals (or over a number field) has only a finite number of rational points: via l-adic representations in Chapter IV, and via diophantine approximations in Chapter IX. But in this case, Mordell's conjecture is now Faltings' theorem.

Parts of the subject are more accessible than others because they require less knowledge for their understanding. To increase accessibility of some parts, I have reproduced some definitions from basic algebraic geometry. This is especially true of the first chapter, dealing with qualitative questions. If substantially more knowledge was required for some results, then I did not try to reproduce such definitions, but I just used whatever language was necessary. Obviously decisions as to where to stop in the backward tree of definitions depend on personal judgments, influenced by several people who have commented on the manuscript before publication.

The question also arose where to stop in the direction of diophantine approximations. I decided not to include results of the last few years centering around the explicit Hilbert Nullstellensatz, notably by Brownawell, and related bounds for the degrees of polynomials vanishing on certain subsets of group varieties, as developed by those who needed such estimates in the theory of transcendental numbers. My not including these results does not imply that I regard them as less important than some results I have included. It simply means that at the moment, I feel they would fit more appropriately in a volume devoted to diophantine approximations or computational algebraic geometry.

I have included several connections of diophantine geometry with other parts of mathematics, such as PDE and Laplacians, complex analysis, and differential geometry. A grand unification is going on, with multiple connections between these fields.

New Haven Serge Lang
Summer 1990

Acknowledgment

I want to thank the numerous people who have made suggestions and corrections when I circulated the manuscript in draft, especially Chai, Coleman, Colliot-Thélène, Gross, Parshin and Vojta. I also thank Chai and Colliot-Thélène for their help with the proofreading.

S.L.

Contents

Notation

Some symbols will be used throughout systematically, and have a more or less universal meaning. I list a few of these.

F^{a} denotes the algebraic closure of a field F. I am trying to replace the older notation $\bar{F}$, since the bar is used for reduction mod a prime, for complex conjugates, and whatnot. Also the notation F^{a} is in line with F^{s} or F^{nr} for the separable closure, or the unramified closure, etc.

# denotes number of elements, so $\#(S)$ denotes the number of elements of a set S.

$\ll$ is used synonymously with the big Oh notation. If f, g are two real functions with g positive, then $f \ll g$ means that $f(x) = O(g(x))$. Then $f \gg\ll g$ means $f \ll g$ and $g \ll f$.

$A[\varphi]$ means the kernel of a homomorphism φ when A is an abelian group.

$A[m]$ is the kernel of multiplication by an integer m.

Line sheaf is what is sometimes called an invertible sheaf. The French have been using the expression "faisceau en droites" for quite some time, and there is no reason to lag behind in English.

Vector sheaf will, I hope, replace locally free sheaf of finite rank, both because it is shorter, and because the terminology becomes functorial with respect to the ideas. Also I object to using the same expression vector bundle for the bundle *and* for its sheaf of sections. I am fighting an uphill battle on this, but again the French have been using faisceau vectoriel, so why not use the expression in English, functorially with respect to linguistics?

CHAPTER I

Some Qualitative Diophantine Statements

The basic purpose of this chapter is to list systematically fundamental theorems concerning the nature of sets of rational points, as well as conjectures which make the theory more coherent. We use a fairly limited language from algebraic geometry, and hence for the convenience of those readers whose background is foreign to algebraic geometry, I have started with a section reproducing the basic definitions which we shall use.

Most cases treated in this chapter are those when the set of rational points is "as small as possible". One of the purposes is to describe what this means. "Small" may mean finite, it may mean thinly distributed, or if there is a group structure it may mean finitely generated. As much as possible, we try to characterize those situations when the set of rational points is small by algebraic geometric conditions. In Chapter VIII we relate these algebraic conditions to others which arise from one imbedding of the ground field into the complex numbers, and from complex analysis or differential geometry applied to the complex points of the variety after such an imbedding.

We shall also try to describe conjecturally qualitative conditions under which there exist many rational points. These conditions seem to have to do with group structures in various ways. The qualitative statements of this first chapter will be complemented by quantitative statements in the next chapter, both in the form of theorems and of conjectures.

The first section of Chapter II is extremely elementary, and many readers might want to read it first. It shows the sort of fundamental results one wishes to obtain, admitting very simple statements, but for which no known proofs are known today. The elaborate machinery being built up strives partly to prove such results.

I, §1. BASIC GEOMETRIC NOTIONS

For the convenience of the reader we shall give definitions of the simple, basic notions of algebraic geometry which we need for this chapter. A reader acquainted with these notions may skip this part. It just happens that we need very few notions, and so a totally uninformed reader might still benefit if provided with these basic definitions.

Let k be a field. Consider the polynomial ring in n variables $k[z_1, \ldots, z_n]$. Let I be an ideal in this ring, generated by a finite number of polynomials $g_1, \ldots, g_m$. Assume that $g_1, \ldots, g_m$ generate a prime ideal in the ring $k^{a}[z_1, \ldots, z_n]$ over the algebraic closure of k. The set of zeros of I is called an **affine variety** X. The variety defined by the zero ideal is all of affine space $\mathbf{A}^n$. If k' is a field containing k, the set of zeros of I with coordinates $z_1, \ldots, z_n \in k'$ is called the set of **rational points of** X **in** k', and is denoted by $X(k')$. It is equal to the set of solutions of the finite number of equations

$$g_j(z_1, \ldots, z_n) = 0 \qquad \text{with} \quad j = 1, \ldots, m$$

and $z_i \in k'$ for all $i = 1, \ldots, n$.

The condition that the polynomials generate a prime ideal is to insure what is called the irreducibility of the variety. Under our condition, it is not possible to express a variety as the finite union of proper subvarieties.

By pasting together a finite number of affine varieties in a suitable way one obtains the general notion of a variety. To avoid a foundational discussion here, we shall limit ourselves ad hoc to the three types of varieties which we shall consider: affine, projective, and quasi projective, defined below. But for those acquainted with the scheme foundations of algebraic geometry, a **variety** is a scheme over a field k, reduced, integral, separated and of finite type, and such that these properties are preserved under arbitrary extensions of the ground field k.

Let $\mathbf{P}^n$ denote projective n-space. If F is a field, then $\mathbf{P}^n(F)$ denotes the set of points of $\mathbf{P}^n$ over F. Thus $\mathbf{P}^n(F)$ consists of equivalence classes of $(n+1)$-tuples

$$P = (x_0, \ldots, x_n) \qquad \text{with} \quad x_j \in F, \text{ not all } x_j = 0,$$

where two such $(n+1)$-tuples $(x_0, \ldots, x_n)$ and $(y_0, \ldots, y_n)$ are equivalent if and only if there exists $c \in F$, $c \neq 0$ such that

$$(y_0, \ldots, y_n) = (cx_0, \ldots, cx_n).$$

By a **projective variety** X over a field k we mean the set of solutions in a projective space $\mathbf{P}^n$ of a finite number of equations

$$f_j(T_0, \ldots, T_n) = 0 \qquad (j = 1, \ldots, m)$$

such that each f_j is a homogeneous polynomial in $n+1$ variables with coefficients in k, and $f_1, \ldots, f_m$ generate a prime ideal in the polynomial ring $k^{\mathrm{a}}[T_0, \ldots, T_n]$. If k' is a field containing k, by $X(k')$ we mean the set of such zeros having some projective coordinates $(x_0, \ldots, x_n)$ with $x_i \in k'$ for all $i = 0, \ldots, n$. We denote by $k(x)$ (the **residue class field of the point**) the field

$$k(x) = k(x_0, \ldots, x_n)$$

such that at least one of the projective coordinates is equal to 1. It does not matter which such coordinate is selected. If for instance $x_0 \neq 0$, then

$$k(x) = k(x_1/x_0, \ldots, x_n/x_0).$$

We shall give a more intrinsic definition of this field below. We say that $X(k')$ is the set of **rational points of** X **in** k'. The set of points in $X(k^{\mathrm{a}})$ is called the set of **algebraic points** (over k).

We can define the **Zariski topology on** $\mathbf{P}^n$ by prescribing that a **closed set** is a finite union of varieties. A **Zariski open set** is defined to be the complement of a closed set. By a **quasi-projective variety**, we mean the open subset of a projective variety obtained by omitting a closed subset. A closed subset is simply the set of zeros of a finite number of polynomials, or equivalently of some ideal, which need not be a prime ideal. A projective variety is covered by a finite number of affine varieties, as follows.

Let, say, $z_i = T_i/T_0$ $(i = 1, \ldots, n)$ and let

$$g_j(z_1, \ldots, z_n) = f_j(1, z_1, \ldots, z_n).$$

Then the polynomials $g_1, \ldots, g_n$ generate a prime ideal in $k^{\mathrm{a}}[z_1, \ldots, z_n]$, and the set of solutions of the equations

$$g_j(z_1, \ldots, z_n) = 0 \qquad (j = 1, \ldots, m)$$

is an affine variety, which is an open subset of X, denoted by U_0. It consists of those points $(x_0, \ldots, x_n) \in X$ such that $x_0 \neq 0$. Similarly, we could have picked any index instead of 0, say j, and let

$$z_i^{(j)} = T_i/T_j \qquad \text{for} \quad i = 0, \ldots, n \text{ and } i \neq j.$$

Thus the set of points $(x_0, \ldots, x_n)$ such that $x_j \neq 0$ is an affine open subset of X denoted by U_j. The projective variety X is covered by the open sets $U_0, \ldots, U_n$.

By a **subvariety** of a variety X we shall always means a closed subvariety unless otherwise specified. Consider a maximal chain of subvarieties

$$Y_0 \subset Y_1 \subset \cdots \subset Y_r = X,$$

where Y_0 is a point and $Y_i \neq Y_{i+1}$ for all i. Then all such chains have the same number of elements r, and r is called the **dimension** of X. If k is a subfield of the complex numbers, then $X(\mathbf{C})$ is a complex analytic space of complex analytic dimension r. A projective variety of dimension r is sometimes called an ***r*-fold**.

A **hypersurface** is a subvariety of $\mathbf{P}^n$ of codimension 1, defined by one equation $f(T_0, \ldots, T_n) = 0$. The degree of f is called the **degree** of the hypersurface. If X is a subvariety of $\mathbf{P}^n$ of dimension $n - r$, defined by r equations $f_j = 0$ $(j = 1, \ldots, r)$, then we say that X is a **complete intersection.**

A **curve** is a variety of dimension 1. A **surface** is a variety of dimension 2. In the course of a discussion, one may wish to assume that a curve or a surface is projective, or satisfies additional conditions such as being non-singular (to be defined below), in which case such conditions will be specified.

Let Z be an affine variety in affine space $\mathbf{A}^n$, with coordinates $(z_1, \ldots, z_n)$, and defined over a field k. Let $P = (a_1, \ldots, a_n)$ be a point of Z. Suppose k algebraically closed and $a_i \in k$ for all i. Let

$$g_j = 0 \qquad (j = 1, \ldots, m)$$

be a set of defining equations for Z. We say that the point P is **simple** if the matrix $(D_i g_j(P))$ has rank $n - r$, where r is the dimension of Z. We have used D_i for the partial derivative $\partial/\partial z_i$. We say that Z is **non-singular** if every point on Z is simple. A **projective variety** is called **non-singular** if all the affine open sets $U_0, \ldots, U_n$ above are non-singular. If X is a variety defined over the complex numbers, then X is non-singular if and only the set of complex points $X(\mathbf{C})$ is a complex manifold.

Let X be an affine variety, defined by an ideal I in $k[z_1, \ldots, z_n]$. The ring $R = k[z_1, \ldots, z_n]/I$ is called the **affine coordinate ring** of X, or simply the **affine ring** of X. This ring has no divisors of zero, and its quotient field is called the **function field** of X over k. An element of the function field is called a **rational function** on X. A rational function on X is therefore the quotient of two polynomial functions on X, such that the denominator does not vanish identically on X. The function field is denoted by $k(X)$.

Let X be a projective variety. Then the function fields $k(U_0), \ldots, k(U_n)$ are all equal, and are generated by the restrictions to X of the quotients T_i/T_j (for all i, j such that T_j is not identically 0 on X). The **function field** of X over k is defined to be $k(U_i)$ (for any i), and is denoted by $k(X)$. A rational function can also be expressed as a quotient of two homogeneous polynomial functions $f_1(T_0, \ldots, T_n)/f_2(T_0, \ldots, T_n)$ where f_1, f_2 have the same degree.

Let P be a point of X. We then have the **local ring of regular functions** $\mathcal{O}_P$ at P, which is defined to be the set of all rational functions φ, expressible as a quotient $\varphi = f/g$, where f, g are polynomial functions on X and $g(P) \neq 0$. This local ring has a unique maximal ideal $\mathcal{M}_P$, consisting of quotients as above such that $f(P) = 0$. The residue class field at P is defined to be

$$k(P) = \mathcal{O}_P/\mathcal{M}_P.$$

A variety is said to be **normal** if the local ring of every point is integrally closed. A non-singular variety is normal.

Let X, Y be varieties, defined over a field k. By a **morphism**

$$f\colon X \to Y$$

defined over k we mean a map which is given locally in the neighborhood of each point by polynomial functions. An **isomorphism** is a morphism which has an inverse, i.e. a morphism $g\colon Y \to X$ such that

$$f \circ g = \mathrm{id}_X \quad \text{and} \quad g \circ f = \mathrm{id}_Y.$$

We say that f is an **imbedding** if f induces an isomorphism of X with a subvariety of Y.

A **rational map** $f\colon X \to Y$ is a morphism on a non-empty Zariski open subset U of X. If V is a Zariski open subset of X, and $g\colon V \to Y$ is a morphism which is equal to f on $U \cap V$, then g is uniquely determined. Thus we think of a rational map as being extended to a morphism on a maximal Zariski open subset of X. A **birational map** is a rational map which has a rational inverse. If f is a birational map, then f induces an isomorphism of the function fields. Two varieties X, Y are said to be **birationally equivalent** if there exists a birational map between them. If needed, we specify the field over which rational maps or birational maps are defined. For instance, there may be a variety over a field k which is isomorphic or birationally equivalent to projective space $\mathbf{P}^m$ over an extension of k, but not over k itself. Example: the curve defined by the equation in $\mathbf{P}^2$

$$x_0^2 + x_1^2 + x_2^2 = 0.$$

Let $f\colon X \to Y$ be a rational map, defined over the field k. We say that f is **generically surjective** if the image of a non-empty Zariski open subset of X under f contains a Zariski open subset of Y. In this case, f induces an injection of function fields

$$k(Y) \hookrightarrow k(X).$$

A variety is said to be **rational** (resp. **unirational**) if it is birationally equivalent to (resp. a rational image of) projective space.

Next we describe divisors on a variety. There are two kinds.

A **Weil divisor** is an element of the free abelian group generated by the subvarieties of codimension 1. A Weil divisor can therefore be written as a linear combination

$$D = \sum n_i D_i$$

where D_i is a subvariety of codimension 1, and $n_i \in \mathbf{Z}$. If all $n_i \geqq 0$ then D is called **effective**.

A **Cartier divisor** is defined as follows. We consider pairs (U, φ) consisting of a Zariski open set U and a rational function φ on X. We say that two such pairs are equivalent, and write $(U, \varphi) \sim (V, \psi)$ if the rational function $\varphi\psi^{-1}$ is a unit in the local ring $\mathcal{O}_P$ for every $P \in U \cap V$. In other words, both $\varphi\psi^{-1}$ and $\varphi^{-1}\psi$ are regular functions at all points of $U \cap V$. A maximal family of equivalent pairs whose open sets cover X is defined to be a **Cartier divisor**. A pair (U, φ) is said to **represent the divisor locally**, or on the open set U. The Cartier divisor is said to be **effective** if for all representing pairs (U, φ) the rational function φ is regular at all points of U. We then view the Cartier divisor as a hypersurface on X, defined locally on U by the equation $\varphi = 0$. The Cartier divisors form a group. Indeed, if Cartier divisors are represented locally by (U, φ) and (U, φ') respectively, then their sum is represented by $(U, \varphi\varphi')$.

It is a basic fact that if X is non-singular then the groups of Weil divisors and Cartier divisors are isomorphic in a natural way.

Let φ be a non-zero rational function. Then φ defines a Cartier divisor denoted by (φ), represented by the pairs (U, φ) for all open sets U. Such Cartier divisors are said to be **rationally** or **linearly equivalent** to 0. The factor group of all Cartier divisors modulo the group of divisors of functions is called the **Cartier divisor class group** or the **Picard group** $\operatorname{Pic}(X)$. (See [Ha 77], Chapter II, Proposition 6.15.)

One can also define the notion of linearly equivalent to 0 for Weil divisors. Let W be a subvariety of X of codimension 1. Let $\mathcal{O}_W$ be the local ring of rational functions on X which are defined at W. If f is a rational function on X which lies in $\mathcal{O}_W$, $f \neq 0$, then we define the **order** of f at W to be

$$\operatorname{ord}_W(f) = \text{length of the } \mathcal{O}_W\text{-module } \mathcal{O}_W/f\mathcal{O}_W.$$

The order function extends to a homomorphism of the group of non-zero rational functions on X into $\mathbf{Z}$. To each rational function we can associate its **divisor**

$$(f) = \sum \operatorname{ord}_W(f)(W).$$

The subgroup of the Weil divisor group consisting of the divisors of rational functions defines the group of divisors **rationally equivalent** to 0, and the factor group is called the **Chow group** $\mathrm{CH}^1(X)$.

It is a pain to have to deal with both groups. When dealing with the Chow group, we shall usually assume that the variety is complete and non-singular in codimension 1. For simplicity, we shall now state some properties of divisor classes for the Cartier divisor class group. Analogous properties also apply to the Chow group. One reason why the Chow group is important for its own sake is that one can form similar groups with subvarieties of higher codimension, and these are interesting for their own sake. See Fulton's book *Intersection Theory.*

There is a natural homomorphism from Cartier divisors to Weil divisors, inducing a homomorphism

$$\mathrm{Pic}(X) \to \mathrm{CH}^1(X),$$

which is injective if X is normal, and an isomorphism if the variety X is non-singular.

Divisors also satisfy certain positivity properties. We have already defined effective divisors. A **divisor class** c is called **effective** if it contains an effective divisor. But there is a stronger property which is relevant. A **divisor** D on X is called **very ample** if there exists a projective imbedding

$$f: X \to \mathbf{P}^m$$

such that D is linearly equivalent to $f^{-1}(H)$ for some hyperplane H of $\mathbf{P}^m$. A **divisor class** c is called **very ample** if it contains a very ample divisor. We call a divisor D **ample** if there exists a positive integer q such that qD is very ample, and similarly for the definition of an ample divisor class. Equivalently, a divisor class c is ample if and only if there exists a positive integer q such that qc is very ample. We have a basic property:

Proposition 1.1. *Let X be a projective variety. Given a divisor D and an ample divisor E, there exists a positive integer n such that $D + nE$ is ample, or even very ample. In particular, every divisor D is linearly equivalent*

$$D \sim E_1 - E_2$$

where E_1, E_2 are very ample.

We view ampleness as a property of "positivity". We shall see in Chapter VIII that this property has an equivalent formulation in terms of differential geometry, and in Chapter II we shall see how it gives rise to positivity properties of heights.

By the **support** of a Cartier divisor D we mean the set of points P

such that if D is represented by (U, φ) in a neighborhood of P, then φ is not a unit in the local ring $\mathcal{O}_P$. The support of D is denoted by supp(D).

A morphism $f: X \to Y$ induces an inverse mapping

$$f^*: \operatorname{Pic}(Y) \to \operatorname{Pic}(X).$$

Indeed, let D be a Cartier divisor on Y, and suppose $f(X)$ is not contained in the support of D. Suppose D is represented by (V, ψ). Then $(f^{-1}(V), \psi \circ f)$ represents the inverse image $f^{-1}D$, which is a Cartier divisor on X. This inverse image defines the inverse image of the divisor class, and thus defines our mapping f^*.

Example. Let $X = \mathbf{P}^n$ be projective space, and let $T_0, \ldots, T_n$ be the projective variables. The equation $T_0 = 0$ defines a hyperplane in $\mathbf{P}^n$, and the complement of this hyperplane is the affine open set which we denoted by U_0. On U_i with $i \neq 0$, the hyperplane is represented by the rational function T_0/T_i. Instead of the index 0, we could have selected any other index, of course. More generally, let $a_0, \ldots, a_n$ be elements of k not all 0. The equation

$$f(T) = a_0 T_0 + \cdots + a_n T_n = 0$$

defines a hyperplane. On U_i, this hyperplane is represented by the rational function

$$\frac{f(T)}{T_i} = a_0(T_0/T_i) + \cdots + a_n(T_n/T_i).$$

Let X be a projective non-singular variety defined over an algebraically closed field k. Let D be a divisor on X. We let

$H^0(X, D) = k$-vector space of rational functions $\varphi \in k(X)$ such that

$$(\varphi) \geqq -D.$$

In other words, $(\varphi) = E - D$ where E is an effective divisor. Let $\{\varphi_0, \ldots, \varphi_N\}$ be a basis of $H^0(X, D)$. If $P \in X(k)$ is a point such that $\varphi_j \in \mathcal{O}_P$ for all j, and for some j we have $\varphi_j(P) \neq 0$ then

$$(\varphi_0(P), \ldots, \varphi_N(P))$$

is viewed as a point in projective space $\mathbf{P}^N(k)$, and we view the association

$$f: P \mapsto (\varphi_0(P), \ldots, \varphi_N(P))$$

as a map, which is defined on a non-empty Zariski open subset U of X.

Thus we obtain a morphism

$$f\colon U \to \mathbf{P}^N.$$

Similarly, for each positive multiple mD of D, using a basis for $H^0(X, mD)$, we obtain a morphism

$$f_m\colon \text{a non-empty Zariski open subset of } X \to \mathbf{P}^{N(m)}.$$

If there exists some positive integer m such that f_m is an imbedding of some non-empty Zariski open subset of X into a locally closed subset of $\mathbf{P}^{N(m)}$, then we say that D is **pseudo ample**. More generally, a divisor D_1 is defined to be **pseudo ample** if and only if D_1 is linear equivalent to a divisor D which is pseudo ample in the above sense. Thus the property of being pseudo ample is a property of divisor classes. It is a result of Kodaira that:

On a non-singular projective variety, a divisor D is pseudo ample if and only if there exists some positive integer m such that

$$mD \sim E + Z$$

where E is ample and Z is effective.

I, §2. THE CANONICAL CLASS AND THE GENUS

We shall discuss a divisor class which plays a particularly important role. We first deal with varieties of dimension 1, and then we deal with the general case.

Curves

We define a **curve** to be a projective variety of dimension 1. Let X be a non-singular curve over k. Then divisors can be viewed as Weil divisors, and a subvariety of codimension 1 is a point. Hence a divisor D can be expressed as a linear combination of points

$$D = \sum_{i=1}^{r} m_i(P_i) \qquad \text{with} \quad m_i \in \mathbf{Z}, \quad \text{and} \quad P_i \in X(k^{\mathrm{a}}).$$

We define the **degree** of the divisor D to be

$$\deg D = \sum m_i.$$

Suppose for simplicity that k is algebraically closed. Let $y \in k(X)$ be a rational function, and $y \neq 0$. Let $P \in X(k)$. Let $\mathcal{O}_P$ be the local ring with maximal ideal P. Then $\mathcal{M}_P$ is a principal ideal, generated by one element t, which is called a **local parameter** at P. Every element $y \neq 0$ of $k(X)$ has an expression

$$y = t^r u \qquad \text{where } u \text{ is a unit in } \mathcal{O}_P.$$

We define $\mathrm{ord}_P(y) = r$. The function $y \mapsto v_p(y) = \mathrm{ord}_P(y)$ defines an absolute value on $k(X)$. The completion of $\mathcal{O}_P$ can be identified with the power series ring $k[[t]]$. Let $F = k(X)$. We may view F as a subfield of the quotient field of $k[[t]]$, denoted by $k((t))$. In this quotient field, y has a power series expansion

$$y = a_r t^r + \text{higher terms}, \qquad \text{with} \quad a_r \in k, \quad a_r \neq 0.$$

To each rational function y as above we can associate a Weil divisor

$$(y) = \sum_P \mathrm{ord}_P(y)(P).$$

It is a fact that

$$\deg(y) = 0.$$

Hence the degree is actually a function on divisor classes, i.e. on $\mathrm{CH}^1(X)$.

Let x, $y \in k(X)$. A differential form $y\,dx$ will be called a **rational differential form**. Let P be a point in $X(k)$. In terms of a local parameter t, we write

$$y\,dx = y(t)\frac{dx}{dt}\,dt,$$

where y, x are expressible as power series in t. Then we define

$$\mathrm{ord}_P(y\,dx) = \text{order of the power series } y(t)\frac{dx}{dt}.$$

We can associate a divisor to the differential form $y\,dx$ by letting

$$(y\,dx) = \sum_P \mathrm{ord}_P(y\,dx)(P).$$

Since every rational differential form is of type $uy\,dx$ for some rational function u, it follows that the degrees of non-zero differential forms are all equal. One possible definition of the **genus** of X is by the formula

$$\deg(y\,dx) = 2g - 2.$$

Furthermore, the divisors of rational differential forms constitute a class in $\mathrm{CH}^1(X)$, called the **canonical class**. Thus we may say that:

The degree of the canonical class is $2g - 2$.

A differential form ω is said to be of the **first kind** if $\mathrm{ord}_P(\omega) \geqq 0$ for all P. The differential forms of first kind form a vector space over k, denoted by $\Omega^1(X)$. The following property also characterizes the genus.

The dimension of $\Omega^1(X)$ *is equal to* g.

The genus can also be characterized topologically over the complex numbers. Suppose $k = \mathbf{C}$. Then $X(\mathbf{C})$ is a compact complex manifold of dimension 1, also called a compact Riemann surface. The genus g is equal to the number of holes in the surface. It also satisfies the formula

$$2g = \operatorname{rank} H_1(X(\mathbf{C}), \mathbf{Z}),$$

where $H_1(X(\mathbf{C}), \mathbf{Z})$ is the first topological homology group, which is a free abelian group over $\mathbf{Z}$.

Finally, we want to be able to compute the genus when the curve is given by an equation. We shall give the value only in the non-singular case. Let the curve X be defined by the homogeneous polynomial equation

$$f(T_0, T_1, T_2) = 0,$$

of degree n in the projective plane $\mathbf{P}^2$, over the algebraically closed field k. Suppose that X is non-singular. Then

$$\boxed{\text{genus of } X = \frac{(n-1)(n-2)}{2}.}$$

For instance, the Fermat curve

$$X_0^n + X_1^n + X_2^n = 0$$

over a field of characteristic p *with* $p \nmid n$ has genus $(n-1)(n-2)/2$. This genus is $\geqq 2$ when $n \geqq 4$.

In general, on a non-singular curve X, *a divisor class* c *is ample if and only if*

$$\deg(c) > 0.$$

Therefore, the canonical class is ample if and only if $g \geqq 2$.

For $g = 1$, the canonical class is 0. For $g = 0$, the canonical class has degree -2.

Examples. Let X be a projective non-singular curve of genus 0 over a field k, not necessarily algebraically closed. Then X is isomorphic to $\mathbf{P}^1$ over k if and only if X has a rational point in k.

Let X be a projective non-singular curve of genus 1 over a field k of characteristic $\neq 2$ or 3. Then X has a rational point in k if and only if X is isomorphic to a curve defined by an equation

$$y^2 = 4x^3 - g_2 x - g_3 \qquad \text{with} \quad g_2, g_3 \in k, \quad g_2^3 - 27g_3^2 \neq 0.$$

A curve of genus 1 with a rational point is called an **elliptic curve over** k.

We now have defined enough notions to pass to diophantine applications. We shall deal with the following kinds of fields:

A number field, which by definition is a finite extension of $\mathbf{Q}$.

A function field, which is defined as the function field of a variety, over a field k. Such fields can be characterized as follows. An extension F of k is a function field over k if and only if F is finitely generated over k; every element of F algebraic over k lies in k; and there exist algebraically independent elements $t_1, \ldots, t_r$ in F over k such that F is a finite separable extension of $k(t_1, \ldots, t_r)$. Under these circumstances, we call k the **constant field**.

If F is a finitely generated field over the prime field, then F is a function field, whose constant field k is the set of elements in F which are algebraic over the prime field.

Let X_0 be a variety defined over k. Suppose F is the function field of a variety W over k. Then there is a natural bijection directly from the definitions between the set of rational points $X_0(F)$ and the rational maps $W \to X_0$ defined over k. We refer to this situation as the **split case** of a variety over F.

Mordell's conjecture made in 1922 [Mord 22] became **Faltings' theorem** in 1983 [Fa 83].

Theorem 2.1. *Let X be a curve defined over a number field F. Suppose X has genus $\geqq 2$. Then X has only a finite number of rational points in F, that is, $X(F)$ is finite.*

Using specialization techniques dating back to the earlier days of diophantine geometry, one then obtains the following corollary.

Corollary 2.2. *Let X be a curve defined over a field F finitely generated over the rational numbers. Then $X(F)$ is finite.*

Aside from this formulation in what we may call the **absolute case**, there is a **relative** formulation, in what is usually called the **function field case**.

In [La 60a] I conjectured the following analogue of Mordell's conjecture for a curve (assumed non-singular).

Theorem 2.3. *Let X be a curve defined over the function field F over k of characteristic* 0, *and of genus* $\geqq 2$. *Suppose that X has infinitely many rational points in $X(F)$. Then there exists a curve X_0 defined over k, such that X_0 is isomorphic to X over F, and all but a finite number of points in $X(F)$ are images under this isomorphism of points in $X_0(k)$.*

This formulation was proved by Manin [Man 63]. We shall describe certain features of Manin's proof, as well as several other proofs given since, in Chapter VI.

Note as in [La 60a] that the essential difficulty occurs when F has transcendence degree 1 over k, that is, when F is a finite extension of a rational field $k(t)$ with a variable t. Elementary reduction steps reduce the theorem to this case. Indeed, there exists a tower

$$k = F_0 \subset F_1 \subset \cdots \subset F_r = F$$

such that each F_i is a function field over F_{i-1}, of dimension 1. One can then apply induction to the case of dimension 1 to handle the general case.

Remark. If k is a finite field with q elements and X_0 is defined over k, and X_0 has a point in an extension F of k, then iterations of Frobenius on this point yield other points, so exceptional cases have to be excluded in characteristic p. See Chapter 6, §5.

The case of Theorem 2.3 when X is isomorphic to some curve X_0 over k is the split case in which the conclusion may be reformulated in the following geometric form. For a proof see [La 60a], p. 29, and [La 83a], p. 223.

Theorem of de Franchis. *Let X_0 be a curve in characteristic* 0, *of genus* $\geqq 2$. *Let W be an arbitrary variety. Then there is only a finite number of generically surjective rational maps of W onto X_0.*

Higher dimensions

Let X be a projective non-singular variety of dimension n, defined over an algebraically closed field k.

Let W be a subvariety of codimension 1. In particular, W is a divisor on X. Let $P \in X(k)$, and let $\mathcal{O}_P$ be the local ring of P on X, with maximal ideal $\mathcal{M}_P$. Then the hypothesis that X is non-singular implies that $\mathcal{O}_P$ is a unique factorization ring. There exists an element $\varphi \in \mathcal{O}_P$, well defined up to multiplication by a unit in $\mathcal{O}_P$, such that W is defined in a Zariski open neighborhood U of P by the equation $\varphi = 0$. If W does not pass through P, then φ is a unit, and $U \cap W$ is empty. If W passes through P, then φ is an irreducible (or prime) element in $\mathcal{O}_P$. The collection of pairs (U, φ) as above define the **Cartier divisor associated with** W.

Let $y, x_1, \ldots, x_n$ be rational functions on X, so in $k(X)$. A form of type $y\, dx_1 \wedge \cdots \wedge dx_n$ is called a **rational differential form** of top degree on X. Let ω be such a form. Let $t_1, \ldots, t_n$ be **local parameters** at P (that is, generators for the maximal ideal in the local ring at P). In a neighborhood U of P we may write

$$\omega = \psi\, dt_1 \wedge \cdots \wedge dt_n$$

for some rational function ψ. The collection of pairs (U, ψ) defines a Cartier divisor, which is called the **divisor associated with** ω, and is denoted by (ω). All such divisors are in the same linear equivalence class, and again this class is called the canonical class of X. A canonical divisor is sometimes denoted by K, as well as its class, or by K_X if we wish to emphasize the dependence on X.

A differential form of top degree is called **regular at** P if its divisor is represented in a neighborhood of P by a pair (U, ψ) where $\psi \in \mathcal{O}_P$. The differential form is called **regular** if it is regular at every point, or in other words, if its associated divisor is effective. The regular differential forms of top degree form a vector space over k, whose dimension is called the **geometric genus**, and is classically denoted by p_g.

Examples. The canonical class of $\mathbf{P}^m$ itself is given by

$$K_{\mathbf{P}^m} \sim -(m+1)H \text{ for any hyperplane } H \text{ on } \mathbf{P}^m.$$

In particular, for $m = 1$, any two points on $\mathbf{P}^1$ are linearly equivalent, and for any point P on $\mathbf{P}^1$ the canonical class on $\mathbf{P}^1$ is given by

$$K_{\mathbf{P}^1} \sim -2(P).$$

Suppose that X is a non-singular hypersurface in projective space $\mathbf{P}^m$, defined by the equation

$$f(T_0, \ldots, T_m) = 0$$

where f is a homogeneous polynomial of degree d. Let H_X be the restriction to X of a hyperplane which does not vanish identically on X. Then the canonical class on X is given by

$$K_X \sim (d - (m+1))H_X.$$

Thus the canonical class is ample if and only if $d \geqq m + 2$.

The class $-K_X$ is called the **co-canonical** or **anti-canonical class.**

A non-singular projective variety is defined to be:

canonical if the canonical class K_X is ample
very canonical if K_X is very ample
pseudo canonical if K_X is pseudo ample
anti-canonical if $-K_X$ is ample
and so on.

Instead of pseudo canonical, a variety has been called of **general type**, but *with the support of Griffiths*, I am trying to make the terminology functorial with respect to the ideas. (I know I am fighting an uphill battle on this.)

Finally, suppose that X is a projective variety, but possibly singular. We say that X is **pseudo canonical** if X is birationally equivalent to a projective non-singular pseudo canonical variety. In characteristic 0, **resolution of singularities** is known, and due to Hironaka. This means that given X a projective variety, there exists a birational morphism

$$f: X' \to X$$

such that X' is projective and non-singular and f is an isomorphism over the Zariski open subset subset of X consisting of the simple points.

It is an elementary fact of algebraic geometry that if $f: X \to Y$ is a birational map between non-singular projective varieties, then for every positive integer n, $f^*: H^0(Y, nK_Y) \to H^0(X, nK_X)$ is an isomorphism. In particular, K_Y is pseudo ample if and only if K_X is pseudo ample. In analogy with the case of curves, one defines the **geometric genus**

$$p_g(X) = \dim H^0(X, K_X).$$

It is a basic problem of algebraic geometry to determine under which conditions the canonical class is ample, or pseudo ample. We shall relate these conditions with diophantine conditions in the next section.

I, §3. THE SPECIAL SET

For simplicity let us now assume that our fields have characteristic 0.

I shall give a list of conjectures stemming from [La 74] and [La 86]. Let X be a variety defined over an algebraically closed field of characteristic 0. Then X can also be defined over a finitely generated field F_0 over the rational numbers. We say that X is **Mordellic** if $X(F)$ is finite for every finitely generated field F over $\mathbf{Q}$, containing F_0. We ask under what conditions can there be infinitely many rational points of X in some such field F? One can a priori describe such a situation. First, if there is a rational curve in X, i.e. a curve birationally equivalent to $\mathbf{P}^1$, then all the rational points of this curve give rise to rational points on X. Note that we are dealing here with curves (and later subvarieties) defined over some finitely generated extension, i.e. we are dealing with the "geometric" situation. A more general example is given by a **group variety**, that is a variety which is a group such that the composition law and the inverse map are morphisms. If G is a group variety, then from two random rational points in some field F we can construct lots of other points by using the law of composition. Roughly speaking, the conjecture is that these are the only examples. Let us make this conjecture more precise.

Let X be a projective variety. Let us define the **algebraic special set** $\mathrm{Sp}(X)$ to be the Zariski closure of the union of all images of non-constant rational maps $f: G \to X$ of group varieties into X. This special set may be empty, it may be part of X, or it may be the whole of X. Note that the maps f may be defined over finitely generated extensions, i.e. the special set is defined geometrically. I conjectured:

3.1. *The complement of the special set is Mordellic.*

3.2. *A projective variety is Mordellic if and only if the special set is empty, i.e. if and only if every rational map of a group variety into X is constant.*

Note that the affine line $\mathbf{A}^1$, or the multiplicative group $\mathbf{G}_m$, are birationally equivalent to $\mathbf{P}^1$, so the presence of rational curves in X can be viewed from the point of view that these lines are rational images of group varieties. A group variety which is projective is called an **abelian variety**. We shall study their diophantine properties more closely later. Other examples of group varieties are given by linear group varieties, i.e. subgroups of the general linear group which are Zariski closed subsets. A general structure theorem due to Chevalley states that the only group varieties are group extensions of an abelian variety by a linear group; and the function field of a linear group is unirational but rational over an algebraically closed field. Hence:

3.3. *The special set* $\mathrm{Sp}(X)$ *is the Zariski closure of the union of all images of non-constant rational maps of* $\mathbf{P}^1$ *and abelian varieties into X.*

Since $\mathbf{P}^1$ itself is a rational image of an abelian variety of dimension 1, we may also state:

3.4. *The special set* $\mathrm{Sp}(X)$ *is the Zariski closure of the union of the images of all non-constant rational maps of abelian varieties into X.*

We define a projective variety to be **algebraically hyperbolic** if and only if the special set is empty. We make this definition to fit conjecturally with the theory of hyperbolicity over the complex numbers. See Chapter V. Examples of hyperbolic projective varieties will be given in Chapter V. Analytic hyperbolicity implies algebraic hyperbolicity, so these examples are also examples of algebraically hyperbolic varieties, and hence conjecturally of Mordellic varieties.

We define a projective variety to be **special** if $\mathrm{Sp}(X) = X$, that is, if the special set is the whole variety. A variety which is not special can be called **general**. This fits older terminology (**general type**) in light of Conjecture 3.5 below.

There are the two extreme cases: when the special set is empty, i.e. the variety is algebraically hyperbolic, and when the special set is the whole variety, i.e. the variety is special. One wants a classification of both types of varieties, which amounts to problems principally in algebraic geometry, although already a diophantine flavor intervenes because one is led to consider generic fiber spaces, where there may not exist a rational section. We shall mention below some specific examples.

One basic conjecture states:

3.5. *The special set* $\mathrm{Sp}(X)$ *is a proper subset if and only if* X *is pseudo canonical.*

As a result, one gets the conjecture:

3.6. *The following conditions are equivalent for a projective variety* X:
X is algebraically hyperbolic, i.e. the special set is empty.
X is Mordellic.
Every subvariety of X is pseudo canonical.

In light of this conjecture, there are no other examples of Mordellic projective varieties besides hyperbolic ones.

In this context, it is natural to define a variety X to be **pseudo Mordellic** if there exists a proper Zariski closed subset Y of X such that $X - Y$ is Mordellic. Then I conjectured:

3.7. *A projective variety X is pseudo Mordellic if and only if X is pseudo canonical. The Zariski closed subset Y can be taken to be the special set* $\mathrm{Sp}(X)$.

As a consequence of the conjecture, we note that for any finitely generated field F over $\mathbf{Q}$, if X is pseudo Mordellic, then $X(F)$ is not Zariski dense in X. The converse is not true, however. For instance, let C be a curve of genus $\geqq 2$ and let

$$X = C \times \mathbf{P}^1.$$

We have $X = \mathrm{Sp}(X)$, and X is fibered by projective lines. By Faltings' theorem, for every finitely generated field F over $\mathbf{Q}$ the set $X(F)$ is not Zariski dense in X, but X is not pseudo canonical. However, conjecturally:

3.8. *If the special set is empty, then the canonical class is ample.*

The above discussion and conjectures give criteria for the special set to be empty or unequal to the whole projective variety. In the opposite direction, one is interested in those cases when X is special. This leads at first into problems of pure algebraic geometry, independently of diophantine applications, concerning the structure of the special set. Notably, we have the following problems.

Sp 1. If we omit taking the Zariski closure, do we still get the same set?

Sp 2. Are the irreducible components of the special set **generically fibered** by rational images of group varieties?

By this we mean the following. Let X be a special projective variety. The condition **Sp 2** means that there exists a generically surjective rational map $f: X \to Y$ such that the inverse image $f^{-1}(y)$ of a generic point of Y is a subvariety W_y for which there exist a group variety G and a generically surjective rational map $g: G \to W_y$ defined over some finite extension of the field $k(y)$. To get "fibrations" (in the strict sense, with disjoint fibers), one must of course allow for blow ups and the like, to turn rational maps into morphisms.

One can also formulate an alternative for the second question, namely:

Sp 3. Suppose $X = \mathrm{Sp}(X)$ is special. Is there a generically finite rational map from a variety X' onto X such that X' is generically fibered by a rational image of a group variety?

It would still follow under this property that it is not necessary to take the Zariski closure in defining the special set. As a refinement of **Sp 3**, one can also ask for those conditions under which X' would be generically fibered by a group variety rather than a rational image.

Note that we are dealing here with rational fibrations, so only up to birational equivalence. In connection with **Sp 2**, suppose $f: X \to Y$ is a rational map whose generic fiber is a rational image of a group variety. Then one asks the general question:

Sp 4. When is there a rational section of f, so that the generic fiber has a rational point over the function field $k(y)$? If there is no rational section, over a number field k, how many fibers have one or more rational points?

We shall discuss several examples below and in Chapter X, §2.

As refinements of **Sp 2** and **Sp 3**, I would raise the possibility of finding "good" models, possibly not complete, for the complete X, whose fibers may be homogeneous spaces for linear groups, i.e. without generic sections. As we shall see later, such models can be found for abelian varieties (Néron models), and the question is to what extent there exists an analogous theory for non-complete special group varieties.

The question arises when the canonical class is 0 whether there exists a generic fibration as in **Sp 2** and **Sp 3** by rational images of abelian varieties, or abelian varieties themselves. Interesting cases when the canonical class K_X is not pseudo ample arise not only when $K_X = 0$ but also when $-K_X$ contains an effective divisor, or is pseudo ample, or ample. In these cases, Conjecture 3.5 implies that X is special. As $-K_X$ is assumed more and more ample, I would expect that there are more and more rational curves on the variety, where ultimately the special set is covered by rational curves and no abelian varieties are needed. The existence of rational curves has long been a subject of interest to algebraic geometers, and has received significant impetus through the work of Mori [Mori 82]. Such algebraic geometers work over an algebraically closed field, and there is of course the diophantine question whether rational curves are defined over a given field of definition for the variety or hypersurface. But still working geometrically we have the following theorem of [Mori 82], see also [ClKM 88], §1, Theorem 1.8.

Theorem 3.9a. *Let X be a projective non-singular variety and assume that $-K_X$ is ample. Then through every point of X there passes a rational curve in X.*

In particular, under the hypothesis of Theorem 3.9, X is special, but we note that only rational images of $\mathbf{P}^1$ are needed to fill out the special set. More generally I conjectured:

Theorem 3.9b. *Let X be a projective non-singular variety and assume that $-K_X$ is pseudo ample. Then X is special, and is equal to the union of rational curves in X.*

I asked Todorov if Mori's theorem would still be valid under the weaker hypothesis that the anti-canonical class is pseudo ample, and he told me that Mori's proof shows that in this case, there exists a rational curve passing through every point except possibly where the rational map defined by a large multiple of $-K_X$ is not an imbedding, which implies Theorem 3.9b as a corollary. These considerations fit well with those of Chapter X. In addition, **3.9a** and **3.9b** raise the question whether a generically finite covering of X is generically fibered by unirational varieties. Here the role of non-complete linear groups is not entirely clear. They intervene in the context of Chapter X, besides intervening in the Néron models of abelian varieties which will be defined later. In both cases, they reflect the existence of degenerate fibers. This suggests the existence of a theory of non-complete special models of special varieties which remains to be elaborated. In particular, the following questions also arise:

3.10. Suppose that $-K_X$ is pseudo ample. (a) Is **Sp 3** then necessarily satisfied with a linear group variety? (b) Suppose that X is defined over a field k and has a k-rational point. When is X unirational (resp. rational) over k?

In this direction, there is a condition which is much stronger than having $-K_X$ ample. Indeed one can define the notion of **ampleness** for a vector sheaf $\mathscr{E}$ as follows. Let $\mathbf{P}\mathscr{E}$ be the associated projective variety and let $\mathscr{L}$ be the corresponding line sheaf of hyperplanes. One possible characterization of $\mathscr{E}$ being **ample** is that $\mathscr{L}$ is ample. One of Mori's theorems proved a conjecture of Hartshorne:

Theorem 3.11. *Let X be a non-singular projective variety and assume that its tangent sheaf is ample. Then X is isomorphic to projective space.*

This result is valid over the algebraic closure of a field of definition of X. Over a given field of definition, a variety may not have any rational point, or may be only unirational. We shall discuss other examples in Chapter X.

There is another notion which has currently been used by algebraic geometers to describe when a variety is generically fibered by rational curves. Indeed, a variety X of dimension r over an algebraically closed field k is said to be **uniruled** if there exists an $(r-1)$-dimensional variety W over k and a generically surjective rational map $f: \mathbf{P}^1 \times W \to X$. For important results when threefolds are uniruled, having to do with negativity properties of the canonical class, in addition to Mori's paper already cited see Miyaoka–Mori [MiyM 86] and Miyaoka [Miy 88], Postscript Theorem, p. 332, which yield the following result among others.

Theorem 3.12. *Let X be a non-singular projective threefold (characteristic 0). The following three conditions are equivalent.* (a) *Through every point of X there passes a rational curve in X.* (b) *X is uniruled.* (c) *We have $H^0(X, mK_X) = 0$ for all $m > 0$.*

Note that $-K_X$ pseudo ample implies $H^0(X, mK_X) = 0$ for all $m > 0$, by the Kodaira criterion for pseudo ampleness. For further results see also Batyrev [Bat 90], and for a general exposition of Mori's program, see Kollár [Koll 89].

Extending the classical terminology of uniruled, I propose to define a variety X to be **unigrouped** if there exists a variety X' as in condition **Sp 3**. We shall now consider several significant examples illustrating the **Sp** conditions, among other things.

Example 1 (Subvarieties of abelian varieties). In this case the structure of the special set is known, and the answers to **Sp 1** and **Sp 2** are yes in both cases. This example will be discussed at length in §6. See also Chapter VIII, §1.

Example 2. Let X be a projective non-singular surface. One says that X is a **K3 surface** if $K_X = 0$ and if every rational map of X into an abelian variety is constant. If A is an abelian variety of dimension 2 and Z is the quotient of A by the group $\{\pm 1\}$, then a minimal desingularization of Z is a K3 surface, called a **Kummer surface**. If X is a K3 surface, then by a result of Green–Griffiths [GrG 80] and Bogomolov–Mumford completed by Mori–Mukai [MoM 83], Appendix, a generically finite covering X' of X has a generic fibration by curves of genus 1, so that X is unigrouped, and in particular X is special. Sometimes there exists a rational section and sometimes not. When such a section does not exist, over number fields, the problem arises how many fibers have rational points, or a point of infinite order on the fibral elliptic curve. We shall see an example with the Fermat surface below, and similar questions arise in the higher dimensional case of Fermat hypersurfaces, or in the case of the Châtelet surface of Chapter X, §2.

The next two cases deal with generic hypersurfaces.

Example 3. Let X be the **generic hypersurface** of degree d in $\mathbf{P}^n$, and suppose $d \geqq n + 2$, so that the canonical class is ample. By **generic** we mean that the polynomial defining X has algebraically independent coefficients over $\mathbf{Q}$. Then conjecturally the special set is empty.

The analogous conjecture goes for the generic complete intersection. These are algebraic formulations of a conjecture of Kobayashi in the complex analytic case. See Example 1.5 of Chapter VIII. Note that in light of Conjecture 3.6 these generic complete intersections would be Mordellic.

Example 4. Let X be the generic hypersurface of degree 5 in $\mathbf{P}^4$. One says that X is the **generic quintic threefold** in $\mathbf{P}^4$. Then $K_X = 0$. Following a construction of Griffiths, Clemens ([Cl 83], [Cl 84]) proved the existence of infinitely many rational curves which are Zariski dense, homologically equivalent, but which are linearly independent modulo algebraic equivalence. Hence in this case we have $\mathrm{Sp}(X) = X$, in other words X is special. It is then a problem to determine whether X satisfies conditions **Sp 2** or **Sp 3** above, especially whether a generically finite covering of X is generically fibered by elliptic curves, or K3 surfaces, or by rational images of abelian surfaces, if not by abelian surfaces themselves. In such a case, the Clemens curves might then be interpreted as sections over rational curves, thus explaining their independence in more geometric terms. The existence of such fibrations in the case of subvarieties of abelian varieties can be taken as an indication that a positive answer may exist in general, but the evidence at the moment is still too scarce to convince everyone that the answer will always be positive. For more on the quintic threefold, see [ClKM 88], §22 and Remark 5.5.

The above examples conjecturally illustrate some general principles on some generic hypersurfaces. Roughly speaking, as the degree increases (so the canonical class becomes more ample), the variety becomes less and less rational, and fibrations of the special set if they exist involve abelian varieties, whereas for lower degrees, these fibrations may involve only linear groups and rational or unirational fibers. Changes of behavior occur especially for $d = n + 2$, $n + 1$, and n. The less the canonical class is ample, the more a variety has a tendency to contain rational curves. For instance:

Let X be a hypersurface of degree d in $\mathbf{P}^n$. If $d \leqq n - 1$, then X contains a line through every point.

This result is classical and easy. For the argument, see [La 86], p. 196.

In all the above, it is a problem to determine what happens on Zariski open subsets rather than generically. For some examples in other contexts, see Chapter VIII, §1 for the Brody–Green perturbation of the Fermat hypersurface, and Chapter X, §2 for the Châtelet surface. Here we now consider:

Example 5 (The Fermat hypersurface). Since this hypersurface

$$T_0^d + \cdots + T_n^d = 0 \qquad \text{or} \qquad T_1^d + \cdots + T_n^d = T_0^d$$

contains lines, we see that the condition that X has ample canonical class does not imply that X is Mordellic or that the special set is empty.

Euler was already concerned with the problem of finding rational curves, that is, solving the Fermat equation with polynomials. Swinnerton-Dyer [SWD 52] gives explicit examples of rational curves over the rationals, on

$$T_0^5 + \cdots + T_5^5 = 0.$$

Here X has degree $d = n = 5$, and so the anti-canonical class is very ample. Swinnerton-Dyer says: "It is very likely that there is a solution in four parameters, or at least that there are an infinity of solutions in three parameters, but I see no prospect of making further progress by the methods of this paper." In general, I conjectured:

3.13. *For the Fermat hypersurface if $d = n$, then the rational curves are Zariski dense, and the Fermat hypersurface is unirational over $\mathbf{Q}$.*

Of course one must take either d odd or express the Fermat equation in an indefinite form.

Example: When $d = n = 3$, the Fermat hypersurface has Ramanujan's taxicab rational point (1729 is the sum of two cubes in two different ways: 9, 10 and 12, 1). Furthermore, the conjecture is true in this case, i.e. for $d = n = 3$, the Fermat surface is a rational image of $\mathbf{P}^2$ over $\mathbf{Q}$, by using Theorem 12.11 of Manin's book [Man 74]. But so far there are no systematic results known for the general Fermat hypersurfaces from the present point of view of algebraic geometry, for the existence of rational curves, both geometrically and over $\mathbf{Q}$, and for the possibility of their being rational images of projective space for low degrees compared to n.

The Fermat equation is even more subtle for $d = n + 1$, when one expects fewer solutions. Euler had a false intuition when he guessed that there would be no non-trivial rational solutions. First, Lander and Parkin [LandP 66] found the solution in degree 5:

$$27^5 + 84^5 + 110^5 + 133^5 = 144^5.$$

Then Elkies [El 88] found infinitely many solutions in degree 4, including

$$2682440^4 + 15365639^4 + 18796760^4 = 20615673^4.$$

He was led to this solution by a mixture of theory and computer search. The point is that for the degree $d = n + 1$ there is no expectation that the Fermat hypersurface is unirational. Rather, *it is fibered by curves of genus* 1, *and the question is when a fiber has a rational point.* Elkies found theoretically that in many cases there could not be a rational point, and in one remaining case, he knew how to make the computer deliver. Furthermore, he proved that infinitely many fibers have at least two rational points; one of them can be taken as an origin, and the other one has infinite order on the fibral elliptic curve. This leaves open the problem of giving an asymptotic estimate for the number of rational points on the base curve of height bounded by $B \to \infty$, such that the fiber above those points is an elliptic curve with a rational point of infinite order.

The fibration of the Fermat surface by elliptic curves over $\mathbf{C}$ is classical, perhaps dating back to Gauss. Over $\mathbf{Q}$, as far as I know, a fibration comes from Demjanenko [Dem 74], and it is the one used by Elkies. When written in the form

$$x_0^4 + x_1^4 = x_2^4 + x_3^4,$$

there are other fibrations, related to modular curves. The Fermat surface can also be viewed as an example of a K3 surface. For various points of view, see also Mumford [Mu 83], p. 55 and Shioda [Shio 73], §4.

Let us now discuss the function field case, analogous to Theorem 2.3 for curves. The following analogue of de Franchis' theorem was proved by Kobayashi–Ochiai [KoO 75], motivated by my conjecture that a hyperbolic projective variety is Mordellic, see Chapter VIII, §4.

Theorem 3.14 (Kobayashi–Ochiai). *Let W be a variety, and let X_0 be a pseudo-canonical projective variety. Then there is only a finite number of generically surjective rational maps of W onto X_0.*

Theorem 3.14 is the **split** case. On this subject see also [DesM 78]. More generally, it has been conjectured that:

3.15. *Given a variety W, there is only a finite number of isomorphism classes of pseudo-canonical varieties X_0 for which there exists a generically surjective rational map of W onto X_0.*

For a partial result in this direction, see Maehara [Mae 83], who investigates algebraic families of pseudo-canonical varieties and rational maps.

In the non-split case, we only have a conjecture.

3.16. *Let X be a pseudo-canonical projective variety defined over a function field F with constant field k. Suppose that the set of rational points $X(F)$ is Zariski dense in X. Then there exists a variety X_0 defined over k such that X is birationally equivalent to X_0 over F. All the rational points in $X(F)$ outside some proper Zariski closed subset of X are images of points in $X_0(k)$ under the birational isomorphism.*

As in the case of curves, the problem is to bound the degrees of sections, so that they lie in a finite number of families. Again see [Mae 83]. It then follows that there exists a parameter variety T and a generically surjective rational map

$$f: T \times W \to X.$$

From this one wants to split X birationally. In [La 86] I stated the following self-contained version as a conjecture.

Theorem 3.17. *Let $\pi: X \to W$ be a generically surjective rational map, whose generic fiber is geometrically irreducible and pseudo canonical. Assume that there exists a variety T and a generically surjective rational map*

$$f: T \times W \to X.$$

Then X is birationally equivalent to a product $X_0 \times W$.

Viehweg pointed out to me that this statement is essentially proved in [Mae 83] over an algebraically closed field of characteristic 0.

The interplay between the diophantine problems and algebraic geometry is reflected in the history surrounding Theorem 3.14. My conjecture that hyperbolic projective varieties are Mordellic led Kobayashi–Ochiai to their theorem about pseudo-canonical varieties in the split case, and this theorem in turn made me conjecture that pseudo-canonical varieties are pseudo Mordellic, thus coming to the conjecture that a variety is Mordellic (or hyperbolic) if and only if every subvariety is pseudo canonical, and coming to the definition of the special set.

For more information on the topics of this section, readers might look up my survey [La 86]. For quantitative formulations, see Vojta's conjectures in Chapter II, §4.

I, §4. ABELIAN VARIETIES

An **abelian variety** is a projective non-singular variety which is at the same time a group such that the law of composition and inverse are morphisms. Over the complex numbers, abelian varieties are thus compact complex Lie groups, and are thus commutative groups. Weil originally developed the theory algebraically, although the fact that abelian varieties are commutative in all characteristics is due to Chevalley.

Example. Let A be an abelian variety of dimension 1, and suppose A is defined over a field k of characteristic $\neq 2, 3$. Then A can be defined by an affine equation in Weierstrass form

$$y^2 = 4x^3 - g_2 x - g_3 \qquad \text{with} \quad g_2, g_3 \in k$$

and $\Delta = g_2^3 - 27g_3^2 \neq 0$. The corresponding projective curve is (isomorphic to) A, whose points $A(k)$ consist of the solutions of the affine equation with $x, y \in k$ together with the point at infinity in $\mathbf{P}^2$. If $k = \mathbf{C}$ is the complex numbers, then $A(\mathbf{C})$ can be parametrized by the Weierstrass functions with respect to some lattice Λ. In other words, there exists a lattice, with basis ω_1, ω_2 over $\mathbf{Z}$, such that the map

$$z \mapsto (\wp(z), \wp'(z))$$

is an isomorphism of $\mathbf{C}/\Lambda$ with $A(\mathbf{C})$. The function $\wp$ is the Weierstrass function, defined by

$$\wp(z) = \sum_{\omega \neq 0} \left(\frac{1}{(z-\omega)^2} - \frac{1}{\omega^2} \right).$$

The coefficients of the equation are given by

$$g_2 = 60 \sum_{\omega \neq 0} \omega^{-4} \qquad \text{and} \qquad g_3 = 140 \sum_{\omega \neq 0} \omega^{-6}.$$

The sums over $\omega \neq 0$ are taken over all elements of the lattice $\neq 0$. In the above parametrization, the lattice points map to the point at infinity on $\mathbf{P}^2$.

Abelian varieties of dimension 1 over a field k are precisely the curves of genus 1 together with a rational point, which is taken as the origin on the abelian variety for the group law.

In higher dimension, it is much more difficult to write down equations for abelian varieties. See Mumford [Mu 66] and Manin–Zarhin [MaZ 72].

Let A, B be abelian varieties over a field k. By $\mathrm{Hom}_k(A, B)$ we mean the **homomorphisms of** A into B which are algebraic, i.e. the morphisms of A into B which are also group homomorphisms, and are defined over k. A basic theorem states that $\mathrm{Hom}_k(A, B)$ is a free abelian group, finitely generated. We sometimes omit the reference to k in the notation, especially if k is algebraically closed.

Abelian varieties are of interest intrinsically, for themselves, and also because they affect the theory of other varieties in various ways. One of these ways is described in §5.

We are interested in the structure of the group of rational points of an abelian variety over various fields.

As usual, we consider the two important cases when F is a number field and F is a function field.

Theorem 4.1 (Mordell–Weil theorem). *Let A be an abelian variety defined over a number field F. Then $A(F)$ is a finitely generated abelian group.*

When $F = \mathbf{Q}$ and $\dim A = 1$, so when A is a curve of genus 1, the finite generation of $A(\mathbf{Q})$ was conjectured by Poincaré and proved by Mordell in 1921 [Mo 21]. Weil extended Mordell's theorem to number fields and arbitrary dimension [We 28]. Néron [Ne 52] extended the theorem to finitely generated fields over $\mathbf{Q}$.

Next we handle the function field case. Let k be a field and let F be a finitely generated extension of k, such that F is the function field of a variety defined over k. Let A be an abelian variety defined over F. By an **F/k-trace** of A we mean a pair (B, τ) consisting of an abelian variety B defined over k, and a homomorphism

$$\tau\colon B \to A$$

defined over F, such that (B, τ) satisfies the universal mapping property for such pairs. In other words, if (C, α) consists of an abelian variety C over k, and a homomorphism $\alpha: C \to A$ over F, then there exists a unique homomorphism $\alpha_*: C \to B$ over k such that the following diagram commutes.

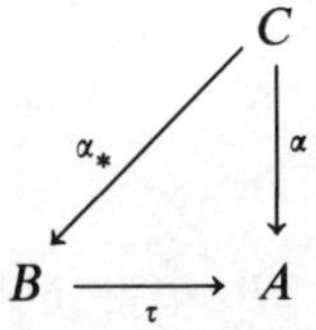

Chow defined and proved the existence of the F/k-trace. He also proved that the homomorphism τ is injective. See [La 59].

The analogue of the Mordell–Weil theorem was then formulated and proved in the function field case as follows [LN 59].

Theorem 4.2 (Lang–Néron theorem). *Let F be the function field of a variety over k. Let A be an abelian variety defined over F, and let (B, τ) be its F/k-trace. Then $A(F)/\tau(B(k))$ is finitely generated.*

Example. Consider for example the case when $\dim A = 1$. Suppose A is defined by the Weierstrass equation as recalled at the beginning of this section. Let as usual

$$j = 1728 g_2^3/\Delta.$$

We assume the characteristic is $\neq 2, 3$. Suppose A is defined over the function field F with constant field k. Assume that the F/k-trace is 0. There may be two cases, when $j \in k$ or when j is transcendental over k. In both cases, the Lang–Néron theorem guarantees that $A(F)$ is finitely generated.

Corollary 4.3. *Let F be finitely generated over the prime field. Let A be an abelian variety defined over F. Then $A(F)$ is finitely generated.*

This corollary is the absolute version of Theorem 4.2, and follows since the set of points of a variety in a finite field (the constant field) is finite, no matter what the variety, or in characteristic 0, by using Theorem 4.1.

Questions arise as to the rank and torsion of the group $A(F)$. First consider the generic case. Abelian varieties distribute themselves in algebraic families, of which the generic members are defined over a function field F over the complex numbers. Shioda in dimension 1 (for elliptic curves) [Shio 72] and Silverberg in higher dimension [Slbg 85] have shown that if A is the generic member of such families, then $A(F)$ is finite. Torsion for elliptic curves over a base of dimension 1 has been studied extensively, and I mention only the latest paper known to me giving fairly general results by Miranda–Persson [MirP 89].

Over the rational numbers, or over number fields, the situation involves a great deal of arithmetic, some of which will be mentioned in other chapters. Here, we mention only two qualitative conjectures, in line with the general topics which have been discussed.

Conjecture 4.4. *Over the rational numbers* $\mathbf{Q}$, *there exist elliptic curves* A *(abelian varieties of dimension* 1) *such that* $A(\mathbf{Q})$ *has arbitrarily high rank.*

No example of such elliptic curves is known today. Shafarevich–Tate have given such examples for elliptic curves defined over function fields over a finite field [ShT 67]. For rank 10 see [Ne 52] and rank 14 see [Me 86]. One problem is to give a quantitative measure, or probabilistic description, of those which have one rank or the other, and an asymptotic estimate of how many have a given positive integer as rank. The problem is also connected with the Birch–Swinnerton-Dyer conjecture, which relates the rank to certain aspects of a zeta function associated with the curve, and which we shall discuss in Chapter III. For some current partial results on the rank, see Goldfeld [Go 79], and for computations giving relatively high frequency of rank > 1, see Zagier–Kramarz [ZaK 87] and Brumer–McGuinness [BruM 90].

Aside from the rank, one also wants to describe the torsion group, for individual abelian varieties, and also uniformly for families. The general expectation lies in:

Conjecture 4.5. *Given a number field* F, *and a positive integer* d, *there exists a constant* $C(F, d)$ *such that for all abelian varieties* A *of dimension* d *defined over* F, *the order* $\# A(F)_{\mathrm{tor}}$ *of the torsion group is bounded by* $C(F, d)$.

Furthermore, Kamienny [Kam 90] has shown for $d = 1$ and $n = 2$ that the following stronger uniformity is true: *There exists a constant* $C(n, d)$ *such that for all abelian varieties* A *of dimension* d *over a number field* F *of degree* n *the order of the torsion group* $\# A(F)_{\mathrm{tor}}$ *is bounded by* $C(n, d)$.

For elliptic curves over the rationals, Mazur has proved that the order of the torsion group is bounded by 16, developing in the process an extensive theory on modular curves [Maz 77] and [Maz 78] which we shall mention in Chapter V. Over number fields, some results have been obtained by Kubert [Ku 76], [Ku 79]. Mazur has conjectured, or more cautiously, raised the question whether the following is true.

Question 4.6. *Given a number field* F, *there is an integer* $N_1(F)$ *and a finite number of values* $j_1, \ldots, j_{c(F)}$ *such that if* A *is an elliptic curve over* F *with a cyclic subgroup* C *of order* $N \geq N_1(F)$, *and* C *is stable under the Galois group* G_F, *then* $j(A)$ *is equal to one of the* $j_1, \ldots, j_{c(F)}$.

We mention here that a certain diophantine conjecture, the *abc* conjecture (stated in Chapter II, §1) implies Conjecture 4.5 by an argument due to Frey (see [Fr 87a, b], [La 90], and also [HiS 88]). It also has something to do, but less clearly, with the above question of Mazur.

Questions also arise as to the behavior of the rank and torsion in infinite extensions. For instance:

Theorem 4.7. *Let A be an abelian variety defined over a number field F. Let $\boldsymbol{\mu}$ denote the group of all roots of unity in the algebraic numbers. Then the group of torsion points*

$$A(F(\boldsymbol{\mu}))_{\text{tor}}$$

is finite.

This was proved by Ribet (see the appendix of [KaL 82]). We also have:

Theorem 4.8 (Zarhin [Zar 87]). *Let A be a simple abelian variety over a number field F. Then $A(F^{\text{ab}})_{\text{tor}}$ is finite if and only if A does not have complex multiplication over F.*

For the convenience of the reader, we recall the definition that A has **complex multiplication**, or has **CM type** over a field F, if and only if $\text{End}_F(A)$ contains a semisimple commutative $\mathbf{Q}$-algebra of dimension $2 \dim A$. The above theorem of Zarhin comes from other theorems concerned with non-abelian representations of the Galois group, for which I refer to his paper. Zarhin also has results for the finiteness of torsion points in non-abelian extensions, for instance:

Theorem 4.9 ([Zar 89]). *Let l be a prime number and let L be an infinite Galois extension of F such that* $\text{Gal}(L/F)$ *is a compact l-adic Lie group. Let A be an abelian variety over F. If $A[p] \cap A(L)$ is $\neq 0$ for infinitely many primes p, and A is simple over L, then A has* CM *type over L.*

Mazur [Maz 72] has related the question of points of abelian varieties in certain cyclotomic extensions with the arithmetic of such extensions, and we shall state his main conjecture. Let F be a number field. Let p be an odd prime (for simplicity). Let F_n be the cyclic extension of F consisting of the largest subfield of $F(\boldsymbol{\mu}_{p^n})$ of p-power degree. Let

$$F_\infty = \bigcup F_n.$$

Then F_∞ is called the **cyclotomic $\mathbf{Z}_p$-extension** of F. Mazur raises the possibility that the following statement is true:

Let A be an abelian variety over F. Then $A(F_\infty)$ is finitely generated.

For a survey of results and conjectures connecting the rank behavior of the Mordell–Weil group in towers of number fields with Iwasawa type theory and modular curves, see Mazur [Maz 83]. For other results concerning points in extensions whose Galois group is isomorphic to the p-adic integers $\mathbf{Z}_p$, see for instance Wingberg [Win 87].

I, §5. ALGEBRAIC EQUIVALENCE AND THE NÉRON–SEVERI GROUP

There is still another important possible relation between divisor classes. Let X be a projective variety, non-singular in codimension 1, and let Y be a non-singular variety. Let c be a divisor class on $X \times Y$. If x is a simple point of X we write

$$c(x) = \text{restriction of } c \text{ to } \{x\} \times Y \text{ identified with } Y.$$

Similarly, for a point y of Y we write

$${}^t c(y) = \text{restriction of } c \text{ to } X \times \{y\} \text{ identified with } X.$$

The superscript t indicates a transpose, namely ${}^t c$ is the transpose of c on $Y \times X$. The group generated by all classes of the form ${}^t c(y_1) - {}^t c(y_2)$ for all pairs of points $y_1, y_2 \in Y$, and all classes c on products $X \times Y$, will be said to be the group of classes **algebraically equivalent to** 0. This subgroup of $\mathrm{CH}^1(X)$ is denoted by $\mathrm{CH}^1_0(X)$, and is also called the **connected component** of $\mathrm{CH}^1(X)$, for reasons which we shall explain later in this section. The factor group

$$\mathrm{NS}(X) = \mathrm{CH}^1(X)/\mathrm{CH}^1_0(X)$$

is called the **Néron–Severi group**.

Theorem 5.1 (Néron [Ne 52]). *The Néron–Severi group is finitely generated.*

The history of this theorem is interesting. Severi had the intuition that there was some similarity between his conjecture that $\mathrm{NS}(X)$ is finitely generated, and the Mordell–Weil theorem that $A(F)$ is finitely generated for number field F. Néron made this similarity more precise when he proved the theorem, and an even clearer connection was established by Lang–Néron, who showed how to inject the Néron–Severi group in a group of rational points of an abelian variety over a function field. To do this we have to give some definitions.

Let X be any non-singular variety, defined say over a field k, and with a rational point $P \in X(k)$. There exists an abelian variety A over k and a morphism

$$f: X \to A$$

such that $f(P) = 0$, satisfying the universal mapping property for morphisms of X into abelian varieties. In other words, if $\varphi: X \to B$ is a morphism of X into an abelian variety B, then there exists a unique homomorphism $f_*: A \to B$ and a point $b \in B$ such that the following diagram commutes.

$$\begin{array}{ccc} X & \xrightarrow{f} & A \\ & {\scriptstyle\varphi}\searrow \quad \swarrow{\scriptstyle f_* + b} & \\ & B & \end{array}$$

Of course, if $\varphi(P) = 0$ then $b = 0$. The abelian variety A is uniquely determined up to an isomorphism, and is called the **Albanese variety** of X. If k' is an extension of k, then A is also the Albanese variety of X over k', so one usually does not need to mention a field of definition for the Albanese variety. Note that the existence of a simple rational point, or some sort of condition is needed on the variety X. For instance, there may be a projective curve of genus 1, defined over a field k and having no rational point. Over any extension of k where this curve acquires a rational point, we may identify the curve with its Albanese variety, but over the field k itself, the curve does not admit an isomorphism with an abelian variety.

The morphism $f: X \to A$ can be extended. For any field k' containing k, define the group of **0-cycles** $\mathscr{Z}(X(k'))$ to be the free abelian group generated by the points in $X(k')$. A zero cycle $\mathfrak{a}$ can then be expressed as a formal linear combination

$$\mathfrak{a} = \sum n_i(P_i) \quad \text{with} \quad P_i \in X(k') \quad \text{and} \quad n_i \in \mathbf{Z}.$$

We define

$$S_f(\mathfrak{a}) = \sum n_i f(P_i),$$

where the sum on the right-hand side is taken on A. Then

$$S_f: \mathscr{Z}(X) \to A$$

is a homomorphism. Let $\mathscr{Z}_0(X)$ be the subgroup of 0-cycles of degree 0, that is those cycles such that $\sum n_i = 0$. Then the image $S_f(\mathfrak{a})$ is independent of the map f, which was determined only up to a translation, whenever $\mathfrak{a}$ is of degree 0. As a result, one can show that even if X does not admit a rational point over k, there still exists an abelian variety A

over k and a homomorphism (the sum)

$$S\colon \mathscr{Z}_0(X(k')) \to A(k')$$

for every field k' containing k, such that over any field k' where X has a rational point, $A_{k'}$ is the Albanese variety of X, and $S = S_f$. We again call A the **Albanese variety** of X, and we call S the **Albanese homomorphism on the 0-cycles of degree** 0.

If X is a curve, then its Albanese variety is called the **Jacobian**. A canonical map of X into its Jacobian is an imbedding. If X and J are defined over a field k, then one way to approach the study of the rational points $X(k)$ is via its imbedding in $J(k)$.

It will now be important to deal with fields of rationality, so suppose the projective variety X is defined over a field k. By $\mathrm{CH}^1(X, k)$ we mean the group of divisor classes on X, defined over k.

Theorem 5.2 (Lang–Néron [LN 59]). *Let X be a projective variety, non-singular in codimension* 1, *and defined over an algebraically closed field k. Let L_u be a linear variety defined by linear polynomials with algebraically independent coefficients u, and of dimension such that the intersection $X \,.\, L_u$ is a non-singular curve C_u defined over the function field $k(u)$ (purely transcendental over k). Let J_u be the Jacobian of C_u, defined over $k(u)$, and let*

$$S\colon \mathscr{Z}_0(C_u) \to J_u$$

be the Albanese homomorphism. Let (B, τ) be the $k(u)/k$-trace of J_u. Let $\mathscr{C}$ be the subgroup of $\mathrm{CH}^1(X, k)$ consisting of those divisor classes c whose restrictions $c \,.\, C_u$ to C_u have degree 0. *Then*

$$\mathscr{C} \supset \mathrm{CH}^1_0(X, k) \qquad \textit{and} \qquad \mathrm{CH}^1(X, k)/\mathscr{C} \approx \mathbf{Z}.$$

The projective imbedding of X can be chosen originally so that the map

$$c \mapsto S(c \,.\, C_u) \qquad \textit{for} \quad c \in \mathscr{C}$$

induces an injective homomorphism $\mathscr{C} \hookrightarrow J_u(k(u))$, and also an injective homomorphism

$$\mathscr{C}/\mathrm{CH}^1_0(X, k) \hookrightarrow J_u(k(u))/\tau B(k).$$

That $\mathrm{NS}(X, k)$ is finitely generated is then a consequence of the Lang–Néron Theorem 4.2.

Observe how a geometric object, the Néron-Severi group, is reduced to a diophantine object, the rational points of an abelian variety in some function field. Conversely, we shall see instances when geometric objects are associated to rational points, in the case of curves.

Remark. In the statement of the theorem, after a suitable choice of projective imbedding, we get an injection of $\mathrm{NS}(X, k)$ into $J_u(k(u))$. Actually, *given* a projective imbedding, the kernel of the homomorphism $c \mapsto S(c \,.\, C_u)$ is always finitely generated, according to basic criteria of algebraic equivalence [We 54]. This suffices for the proof that $\mathrm{NS}(X, k)$ is finitely generated, which proceeds in the same way.

Having described the Néron–Severi group, we next describe the subgroup $\mathrm{CH}_0^1(X)$ by showing how it can be given the structure of an algebraic group, and in fact an abelian variety. This explains why we call it a **connected component**.

Theorem 5.3. *Let X be a projective variety, non-singular in codimension 1, and defined over a field k. Let $P \in X(k)$ be a simple rational point. Then there exists an abelian variety $A' = A'(X)$, and a class $c \in \mathrm{CH}^1(X \times A')$ such that*

$$ {}^t c(0) = 0, \qquad c(P) = 0, $$

and for any field k' containing k the map

$$ a' \mapsto {}^t c(a') \qquad \text{for} \quad a' \in A'(k') $$

gives an isomorphism $A'(k') \xrightarrow{\approx} \mathrm{CH}_0^1(X, k')$. The abelian variety and c are uniquely determined up to an isomorphism over k.

We call the pair (A', c) the **Picard variety** of X over k. It is a theorem that for every extension k' of k, the base change $(A'_{k'}, c_{k'})$ is also the Picard variety of X over k'. The class c is called the **Poincaré class**.

Let A be the Albanese variety of X and let $f: X \to A$ be a canonical map, determined up to a translation. If (A', δ) is the Picard variety of A, then we can form its pull back to get the Picard variety of X. Indeed, we have a morphism

$$ f \times \mathrm{id}: X \times A' \to A \times A', $$

and the pull back $(f \times \mathrm{id})^*(\delta)$ is the Poincaré class making A' also the Picard variety of X.

The Picard variety of an abelian variety is also called the **dual variety**.

It is a theorem that if (A', δ) is the dual of A then $(A, {}^t\delta)$ is the dual of A'. This property is called the **biduality** of abelian varieties.

Putting together the finite generation of the Néron–Severi group and the finite generation of the Mordell–Weil group, we find:

Theorem 5.4. *Let X be a projective variety, non-singular in codimension 1, defined over a field F finitely generated over the prime field. Then $\mathrm{CH}^1(X, F)$ is finitely generated.*

Remark 5.5. One can define **Chow groups** $\mathrm{CH}^m(X, k)$ for higher co-dimension m, and one can define the notion of algebraic equivalence, as well as cohomological equivalence. Thus one obtains factor groups analogous to the Néron–Severi group. The example of Clemens ([Cl 83], [Cl 84] and Example 4 of §3) shows that over an algebraically closed field, this group is not necessarily finitely generated. It is still conjectured that $\mathrm{CH}^m(X, F)$ is finitely generated if F is finitely generated over $\mathbf{Q}$. In fact there are even much deeper conjectures of Beilinson and Bloch connecting the rank with orders of poles of zeta functions in the manner of the Birch–Swinnerton-Dyer conjecture and the theory of heights. See for instance Beilinson [Be 85].

We shall use other properties of the Picard variety later in several contexts, so we recall them there. By a **polarized abelian variety**, we mean an abelian variety A and an algebraic equivalence class $c \in \mathrm{NS}(A)$ which contains an ample divisor. Such a class is called a **polarization**. A **homomorphism of polarized abelian varieties**

$$f\colon (A, c) \to (A_1, c_1)$$

is a homomorphism of abelian varieties $f\colon A \to A_1$ such that $f^*c_1 = c$. To each polarization we shall associate a special kind of homomorphism of A. We define an **isogeny** $\varphi\colon A \to B$ of abelian varieties to be a homomorphism which is surjective and has finite kernel. Then A, B have the same dimension.

Proposition 5.6. *Given a class $c \in \mathrm{CH}^1(A)$ and $a \in A$, let c_a be the translation of c by a. Let A' be the dual variety. The map*

$$\varphi_c\colon A \to A' \quad \textit{satisfying } \varphi_c(a) = \textit{element } a' \textit{ such that} \quad {}^t\delta(a') = c_a - c$$

is a homomorphism of A into A', depending only on the algebraic equivalence class of c. The association

$$c \mapsto \varphi_c$$

induces an injective homomorphism

$$\mathrm{NS}(A) \hookrightarrow \mathrm{Hom}(A, A').$$

If c is ample, then φ_c is an isogeny.

By abuse of notation, one sometimes writes

$$\varphi_c(a) = c_a - c.$$

In general, let $\varphi: A \to B$ be an isogeny. Then φ is a finite covering, whose degree is called the **degree of the isogeny**. If c is a polarization, then the degree of φ_c is called the **degree of the polarization**. A polarization of degree 1 is a polarization c such that φ_c is an isomorphism, and is called a **principal polarization**.

I, §6. SUBVARIETIES OF ABELIAN AND SEMIABELIAN VARIETIES

Such varieties provide a class of examples for diophantine problems holding special interest, since all curves of genus $\geqq 1$ belong to this class. A basic theorem describes their algebraic structure. An important characterization was given by Ueno [Ue 73], see also Iitaka ([Ii 76], [Ii 77]), who proved:

Let X be a subvariety of an abelian variety over an algebraically closed field. Then X is pseudo canonical if and only if the group of translations which preserve X is finite.

Then we have quite generally, [Ue 73], Theorem 3.10:

Ueno's theorem. *Let X be a subvariety of an abelian variety A, and let B be the connected component of the group of translations preserving X. Then the quotient $f: X \to X/B = Y$ is a morphism, whose image is a pseudo-canonical subvariety of the abelian quotient A/B, and whose fibers are translations of B. In particular, if X does not contain any translates of abelian subvarieties of dimension $\geqq 1$, then X is pseudo canonical.*

For a proof, see also Iitaka [Ii 82], Theorem 10.13, and [Mori 87], Theorem 3.7. We call f the **Ueno fibration of** X.

The variety Y in Ueno's theorem is also a subvariety of an abelian variety. Hence to study the full structure of subvarieties of A and their rational points, we are reduced to pseudo-canonical subvarieties, in which case we have:

Kawamata's structure theorem ([Kaw 80]). *Let X be a pseudo-canonical subvariety of an abelian variety A in characteristic 0. Then there exists a finite number of proper subvarieties Z_i with Ueno fibrations $f_i: Z_i \to Y_i$ whose fibers have dimension $\geqq 1$, such that every translate of an abelian subvariety of A of dimension $\geqq 1$ contained in X is actually contained in the union of the subvarieties Z_i.*

Note that the set of Z_i is empty if and only if X does not contain any translation of an abelian subvariety of dimension $\geqq 1$.

Although the above version of Kawamata's theorem is not stated that way in the given references, I am indebted to Lu for pointing out that these references actually prove the structure theorem as stated. Ochiai [Och 77] made a substantial contribution besides Ueno, but it was actually Kawamata who finally proved the existence of the fibrations by abelian subvarieties, so we call the union of the subvarieties Z_i the **Ueno–Kawamata fibrations in** X when X is pseudo canonical. We now see:

For every subvariety X of an abelian variety A, the Ueno–Kawamata fibrations in X constitute the special set defined in §3.

So in the case of subvarieties of abelian varieties, we have a clear description of this special set.

For an extension of the above results to semiabelian varieties, see Noguchi [No 81a]. The structure theorems constitute the geometric analogue of my conjecture over finitely generated fields [La 60a]:

Conjecture 6.1. *Let X be a subvariety of an abelian variety over a field F finitely generated over* $\mathbf{Q}$. *Then X contains a finite number of translations of abelian subvarieties which contain all but a finite number of points of $X(F)$.*

In light of the determination of the exceptional set, this conjecture corresponds to the general conjecture of §4 applied to subvarieties of abelian varieties. By Kawamata's structure theorem, to prove Conjecture 6.1 it suffices to prove that if X is not the translate of an abelian subvariety of dimension $\geqq 1$, then the set of rational points $X(F)$ is not Zariski dense. The following especially important case from [La 60b] has now been proved.

Theorem 6.2 (Faltings [Fa 90]). *Let X be a subvariety of an abelian variety, and suppose that X does not contain any translation of an abelian subvariety of dimension > 0. Then X is Mordellic.*

In particular, let C be a projective non-singular curve, imbedded in its Jacobian J. In the early days of the theory, I formulated Mordell's conjecture as follows.

Suppose that the genus of C is $\geqq 2$. Let Γ_0 be a finitely generated subgroup of J. Then $C \cap \Gamma_0$ is finite.

No direct proof has been found, and the statement is today a consequence of Faltings' theorem over number fields, combined with the Mordell–Weil and Lang–Néron theorems, as they imply Corollary 4.3. Chabauty [Chab 41] proved the statement when the rank of Γ_0 is smaller than the genus of the curve. Although he works over a number field, his proof makes no use of that fact, and depends only on p-adic analysis.

Such theorems or conjectures apply to somewhat more general group varieties than abelian varieties. By a **linear torus**, or **torus** for short, one means a group variety which is isomorphic over an algebraically closed field to a product of a finite number of multiplicative groups. By a **semiabelian variety**, one means an extension of an abelian variety by a torus. In [La 60a] I proved in characteristic 0:

Let C be a curve in a torus, and let Γ_0 be a finitely generated subgroup of the torus. If $C \cap \Gamma_0$ is infinite then C is the translation of a subtorus.

The method is to pass to coverings and use Hurwitz's genus formula together with an extension of a theorem of Siegel, Chapter IX, Theorem 3.1. In particular, the equation

$$x + y = 1$$

has only a finite number of solutions in Γ_0. One may take Γ_0 to be the group of units in a finitely generated ring over $\mathbf{Z}$, so I called this equation the **unit equation**. For an example in the context of modular units, see [KuL 75]. In general, one sees how it is easier to formulate the statement in characteristic 0, because in characteristic p, one can always raise a solution to the power p, i.e. apply Frobenius, and one gets infinitely many solutions from one of them. Basically, this is the only type of example one can concoct, but we continue to assume characteristic 0, to make the results easier to state. Let F be finitely generated over $\mathbf{Q}$. We know that $A(F)$ is finitely generated. Manin's proof of the analogue of Mordell's conjecture in the function field case led him to ask whether the intersection of a curve with the torsion group of its Jacobian is finite. The same question was raised by Mumford at about the same time. I formulated a conjecture which would cover both the **Manin–Mumford conjecture** and the Mordell conjecture as follows [La 65a]. Let A be a semiabelian variety over $\mathbf{C}$, and let Γ_0 be a finitely generated subgroup of $A(\mathbf{C})$. We define its **division group** Γ to be the group of all points $P \in A(\mathbf{C})$ such that $mP \in \Gamma_0$ for some positive integer m.

Conjecture 6.3. *Let A be a semiabelian variety over $\mathbf{C}$. Let Γ_0 be a finitely generated subgroup of $A(\mathbf{C})$ and let Γ be its division group. Let X be a subvariety of A. Then X contains a finite number of translations of semiabelian subvarieties which contain all but a finite number of points of $X \cap \Gamma$.*

Theorem 6.4. *Conjecture* 6.3 *is a theorem in the following cases.*

(1) *The variety X is a curve in an abelian variety A.* (Raynaud [Ra 83a, b])

(2) *The group* $\Gamma_0 = \{0\}$, *so* $\Gamma = A_{\text{tor}}$ *is the group of torsion points of an abelian variety* A *but* X *is arbitrary.* (Raynaud [Ra 83a, b])
(3) *Again* $\Gamma_0 = \{0\}$ *so* Γ *is the group of torsion points, and* X *is a subvariety of an arbitrary commutative group variety.* (Hindry [Hi 88])

In particular, Raynaud proved the Manin–Mumford conjecture.

Liardet [Li 74], [Li 75] proved the conjecture for curves in toruses. We restate it in more naive terms.

Theorem 6.5. *Let* $f(x, y) = 0$ *be the equation for a curve in the plane, where* f *is an irreducible polynomial over the complex numbers. Let* Γ_0 *be a finitely generated multiplicative group of complex numbers, and let* Γ *be its division group, that is the group of complex numbers* z *such that* $z^m \in \Gamma_0$ *for some positive integer* m. *If there exist infinitely many elements* $x, y \in \Gamma$ *such that* $f(x, y) = 0$, *then* f *is a polynomial of the form*

$$aX^r + bY^s = 0 \quad \text{or} \quad cX^rY^s + d = 0 \quad \text{with } a, b, c, d \in \mathbf{C}.$$

Or briefly put:

If a curve has an infinite intersection with the division group of a finitely generated multiplicative group, then the curve is the translation of a subtorus.

For the intersection with roots of unity my conjecture had been proved by Tate [La 65a]. In higher dimension, for the intersection of a variety with a finitely generated group of units, the conjecture dates back to Chabauty [Chab 38]. It is contained in the next theorem.

Theorem 6.6 (Laurent [Lau 84]). *Let* X *be a subvariety of a torus and let* Γ_0 *be a finitely generated subgroup of the torus. If* $X \cap \Gamma_0$ *is infinite, then* X *contains the translations of a finite number of subtoruses which themselves contain* $X \cap \Gamma_0$.

Hindry [Hin 88] proved his result by a method involving the Galois group of torsion points, stemming from [La 65a], which we describe briefly for curves X on an abelian variety, defined over a field F finitely generated over $\mathbf{Q}$. Let m be an integer $\geqq 2$, and let m_A be multiplication by m on A. As a cycle, $m_A(X) = s \cdot X^{(m)}$ where s is some positive integer and $X^{(m)}$ is the set-theoretic image of X by m_A. Then $m_A(X)$ is algebraically equivalent to m^2X. If $X \neq X^{(m)}$ then $X \cap X^{(m)}$ has at most $m^2(\deg X)^2$ points by a generalization of Bezout's theorem. If $X = X^{(m)}$ then m_A gives an unramified covering of X over itself, of degree m^2, and hence X is of genus 1, so is an abelian subvariety. If X has an infinite intersection with $A(F)_{\mathrm{tor}}$, then we use a Galois theoretic property of torsion points which I conjectured, namely:

(∗) *Let A be an abelian variety defined over F. There exists an integer $c \geqq 1$ with the following property. Let x be a point of period n on A. Let G_n be the multiplicative group of integers prime to n mod n. Let G be the subgroup of G_n consisting of those integers d such that dx is conjugate to x over F, that is there exists an element σ of the Galois group over F such that $dx = \sigma x$. Then*

$$(G_n : G) \leqq c.$$

To apply (∗), suppose that there exist points x_n of period n, $n \to \infty$, lying on X. Let d be a positive integer prime to n. By (∗), there exists an automorphism σ of $F(x_n)$ over F such that $\sigma x_n = d^r x_n$ where r is a positive integer bounded by c. Then

$$\sigma x_n = d^r x_n \in X \cap X^{(d^r)}.$$

Furthermore, if τ is in the group of automorphisms of $F(x_n)$ over F, then

$$\tau\, d^r x_n \in X \cap X^{(d^r)}.$$

If $X \neq X^{(d^r)}$ we obtain the inequalities, using (∗):

$$\frac{\phi(n)}{c} \leqq \text{number of points on } X \cap X^{(d^r)} \leqq d^{2r}(\deg X)^2.$$

We note that $\phi(n) \geqq n^{1/2}$ for sufficiently large n. This immediately gives a contradiction as soon as n is sufficiently large.

Property (∗) was proved by Shimura in the case of complex multiplication, [La 83c] Chapter 4, Theorem 2.4. The general case is not known at this time. A partial result which suffices for the above application has been anounced by Serre, but no proof is yet published. See [Hin 88].

The above results and conjectures concern the absolute case of finiteness for rational points of subvarieties of semiabelian varieties. The relative case for abelian varieties over function fields is due to Raynaud [Ra 83c], namely:

Theorem 6.7. *Let $k = \mathbf{C}$ and let F be the function field of a curve over k. Let X be a subvariety of an abelian variety A, defined over F. Let (B, τ) be the F/k-trace of A. If X does not contain the translation of an abelian subvariety of dimension > 0, then the set of rational points $X(F)$ is finite (modulo the trace $\tau B(k)$).*

The more general case when X may contain a finite number of translates of abelian subvarieties is still unknown as far as I know.

I, §7. HILBERT IRREDUCIBILITY

Let F be a function field over the constant field k. Let X_F be a projective variety over F, and say X_F is non-singular for simplicity. Write $F = k(Y)$ for some parameter variety Y. Then we may view X_F as the generic fiber of a family, namely there exists a morphism

$$\pi: X \to Y$$

such that if η is the generic point of Y then $\pi^{-1}(\eta) = X_\eta = X_F$. Then there exists a non-empty Zariski open set Y_0 of Y such that π is smooth over Y_0. For each $y \in Y_0(k^a)$ we get the fiber X_y called also the **specialization** of X_η (or X_F) over y, depending on our choice of π.

A rational point $P \in X_F(F)$ corresponds to a rational section

$$s_P: Y \to X,$$

and for $y \in Y_0$ the imbedding $\{y\} \subset Y$ induces a point $s_P(y) \in X_y(k(y))$. The map

$$P \mapsto s_P(y)$$

induces a map

$$X_F(F) \to X_y(k(y))$$

called the **specialization map**. If $X = A$ is an abelian variety, then the specialization map

$$A_F(F) \to A_y(k(y))$$

is a homorphism.

Suppose that X_F is a curve of genus $\geqq 1$, and that $\varphi_F: X_F \to J_F$ is a canonical imbedding of X_F into its Jacobian over F. Then we can choose

$$\pi: X \to Y$$

as above, and it is a basic theorem that one can choose the family of Jacobians $\{J_y\}$ for the family $\{X_y\}$ $(y \in Y_0)$ to be compatible with specialization. Then we have an imbedding

$$X_F(F) \subset J_F(F),$$

and the specialization mapping on $X_F(F)$ is induced by the specialization homomorphism on $J_F(F)$.

We ask whether there exist "many" points $y \in Y(k^a)$ such that the specialization homomorphism is injective on $A_F(F)$ for an abelian variety A_F. In the case when $A_F = J_F$ is the Jacobian of a curve, for such y the specialization map

$$X_F(F) \to X_y(k(y))$$

is also injective since $X_F(F) \subset J_F(F)$. Suppose that k is a number field, and that X_F has genus $\geqq 2$. Suppose we know that $X_y(k')$ is finite for every finite extension k' of k. It then follows that $X_F(F)$ is finite, by using points y for which the specialization map is injective.

We have to make more precise what we mean by "many". One way is by means of Hilbert's irreducibility theorem, which we shall now describe.

Let $\mathbf{A}^n$ be affine space over k, with variables $T_1, \ldots, T_n$. Let u be another variable. Let

$$f(T, u) = f(T_1, \ldots, T_n, u) \in k[T, u] = k[T][u]$$

be irreducible, but not necessarily geometrically irreducible, that is, f may become reducible over the algebraic closure k^a. By a **basic Hilbert subset** S_f of $\mathbf{A}^n(k)$ we mean that set of elements $(t_1, \ldots, t_n) \in \mathbf{A}^n(k)$ such that the polynomial $f(t, u) \in k[u]$ has the same degree as f as a polynomial in u, and is irreducible over k. Such a set may of course be empty. By a **Hilbert subset** of $\mathbf{A}^n(k)$ we mean the intersection of a finite number of basic Hilbert subsets with a non-empty Zariski open subset of $\mathbf{A}^n(k)$. We say that k is **Hilbertian** if every Hilbert subset of $\mathbf{A}^n(k)$ (for all n) is not empty. Hilbert's theorem can then be stated in the form:

Theorem 7.1. *Let k be a number field. Then k is Hilbertian.*

The relevance of Hilbert sets to our specialization problem for rational points of an abelian variety over a function field arises as follows. Let

$$\pi: A \to Y$$

be the family whose generic member is a given abelian variety A_F. Let Y_0 be the Zariski open subset where π is smooth. There always exists a non-empty Zariski open subset Y_1 of Y_0 which has a finite morphism into affine space

$$\psi: Y_1 \to \mathbf{A}^n$$

and in fact such that the image of ψ contains a non-empty Zariski open subset of $\mathbf{A}^n$ (where $n = \dim Y$). Thus we obtain the morphism

$$\psi \circ \pi: A \to \mathbf{A}^n$$

whose generic fiber is again A_F. Thus we may view our family of abelian varieties $\{A_y\}$ as parametrized by $\mathbf{A}^n$.

Theorem 7.2 (Néron [Ne 52]). *Then exists a Hilbert subset $S \subset \mathbf{A}^n(k)$ such that for $t \in S$ and $y \in Y_1(k^a)$, $\psi(y) = t$, the specialization homomorphism*

$$A_F(F) \to A_y(k(y))$$

is injective.

In particular, in light of Hilbert's irreducibility theorem, we get:

Corollary 7.3. *Let k be a number field. Let $\pi: X \to Y$ be a family of curves of genus $\geqq 2$ with π defined over k. Let $\psi: Y \to \mathbf{A}^n$ be a generically finite map. Then there exists a Hilbert subset $S \subset \mathbf{A}^n(k)$ (non-empty by Hilbert's theorem) such that for $t \in S$ and $y \in Y(k^a)$ with $\psi(y) = t$ the specialization map*

$$X_F(F) \to X_y(k(y))$$

is injective.

Thus we see that if $X_y(k(y))$ is finite, so is $X_F(F)$.

The proofs of Hilbert's theorem give various quantitative measures of how large such sets are. Cf. [La 83a], Chapter IX, especially Corollary 6.3. On the other hand, a much better estimate for the set of points where the specialization map is injective follows from a theorem of Silverman which will be given in Chapter III, Theorem 2.3. For instance, if $\dim Y = 1$, for all but a finite number of points $t \in \mathbf{A}^1(k)$ the specialization map is injective.

The exposition of [La 83a], Chapter IX, Corollary 6.3, gives a more general scheme theoretic setting for the specialization theorem including reduction modulo a prime. This may be useful in certain applications. See for instance Voloch's theorem at the end of Chapter VI, §5.

I also raised the conjecture that the specialization theorem would apply to non-commutative groups, say subgroups of GL_n, which have only semisimple elements and are finitely generated.

One application of the Hilbert irreducibility theorem which has been realized since Hilbert himself is the possibility of constructing finite Galois extensions of $\mathbf{Q}$ with a given finite group G. If instead of $\mathbf{Q}$ one can construct such an extension over a purely transcendental extension $\mathbf{Q}(t_1, \ldots, t_n)$, then Hilbert's theorem implies that one can specialize the variables in $\mathbf{Q}$ in such a way that one gets the desired extension over $\mathbf{Q}$. The point is that over a field $\mathbf{Q}(t_1, \ldots, t_n)$ there may be various geometric methods which make the construction more natural than over $\mathbf{Q}$. The oldest example is when the independent variables $t_1, \ldots, t_n$ are the elementary symmetric functions of variables $u_1, \ldots, u_n$, thus giving the possibility of constructing extensions whose Galois group is the symmetric group. This point of view was taken by Emmy Noether, and has given rise in recent years to extensive theories by Belyi, Thompson and others.

Another application is to the construction of abelian varieties over $\mathbf{Q}$ with as large a rank as one can manage for the group of rational points $A(\mathbf{Q})$, which was first attempted by Néron [Ne 52], who got rank 10 for elliptic curves by this method.

CHAPTER II

Heights and Rational Points

II, §1. THE HEIGHT FOR RATIONAL NUMBERS AND RATIONAL FUNCTIONS

We wish to determine all solutions of an equation $f(x, y) = 0$, where f is a polynomial with integer coefficients, for instance, and the solutions are in some domain. Part of determining the solutions consists in estimating the size of such solutions, in various ways. For instance if x, y are to be elements of the ring of integers $\mathbf{Z}$, then we can estimate the absolute values $|x|$, $|y|$ or better the maximum $\max(|x|, |y|)$. If x, y are taken to be rational numbers, we estimate the maximum of the absolute values of the numerators and denominators of x and y, written as reduced fractions. Thus we are led to define the size in a fairly general context, involving several variables and more general domains than the integers or rational numbers. The size will be defined technically by a notion called the height. Consider projective space $\mathbf{P}^n$, and first deal with the rational numbers, so we consider $\mathbf{P}^n(\mathbf{Q})$. Let $P \in \mathbf{P}^n(\mathbf{Q})$, and let $(x_0, \ldots, x_n)$ denote projective coordinates of P. Then these projective coordinates x_j can be selected to be integers, and after dividing out by the greatest common divisor, we may assume that $(x_0, \ldots, x_n)$ are relatively prime integers. In other words, $(x_0, \ldots, x_n)$ have no prime factor in common. Then we define the **height** (or **logarithmic height**)

$$h(P) = \log \max_j |x_j|.$$

For example, take $n = 1$. A point in $\mathbf{P}^1(\mathbf{Q}) - \{\infty\}$ can be represented by coordinates $(1, x)$ where x is a rational number. Write $x = x_0/x_1$ where

x_0, x_1 are relatively prime integers. Then

$$h(1, x) = \log \max(|x_0|, |x_1|).$$

The choice of relatively prime integers to represent a point in projective space works well over the rational numbers, but does not work in more general fields, so we have to describe the height in another way, in terms of absolute values other than the ordinary absolute value. We do this fairly generally.

Let F be a field. An **absolute value** v on F is a real valued function

$$x \mapsto |x|_v = |x|$$

satisfying the following properties.

AV 1. We have $|x| \geqq 0$ and $|x| = 0$ if and only if $x = 0$.

AV 2. $|xy| = |x||y|$ for all $x, y \in F$.

AV 3. $|xy| \leqq |x| + |y|$.

If instead of **AV 3** the absolute value satisfies the stronger condition

AV 4. $|x + y| \leqq \max(|x|, |y|)$

then we shall say that it is a **valuation**, or that it is **non-archimedean**.

The absolute value which is such that $|x| = 1$ for all $x \neq 0$ is called **trivial**.

If v is an absolute value, we always denote

$$v(x) = -\log|x|_v.$$

Example 1. Let $F = \mathbf{Q}$ be the rational numbers. There are two types of non-trivial absolute values. The **ordinary absolute value**, which is said to be **at infinity**; and which we denote by v_∞; and for each prime number p the p-adic absolute value v_p which we define as follows. Let x be a rational number $\neq 0$, and write

$$x = p^r a/c \quad \text{with} \quad a, c \in \mathbf{Z} \quad \text{and} \quad p \nmid ac.$$

Then we define the ***p*-adic absolute value**

$$|x|_p = 1/p^r, \qquad \text{where} \quad r = \operatorname{ord}_p(x) = \textbf{order of } x \textbf{ at } p.$$

When we speak of the set of absolute values on $\mathbf{Q}$ we always mean the set of absolute values as above. This set of absolute values satisfies the important relation, called the (Artin–Whaples) **product formula**:

$$\prod_v |x_v| = 1 \qquad \text{for all} \quad x \in \mathbf{Q}, \quad x \neq 0.$$

We may also use additive notation. If $v = v_p$, then

$$v_p(x) = \mathrm{ord}_p(x) \log p.$$

The product formula can be rewritten as a **sum formula**:

$$\sum_v v(x) = 0 \qquad \text{for all} \quad x \in \mathbf{Q}, \quad x \neq 0.$$

It is easy to see that the height can then be defined in terms of the set of all absolute values. Let $P = (x_0, \ldots, x_n)$ be a point in $\mathbf{P}^n(\mathbf{Q})$. Then

$$h(P) = \sum_v \log \max_j |x_j|_v.$$

In that form, the height generalizes. Let F be a field with a set of non-trivial absolute values $\mathbf{V}(F) = \{v\}$. For each $v \in \mathbf{V}(F)$, let m_v be a positive integer. We say that $\mathbf{V}(F)$ satisfies the **product formula with multiplicities** m_v if we have

$$\sum_{v \in \mathbf{V}(F)} m_v v(x) = 0 \qquad \text{for all} \quad x \in F, \quad x \neq 0.$$

We can then define the **height** of a point $P = (x_0, \ldots, x_n) \in \mathbf{P}^n(F)$ by the same formula as before:

$$h(P) = \sum_{v \in \mathbf{V}(F)} m_v \log \max |x_j|_v.$$

By the product formula, the expression on the right-hand side is the same for all equivalent $(n+1)$-tuples representing a point in projective space. We give another fundamental example besides the rational numbers.

Example 2. Let k be a field and let t be a variable over that field. Then we can form the polynomial ring $k[t]$, and its quotient field is the field of rational functions

$$\mathbf{F} = k(t).$$

Let $p = p(t)$ be an irreducible polynomial (of degree $\geqq 1$) with leading coefficient 1 in $k[t]$. For each rational function $\varphi(t) \in k(t)$ we write

$$\varphi(t) = p(t)^r f(t)/g(t)$$

where $f(t)$, $g(t) \in k[t]$ are polynomials and $p \nmid fg$. We may define the absolute value v_p by

$$|\varphi|_p = 1/e^{r \deg p}.$$

On the other hand, we define v_∞ by

$$|\varphi|_\infty = e^{\deg \varphi}.$$

We let $\mathbf{V}$ be the set of absolute values v_∞, v_p for all irreducible polynomials p with leading coefficient 1. Then $\mathbf{V}$ satisfies the product formula.

Suppose the field k is algebraically closed. Then irreducible polynomials of degree $\geqq 1$ and leading coefficient 1 are simply the linear polynomials $t - \alpha$ with $\alpha \in k$.

In general, just as for the rational numbers, a point of $\mathbf{P}^n(\mathbf{F})$ can be represented by coordinates $(f_0, \ldots, f_n)$ where $f_0, \ldots, f_n$ are polynomials which are relatively prime, i.e. have no common irreducible factor. Then the height is given by

$$h(f_0, \ldots, f_n) = \max_j \deg f_j.$$

Suppose that k is a finite field with q elements. Then one can normalize the absolute values v_p as follows. Let

$$k(v_p) = k[t]/(p)$$

be the residue class field. Then $k(v_p)$ is also a finite field, with $q^{\deg p}$ elements. We may define

$$v_p(f) = \operatorname{ord}_p(f) \log \#\big(k(v_p)\big),$$

where $\#$ denotes the number of elements of a set. Similarly, we normalize v_∞ by

$$v_\infty(f) = (\deg f) \log \#(k) = (\deg f) \log q.$$

Thus we multiply all absolute values by $\log \#(k)$. Such a normalization brings the shape of these absolute values in even closer analogy to those defined on the rational numbers.

Example 3. Let X be a projective variety, non-singular in codimension 1, and defined over a field k, which we take to be algebraically closed for simplicity. If W is a subvariety of codimension 1, let $\mathcal{O}_W$ be the subring of $k(X)$ consisting of rational functions which are regular at a generic point of W. Then $\mathcal{O}_W$ is a discrete valuation ring, and consequently if $x \in k(X)$ then we have the order $\operatorname{ord}_W(x)$ of x at this discrete valuation. Furthermore, W is a projective variety, and as such has a projective degree, where

$$\deg(W) = \text{intersection number } (W \,.\, L)$$

where L is a sufficiently general linear variety of projective space, whose codimension is $\dim(W)$. Then we can define an absolute value by

$$v_W(x) = \operatorname{ord}_W(x) \deg(W).$$

The set of all such absolute values as W ranges over all subvarieties of X of codimension 1 satisfies the product formula. We may call this example the **higher dimensional function field case**. The height arising from this product formula may be described also in the following ways. We let $F = k(X)$.

(a) Let $P = (y_0, \ldots, y_n)$ be a point in $\mathbf{P}^n(F)$. Then

$$h_X(P) = \sup_i \deg\,(y_i)_\infty$$

where $(y)_\infty$ is the divisor of poles of a rational function y.

(b) Let $f\colon X \to \mathbf{P}^n$ be the rational map given by the rational functions $(y_0, \ldots, y_n)$. Then

$$h_X(P) = \deg f^{-1}(E)$$

for any sufficiently general hyperplane E in $\mathbf{P}^n$. The degree is the degree in the projective space in which X is imbedded.

The above analogy between rational numbers and rational functions has been one of the most fruitful in mathematics because certain results for the rational numbers can be discovered by this analogy, and in the past, their analogues for rational functions can be proved in an easier way than for the rational numbers. In fact, some of the results proved for rational functions are still unknown for the rational numbers. We shall see several examples of such results.

A significant diophantine example: the *abc* conjecture. This conjecture evolved from the insights of Mason [Mas], Frey [Fr], Szpiro and others. Mason started a trend of thoughts by discovering an entirely new relation among polynomials, in a very original work as follows. Let $f(t)$ be a polynomial with coefficients in an algebraically closed field of characteristic 0. We define

$$n_0(f) = \text{number of distinct zeros of } f.$$

Thus $n_0(f)$ counts the zeros of f by giving each of them multiplicity one.

Theorem 1.1 (Mason's theorem). *Let $a(t)$, $b(t)$, $c(t)$ be relatively prime polynomials not all constant such that $a + b = c$. Then*

$$\max \deg\{a, b, c\} \leqq n_0(abc) - 1.$$

Note that the left-hand side in Mason's inequality is just the height $h(a, b, c)$. In the statement of Mason's theorem, observe that it does not matter whether we assume a, b, c relatively prime in pairs, or without common prime factor for a, b, c. These two possible assumptions are equivalent by the equation $a + b = c$. Also the statement is symmetric in a, b, c and we could have rewritten the equation in the form

$$a + b + c = 0.$$

The proof is quite easy, see also the exposition in [La 90c]. As an application, let us show how Mason's theorem implies:

Fermat's theorem for polynomials. *Let $x(t)$, $y(t)$, $z(t)$ be relatively prime polynomials such that one of them has degree $\geqq 1$, and such that*

$$x(t)^n + y(t)^n = z(t)^n.$$

Then $n \leqq 2$.

Indeed, by Mason's theorem, we get

$$\deg x(t)^n \leqq \deg x(t) + \deg y(t) + \deg z(t) - 1,$$

and similarly replacing x by y and z on the left-hand side. Adding, we find

$$n(\deg x + \deg y + \deg z) \leqq 3(\deg x + \deg y + \deg z) - 3.$$

This yields a contradiction if $n \geqq 3$.

By applying Mason's theorem similarly, we get a theorem of Davenport [Da 65].

Let $f(t)$, $g(t)$ be nonconstant polynomials such that $f(t)^3 - g(t)^2 \neq 0$. Then

$$\deg\big(f(t)^3 - g(t)^2\big) \geqq \tfrac{1}{2} \deg f + 1.$$

Davenport's theorem gives a lower bound for the difference between a cube and a square of polynomials.

Influenced by Mason's theorem, and considerations of Szpiro and Frey, Masser and Oesterlé formulated the *abc* conjecture for integers as follows. Let k be a non-zero integer. Define the **radical** of k to be

$$N_0(k) = \prod_{p \mid k} p$$

i.e. the product of the distinct primes dividing k. Under the analogy between polynomials and integers, n_0 of a polynomial corresponds to $\log N_0$ of an integer. Thus for polynomials we had an inequality formulated additively, whereas for integers, we shall formulate the corresponding inequality multiplicatively.

The *abc* conjecture. *Given $\varepsilon > 0$, there exists a number $C(\varepsilon)$ having the following property. For any non-zero relatively prime integers a, b, c such that $a + b = c$ we have*

$$\max(|a|, |b|, |c|) \leqq C(\varepsilon) N_0(abc)^{1+\varepsilon}.$$

Unlike the polynomial case, it is necessary to have the ε in the formulation of the conjecture, and the constant $C(\varepsilon)$ on the right-hand side. Also the *abc* conjecture is unproved today. The conjecture implies that many or large prime factors of *abc* occur to the first power, and that if some primes occur to high powers, then they have to be compensated by "large primes", or many primes, occurring to the first power.

By a similar argument as for polynomials, one sees that the *abc* conjecture implies the Fermat problem for n sufficiently large. In other words, there exists an integer n_1 such that for all $n \geqq n_1$ the equation

$$x^n + y^n = z^n$$

has only solutions with $x = 0$ or $y = 0$ or $z = 0$ in integers x, y, z. The determination of n_1 depends on the constant $C(\varepsilon)$, or even $C(1)$, taking $\varepsilon = 1$ for definiteness. Today, there is no conjecture as to what $C(1)$, or $C(\varepsilon)$ is like, or how $C(\varepsilon)$ behaves as $\varepsilon \to 0$.

The analogue of Davenport's theorem is Hall's conjecture [Ha 71].

Hall's conjecture. *Let x, y be integers such that $x^3 - y^2 \neq 0$. Then*

$$|x^3 - y^2| \geqq C_1(\varepsilon)|x|^{(1/2)-\varepsilon}.$$

for some constant $C_1(\varepsilon)$, depending only on ε.

This conjecture also follows directly from the *abc* conjecture, as do others of similar type, for expressions of higher degree. Instead of viewing the Hall conjecture as giving a lower bound for a certain expression, we can also view it as giving an upper bound for solutions of the equation

$$x^3 - y^2 = b,$$

where b varies over the integers $\neq 0$. Then the conjecture states that

$$|x| \leqq C_2(\varepsilon)|b|^{2+\varepsilon}.$$

Such equations, cubic equations, already provide difficult unsolved problems. For the most general cubic, one has:

Lang–Stark conjecture. *Consider the equation $y^2 = x^3 + ax + b$ with a, $b \in \mathbf{Z}$ and with $-4a^3 - 27b^2 \neq 0$. Then for x, $y \in \mathbf{Z}$ we have*

$$|x| \leqq C_3(\varepsilon) \max(|a|^3, |b|^2)^{5/3+\varepsilon},$$

except for a finite number of families of cases for which x, y, a, b are polynomials satisfying the above equation. (*See* [La 83b], *Conjecture* 5.) *In particular, there should exist positive numbers k and C such that in all cases, we have*

$$|x| \leqq C \max(|a|^3, |b|^2)^k.$$

Elkies has found one example showing that $k \geqq 2$, and in particular $k > 5/3$, namely:

$$QY^2 = X^3 + AX + B \quad \text{with} \quad Q = 9t^2 - 10t + 3, \quad A = 33,$$
$$B = -18(8t - 1)$$
$$X = 324t^4 - 360t^3 + 216t^2 - 84t + 15$$
$$Y = 36(54t^5 - 60t^4 + 45t^3 - 21t^2 + 6t - 1).$$

Elkies rescales the equation by replacing (A, B, X) by $(4A = 132, 8B, 2X)$, which yields integral points provided $2Q$ is a square. That equation is satisfied by $t = 1$ and thus by infinitely many t, yielding an infinite family of solutions (b, x, y) to

$$y^2 = x^3 + 132x + b \qquad \text{with} \quad x \sim 2^{-25}3^{-4}b^4.$$

The small factor $2^{-25}3^{-4}$ means that one sees the exponent k nearing 2 only for large values of the parameter t.

No other example of such a family is known, and there is no conjecture at this time as to what would constitute all such families. The Lang–Stark conjecture is a consequence of Vojta's conjectures [Vo 87], as in the discussion following Conjecture 5.5.5.1. We shall describe Vojta's conjectures later, in §4.

By the way, the expression $-4a^3 - 27b^2$ is called the **discriminant** D of the polynomial $x^3 + ax + b$. If we have a factorization

$$x^3 + ax + b = (x - \alpha_1)(x - \alpha_2)(x - \alpha_3)$$

in the complex numbers, then

$$D = -4a^3 - 27b^2 = ((\alpha_2 - \alpha_1)(\alpha_3 - \alpha_1)(\alpha_3 - \alpha_2))^2.$$

Thus the condition $D \neq 0$ is equivalent to the condition that the three roots of the polynomial are distinct.

The Szpiro conjecture has to do with this discriminant. We state it in a generalized form.

Generalized Szpiro conjecture. *Fix integers $A, B \neq 0$. Let u, v be relatively prime integers, and let $k = Au^3 + Bv^2 \neq 0$. Then*

$$|u| \leqq C(A, B, \varepsilon)N_0(k)^{2+\varepsilon} \quad \text{and} \quad |v| \leqq C(A, B, \varepsilon)N_0(k)^{3+\varepsilon}.$$

It is an exercise to show that the generalized Szpiro conjecture is equivalent with the *abc* conjecture. To do this one uses Frey's idea, which is to associate with each solution of the equation $a + b = c$ the **Frey polynomial**

$$t(t - a)(t + b).$$

The discriminant of this polynomial is $(abc)^2$. Szpiro actually only made the conjecture

$$|D| \leqq C(\varepsilon)N_0(D)^{6+\varepsilon},$$

where $D = -4a^3 - 27b^2$ is a discriminant $\neq 0$, and a, b are relatively prime. In fact, Szpiro made the conjecture not even quite in this form, but as a function of a more subtle invariant $N(D)$ instead of $N_0(D)$, called the **conductor**. See [Fr]. This conductor is irrelevant for our purposes here.

II, §2. THE HEIGHT IN FINITE EXTENSIONS

Let F be a field with an absolute value v. We may form the completion of F. For instance, let $F = \mathbf{Q}$ and let $v = v_\infty$ be the ordinary absolute value. Then the completion is the field of real numbers $\mathbf{R}$. The construction of $\mathbf{R}$ can be generalized as follows. We let

R = ring of Cauchy sequences of elements of F.

M = maximal ideal of null sequences, i.e. sequences $\{x_n\}$ such that $\lim |x_n|_v = 0$.

Then R/M is a field, to which we can extend the absolute value by continuity. This construction works just as well starting with any field F and any non-trivial absolute value v. The completion is denoted by F_v. It is a fact that the absolute value on the completion extends in a unique way to an absolute value on the algebraic closure of the completion. If K is a field, we denote its algebraic closure by K^a. Thus we obtain an absolute value on F_v^{a}, which may not be complete. We may then form the completion of F_v^{a}, which we denote by $\mathbf{C}_v$. It is a fact that $\mathbf{C}_v$ is algebraically closed.

Example 1. If $v = v_\infty$ on $\mathbf{Q}$, then $\mathbf{Q}_v = \mathbf{R}$ and $\mathbf{C}_v = \mathbf{C}$ is just the field of complex numbers. However, if $v = v_p$ is p-adic, then $\mathbf{Q}_{v_p} = \mathbf{Q}_p$ is called a ***p*-adic field**, and its algebraic closure $\mathbf{Q}_p^{\mathrm{a}}$ is an infinite extension.

Example 2. Let $\mathbf{F} = F_0(t)$ be the field of rational functions, and let $p(t) = t$. Let $v = v_p$. Then the completion F_v is the field of power series $F_0((t))$, consisting of all power series

$$f(t) = \sum_{n=r}^{\infty} a_n t^n$$

where r may be positive or negative integer. If $a_r \neq 0$ then

$$\mathrm{ord}_p(f) = v_p(f) = r.$$

On the other hand, let $v = v_\infty$. Put $u = 1/t$. Then $\mathbf{F}_v = F_0((u)) = F_0((1/t))$ is the field of power series in $1/t$.

Let $\mathbf{F}$ be a field with a set of absolute values $\mathbf{V} = \mathbf{V}(\mathbf{F})$ satisfying the product formula. The two standard examples are the cases when $\mathbf{F} = \mathbf{Q}$ and $\mathbf{F} = k(t)$ as in §1. We shall consider finite extensions of $\mathbf{F}$. A finite extension of $\mathbf{Q}$ is called a **number field**, and a finite extension of $k(t)$ is called a **function field**, or if necessary to make more precise, a **function field in one variable**.

Example 3. This is the higher dimensional version of Example 2, when $\mathbf{F} = k(x_1, \ldots, x_n)$ is the field of rational functions in n variables. We may view $\mathbf{F}$ as the function field of $\mathbf{P}^n$, and the absolute values are in bijection with irreducible homogeneous polynomials in the homogeneous variables $T_0, \ldots, T_n$ such that, for instance, $x_i = T_i/T_0$.

Let F be a finite extension of $\mathbf{F}$. We let $\mathbf{V}(F)$ be the set of absolute values on F which extend those in $\mathbf{V}$. Hence if F is a number field, we let $\mathbf{V}(F)$ be the set of absolute values which extend v_∞ or v_p on $\mathbf{Q}$, for some prime number p. We can describe elements $v \in \mathbf{V}(F)$ as follows. Let v extend the absolute value $v_0 \in \mathbf{V}(\mathbf{F})$. Then the completion F_v is a finite extension of the completion $\mathbf{F}_{v_0}$. We also write $\mathbf{F}_v = \mathbf{F}_{v_0}$ for simplicity. We have the field

$$\mathbf{C}_v = \text{completion of the algebraic closure of } \mathbf{F}_v.$$

There is an imbedding

$$\sigma_v \colon F \to \mathbf{C}_v$$

and v is induced by the absolute value on $\mathbf{C}_v$. Conversely, given such an imbedding $\sigma: F \to \mathbf{C}_v$ we let v_σ be the induced absolute value. Two imbeddings $\sigma_1, \sigma_2: F \to \mathbf{C}_v$ induce the same absolute value on F if and only if there exists an isomorphism τ of F_v (leaving $\mathbf{F}_v$ fixed) such that

$$\sigma_2 = \tau \circ \sigma_1.$$

Example. Suppose F is a finite extension of $\mathbf{Q}$, that is, F is a number field, and v extends the absolute value at infinity. Then v corresponds either to an imbedding of F into the real numbers, or v corresponds to a pair of complex conjugate imbeddings of F into the complex numbers.

For each absolute value v on F we have what we call the **local degree** $[F_v : \mathbf{F}_v]$. We also have the **global degree** $[F : \mathbf{F}]$. These are the degrees of the finite extensions F_v over $\mathbf{F}_v$ and F over $\mathbf{F}$ respectively. In the standard Examples 1, 2 and 3, it is not difficult to prove that if v_0 is an absolute value on $\mathbf{F}$ (i.e. lies in $\mathbf{V}$) then

$$(*) \qquad \sum_{v|v_0} [F_v : \mathbf{F}_v] = [F : \mathbf{F}].$$

The symbol $v|v_0$ means that the restriction of v to $\mathbf{F}$ is v_0, and the sum is taken over all $v \in \mathbf{V}(F)$ such that $v|v_0$. In general, if every absolute value v_0 of V has the property $(*)$, then we say that V is a **proper set of absolute values**. *From now on, we assume that this is the case.*

As a result, it follows that the set of absolute values $\mathbf{V}(F)$ satisfies the product formula with multiplicities $[F_v : \mathbf{F}_v]$, namely

$$\sum_{v \in \mathbf{V}(F)} [F_v : \mathbf{F}_v] v(x) = 0 \qquad \text{for all} \quad x \in F, \quad x \neq 0.$$

We may therefore define the **height** of a point in projective space $\mathbf{P}^n(F)$. Let $P = (x_0, \ldots, x_n)$ with $x_j \in F$, not all $x_j = 0$. We define

$$h(P) = \frac{1}{[F : \mathbf{F}]} \sum_{v \in \mathbf{V}(F)} [F_v : \mathbf{F}_v] \log \max_j |x_j|_v.$$

The factor $1/[F : \mathbf{F}]$ in front has been put there so that the value $h(P)$ is independent of the field F in which the coordinates $x_0, \ldots, x_n$ lie.

Note that if $F = \mathbf{Q}$ and $x_0, \ldots, x_n$ are relatively prime integers, then

$$h(P) = \log \max |x_j|$$

so we get the same height discussed in §1.

In any case, we now view the height

$$h: \mathbf{P}^n(\mathbf{F}^a) \to \mathbf{R}$$

as a real valued function on the set of point in $\mathbf{P}^n$, algebraic over the ground field $\mathbf{F}$. When $\mathbf{F} = \mathbf{Q}$, the height is a function on the set of all algebraic points, i.e. points whose coordinates are algebraic numbers. Aside from the intrinsic interest of knowing about algebraic points, we are interested in algebraic points because sometimes the study of rational points, i.e. points in the rational numbers $\mathbf{Q}$, for a family of equations, can be reduced to the study of algebraic points for a single equation. We shall see examples of this phenomenon later.

Since the expression $[F_v : \mathbf{F}_v]v(x)$ occurs quite frequently, we shall use the notation

$$\|x\|_v = |x|_v^{[F_v : \mathbf{F}_v]}.$$

Then

$$[F_v : \mathbf{F}_v]v(x) = -\log \|x\|_v.$$

Number fields

Suppose F is a number field. It is sometimes useful to deal with the **multiplicative height relative to** F, that is, we define

$$H_F(P) = \exp\big([F : \mathbf{Q}]h(P)\big) = \prod_{v \in \mathbf{V}(F)} \max_j |x_j|_v^{[F_v : \mathbf{Q}_v]}.$$

$$= \prod_{v \in \mathbf{V}(F)} \max_j \|x_j\|_v.$$

Writing the height multiplicatively suggests more directly certain bounds for algebraic numbers. Just as a rational number is a quotient of integers, we have a similar representation for algebraic numbers as follows.

An algebraic number x is said to be an **algebraic integer** if x is a root of a polynomial equation

$$x^n + a_{n-1}x^{n-1} + \cdots + a_0 = 0 \qquad \text{with} \quad a_i \in \mathbf{Z}, \quad n \geqq 1.$$

Given an algebraic number α, there exists an integer $c \in \mathbf{Z}$, $c \neq 0$ such that $c\alpha$ is an algebraic integer. We can take for c the leading coefficient in the irreducible equation for α over $\mathbf{Z}$. The set of algebraic integers in a number field F is a subring, called the **ring of algebraic integers**, generalizing the ring of ordinary integers $\mathbf{Z}$. This ring is denoted by $\mathfrak{o}_F$. It can be shown that $\mathfrak{o}_F$ is a free module over $\mathbf{Z}$, or rank $[F : \mathbf{Q}]$. In other words, letting $d = [F : \mathbf{Q}]$, there exists a basis $\{\alpha_1, \ldots, \alpha_d\}$ of $\mathfrak{o}_F$ over $\mathbf{Z}$.

Let F be a number field. Usually, one denotes by $r_1 = r_1(F)$ the number of real imbeddings of F, and by $r_2 = r_2(F)$ the number of pairs of complex conjugate imbeddings. Then

$$r_1 + 2r_2 = [F : \mathbf{Q}].$$

By a basic theorem of Dedekind, the ideals (non-zero, always) of $\mathfrak{o}_F$ admit unique factorization into prime ideals. Let $\mathfrak{a}$, $\mathfrak{b}$ be ideals of $\mathfrak{o}_F$. We say that $\mathfrak{a}$ is **linearly equivalent** to $\mathfrak{b}$, and write $\mathfrak{a} \sim \mathfrak{b}$ if there exists an element $\alpha \in F$, $\alpha \neq 0$ such that $\mathfrak{b} = \alpha\mathfrak{a}$. It can be easily shown that under multiplication, the ideal classes form a group, called the **ideal class group**. The order of this group is called the **class number** of F, and is denoted by h. (Not to be confused with the height!)

Let $\mathfrak{a}$ be an ideal. We define the **absolute norm** of $\mathfrak{a}$ to be

$$\mathbf{N}\mathfrak{a} = \text{number of elements in the residue class field } \mathfrak{o}_F/\mathfrak{a}.$$

If we have the unique factorization

$$\mathfrak{a} = \prod_{\mathfrak{p}} \mathfrak{p}^{m_{\mathfrak{p}}}$$

then it is a fact that

$$\mathbf{N}\mathfrak{a} = \prod_{p} \mathbf{N}\mathfrak{p}^{m_{\mathfrak{p}}}.$$

Let $x_0, \ldots, x_n$ be elements of $\mathfrak{o}_F$ not all 0, and let $\mathfrak{a}$ be the ideal generated by these elements. Let $P = (x_0, \ldots, x_n)$ be the corresponding point in $\mathbf{P}^n(F)$. Then we have the formula

$$H_F(P) = \mathbf{N}\mathfrak{a}^{-1} \prod_{v \in S_\infty} \max_j \|x_j\|_v,$$

where S_∞ is the set of absolute values on F extending v_∞ on $\mathbf{Q}$.

Let $U = \mathfrak{o}_F^*$ be the group of units. Let S_∞ be the set of absolute values at infinity of F. The map

$$u \mapsto (\ldots, \log \|x\|_v, \ldots)_{v \in S_\infty}$$

is a homomorphism of U into $\mathbf{R}^{r_1+r_2}$ whose image is contained in the hyperplane consisting of those elements such that the sum of the coordinates is 0, by the product formula. Thus the image is contained in a euclidean space of dimension $r = r_1 + r_2 - 1$. Dirichlet's unit theorem states that the image is a lattice in this space, and the **regulator** is the volume of a fundamental domain for this lattice.

Let $\{\alpha_1, \ldots, \alpha_d\}$ be a basis for $\mathfrak{o}_F$ over $\mathbf{Z}$. Let $\sigma_1, \ldots, \sigma_d$ be the distinct imbeddings of F into $\mathbf{C}$. Then the **discriminant** of F is

$$D_F = (\det \sigma_i \alpha_j)^2, \qquad \text{and} \qquad \mathbf{D}_F = |D_F|.$$

We define the (normalized) **logarithmic discriminant** to be

$$d(F) = \frac{1}{[F:\mathbf{Q}]} \log \mathbf{D}_F.$$

The logarithmic discriminant will be useful later, and we immediately list two properties which are used frequently:

if $F_1 \subset F_2$ *then* $d(F_1) \leqq d(F_2)$;
for any number fields F_1, F_2 *we have* $d(F_1 F_2) \leqq d(F_1) + d(F_2)$.

Finally, the **zeta function** of F is defined for complex numbers s with $\operatorname{Re}(s) > 1$ by the formula

$$\zeta_F(s) = \sum_{\mathfrak{a}} \frac{1}{\mathbf{N}\mathfrak{a}^s} = \prod_{\mathfrak{p}} (1 - \mathbf{N}\mathfrak{p}^{-s})^{-1}.$$

The sum is taken over all (non-zero) ideals $\mathfrak{a}$ of $\mathfrak{o}_F$, and the product is taken over all (non-zero) prime ideals $\mathfrak{p}$. The zeta function gives sometimes a convenient analytic garb for some relations between the notions we have defined. Specifically:

Theorem 2.1. *The zeta function has an analytic continuation to a function which is holomorphic on all of* $\mathbf{C}$ *except for a simple pole at* $s = 1$. *The residue at this pole is*

$$\frac{2^{r_1}(2\pi)^{r_2} hR}{w\mathbf{D}^{1/2}}$$

where

$h = h_F$ *is the class number*;
$R = R_F$ *is the regulator*;
$w = w_F$ *is the number of roots of unity in* F;
$\mathbf{D} = \mathbf{D}_F$ *is the absolute value of the discriminant.*

We shall apply further these notations of algebraic number theory to the height. First we have a completely elementary result due to Northcott.

Theorem 2.2. *There is only a finite number of algebraic numbers of bounded height and bounded degree over* $\mathbf{Q}$. *More generally, there is only a finite number of points in projective space* $\mathbf{P}^n(\mathbf{Q}^a)$ *of bounded height and bounded degree.*

The main point of this theorem is that it is uniform in the degree and does not only concern the points in $\mathbf{P}^n(F)$ for some fixed number field F. The idea of the proof is that if we bound all absolute values of an algebraic integer of degree d, then we obtain a bound for the coefficients of an equation for this algebraic integer over $\mathbf{Z}$, and there is only a finite number of ordinary integers in $\mathbf{Z}$ having bounded absolute value. Since the height bounds essentially both the numerator and denominator of an algebraic number, we obtain the theorem for algebraic numbers, not just for algebraic integers.

It is interesting to give an asymptotic estimate for the number of elements of a number field F of height $\leqq B$ for $B \to \infty$. This question can be asked of elements of F, of algebraic integers in $\mathfrak{o}_F$, and of units in $\mathfrak{o}_F$. In each case, the method of proof consists of determining the number of lattice points in a homogeneously expanding domain which has a sufficiently smooth boundary, and using the following basic fact.

Proposition 2.3. *Let W be a subset of $\mathbf{R}^n$, let L be a lattice in $\mathbf{R}^n$, and let $\mathrm{Vol}(L)$ denote the euclidean volume of a fundamental domain for L. Assume that the boundary of W is $(n-1)$-Lipschitz parametrizable. Let*

$$N(t) = N(t, W, L)$$

be the number of lattice points in tW for t real > 0. Then

$$N(t) = \frac{\mathrm{Vol}(W)}{\mathrm{Vol}(L)} t^n + O(t^{n-1}),$$

where the constant implicit in O depends on L, n, and the Lipschitz constants.

For the proof, see [La 64] or [La 70], Chapter VI, §2. The expression **$(n-1)$-Lipschitz parametrizable** means that there exists a finite number of mappings

$$\rho\colon [0, 1]^{n-1} \to \text{Boundary of } W$$

whose images cover the boundary of W, and such that each mapping satisfies a Lipschitz condition. In practice, such mapping exist which are even of Class C^1, that is with continuous partial derivatives.

Theorem 2.4. *Let F be a number field. Let $r = r_1 + r_2 - 1$.*

(i) *The number of algebraic integers $x \in \mathfrak{o}_F$ with height $H_F(x) \leqq B$ is*

$$\gamma_0 B(\log B)^r + O\big(B(\log B)^{r-1}\big)$$

for some constant $\gamma_0 > 0$.

(ii) *The number of units $u \in \mathfrak{o}_F^*$ with $H_F(u) \leqq B$ is*

$$\gamma_0^*(\log B)^r + O(\log B)^{r-1}$$

for some constant $\gamma_0^* > 0$.

Both parts come from a straightforward application of Proposition 2.3, and both constants are easily determined. On the other hand, Schanuel [Sch 64], [Sch 79], has determined the somewhat harder asymptotic behavior of field elements of bounded height in projective space as follows.

Theorem 2.5. *Let $N_F(B)$ be the number of elements $x \in \mathbf{P}^{n-1}(F)$ with height $H_F(x) \leqq B$. Let $d = [F : \mathbf{Q}]$. Then:*

$$N_F(B) = \frac{hR/w}{\zeta_F(n)} \gamma_{F,n} B^n + \begin{cases} O(B \log B) & \text{if } \ d = 1, n = 2 \\ O(B^{n-1/d}) & \text{otherwise.} \end{cases}$$

The constant $\gamma_{F,n}$ has the value

$$\gamma_{F,n} = \left(\frac{2^{r_1}(2\pi)^{r_2}}{\mathbf{D}_F^{1/2}} \right)^n n^r$$

Theorem 2.5 generalizes the classical fact that the number of relatively prime pairs of integers of absolute value $\leqq B$ is

$$\frac{6}{\pi^2} B^2 + O(B \log B).$$

For the setting of Schanuel's counting in a more general (conjectural) context, see Chapter *X*, §3. For the counting of values of binary forms see [May 64], following work of Siegel and Mahler.

II, §3. THE HEIGHT ON VARIETIES AND DIVISOR CLASSES

Throughout this section, we let **F** *be a field with a proper set of absolute values satisfying the product formula. The height of points in* $\mathbf{P}^n(F)$ *for finite extensions* F *of* **F** *is then defined as in* §2.

The fundamental theorem about the relation between heights and divisor classes runs as follows, and is mostly due to Weil [We 28], [We 51].

Theorem 3.1. *Let X be a projective variety, defined over the algebraic closure* $\mathbf{F}^a$. *To each Cartier divisor class $c \in \mathrm{Pic}(X)$ one can associate in one and only one way a function*

$$h_c \colon X(\mathbf{F}^a) \to \mathbf{R},$$

well defined modulo bounded functions (*i.e.* mod $O(1)$), *satisfying the following properties:*

The map $c \mapsto h_c$ *is a homomorphism* mod $O(1)$.
If $f: X \to \mathbf{P}^m$ *is a projective imbedding, and* c *is the class of* $f^{-1}(H)$ *for some hyperplane* H, *then* modulo $O(1)$, *we have*

$$h_c(P) = h(f(P)) = \textit{height of } f(P) \textit{ as described in §2.}$$

This height association satisfies the additional property that if

$$g: X \to Y$$

is a morphism of varieties defined over $\mathbf{F}^a$, *then for* $c \in \operatorname{Pic}(Y)$ *we have*

$$h_{g^*c} = h_c \circ g + O(1).$$

As a matter of notation, if D is a divisor and c its class, we write this class as c_D if we want to make the reference to D explicit, and we also write

$$h_D = h_c.$$

Thus h_D depends only on the Cartier divisor class of D, mod $O(1)$.

Remark. Although I find it convenient here and elsewhere to use the language of divisor classes, we shall also deal with line sheaves, especially in Chapter VI. It is an elementary observation from algebraic geometry (to be recalled more explicitly later) that $\operatorname{Pic}(X)$ is naturally isomorphic to the group of isomorphism classes of line sheaves. Thus if $\mathscr{L}$ is a line sheaf corresponding to the divisor class c, we also write

$$h_{\mathscr{L}} \text{ instead of } h_c$$

for the corresponding height.

The above theorem gives the basic properties of the height in its relation to divisors and the operation of addition, as well as morphisms. We also want to describe the positivity properties of the height.

Observe that the height on projective space as defined in §2 is always $\geqq 0$. By Theorem 3.1, this height is the same as the height h_D, where D is a hyperplane. On any projective variety we have the following theorem.

Theorem 3.2. *If* c *is ample, then* $h_c \geqq -O(1)$, *in other words we can choose* h_c *in its class modulo bounded functions such that* $h_c \geqq 0$. *Furthermore, let* c *be an ample class, and let* c' *be any class. Choose* h_c

such that $h_c \geqq 0$. *Then there exist numbers* $\gamma_1, \gamma_2 > 0$ *such that*

$$|h_{c'}| \leqq \gamma_1 h_c + \gamma_2.$$

Theorem 3.3 is immediate from Theorem 3.1.

Theorem 3.2 translates the strongest positivity property of divisors into a property of the associated height. However, there is also a property corresponding to the weaker notion of effectivity.

Theorem 3.3. *Suppose that* X *is projective and non-singular and that the class* c *contains an effective divisor* D. *Then one can choose* h_c *in its class modulo bounded functions such that*

$$h_c(P) \geqq 0$$

for all $P \in X(\mathbf{F}^a)$, $P \notin \operatorname{supp}(D)$.

Thus, roughly speaking, we may say that the association $c \mapsto h_c$ preserves all the standard operations on divisor classes: the group law, inverse images and positivity. Geometric relations between divisor classes thus give rise to relations between their height functions. We shall see especially significant examples of such relations when dealing with abelian varieties.

The relation of algebraic equivalence gives rise to a height relation when the variety is defined over our fields F. Cf. [La 60a] and [La 83a], Chapter 4, Proposition 3.3, and Chapter 5, Proposition 5.4.

Theorem 3.4. *Let* X *be projective non-singular, defined over* $\mathbf{F}^a$. *Let* $c \in \operatorname{Pic}_0(X)$ *be algebraically equivalent to* 0, *and let* E *be an ample divisor. Then*

$$h_c = o(h_E) \quad \text{on } X(\mathbf{F}^a), \qquad \text{for} \quad h_E \to \infty.$$

In fact, selecting $h_E \geqq 0$, *we have*

$$h_c = O(h_E^{1/2}) + O_{c,E}(1).$$

Example. Let X be a curve. Two divisors on X are algebraically equivalent if and only if they have the same degree. For classes c_1, c_2 which are algebraically equivalent and ample, we have

$$\lim_{h(P)\to\infty} h_{c_1}(P)/h_{c_2}(P) = 1, \qquad \text{for} \quad P \in X(\mathbf{F}^a).$$

The height h denotes the height associated with any ample class. If the height goes to infinity for one ample class, it goes to infinity for every other ample class.

Suppose that c is ample, and that X is defined over a number field F. One can try systematically to give an asymptotic formula for the number of points $N(X, c, B) = N(B)$ in $X(F)$ such that $H_c \leqq B$ for $B \to \infty$. Here we use the exponential height $H_c = \exp h_c$, and h_c should be normalized in a suitable way. Roughly speaking, the following cases emerge:

$N(B) = O(1)$, so there is only a finite number of points.
$N(B)$ grows like $\gamma B^{\alpha}(\log B)^{\alpha'}$ for some constants γ, α, α';
$N(B)$ grows like $\gamma(\log B)^{r}$.

We have seen an example for projective space in §2, with exponential growth. For abelian varieties, we shall find logarithmic growth in the next chapter. Some systematic conjectures due to Manin and others will be discussed in Chapter X, §4.

II, §4. BOUND FOR THE HEIGHT OF ALGEBRAIC POINTS

The finiteness statements for rational points on curves, both in the number field and function field case, are proved by showing that a height is bounded. In the function field case, geometric methods give an explicit bound for the heights of rational points. No such method is known today in the number field case. One difficulty today is that one does not know in an effective way whether there is a rational point or not to start with. Roughly speaking, once one knows the presence of one rational point, one has some methods for bounding the heights of other rational points, and as a result current proofs give an effective (albeit inefficient) bound for the number of rational points. A key ingredient is the following theorem of Mumford [Mu 65].

Theorem 4.1. *Let C be a non-singular curve of genus $\geqq 2$ defined over a finite extension of a field* $\mathbf{F}$ *with a set of proper absolute values satisfying the product formula. Let J be the Jacobian of C and let Γ be a finitely generated subgroup of $J(\mathbf{F}^{a})$. Let $\{P_n\}$ be a sequence of distinct points in $C \cap \Gamma$, ordered by increasing height. Let h be the height associated with an ample class, and assume that the associated quadratic form is positive non-degenerate on Γ. Then there is an integer N and a number $b > 1$ such that for all n we have*

$$h(P_{n+N}) \geqq bh(P_n).$$

Thus Mumford's theorem shows that the points of $C \cap \Gamma$ are thinly distributed, in the sense that their heights grow rapidly. The hypotheses of the theorem apply when C is a curve over a number field F and $\Gamma = J(F)$, by the Mordell–Weil theorem. For instance, in the concrete case of, say, the Fermat curve over the rationals

$$x^d + y^d = z^d$$

with $d \geqq 4$, suppose we have a sequence of solutions

$$P_n = (x_n, y_n, z_n)$$

in relatively prime integers, so that

$$H(P_n) = \max(|x_n|, |y_n|, |z_n|).$$

Then there exist numbers $a_1 > 0$ and $b > 1$ such that

$$\log H(P_n) \geqq a_1 b^n,$$

so

$$H(P_n) \geqq e^{a_1 b^n}$$

grows doubly exponentially.

We stated Mumford's theorem deliberately under quite general hypotheses to show that it is valid without making use of specific arithmetic properties of the ground field. Since only weak hypotheses are used, only a weak statement comes out, albeit a useful one. Indeed, over a number field one has a much stronger result namely the finiteness of the set of rational points. But in characteristic p, using the Frobenius element, Mumford already remarked that the rate of growth for the height that he indicated is best possible.

Note that Faltings' theorem is only vaguely related to Fermat's problem. Faltings' theorem applies to all number fields. Fermat's problem has to do specifically with the rational numbers. Over certain number fields, of course, the Fermat curve will have many other solutions besides the trivial solutions with one of the coordinates equal to 0.

In the function field case, one has the following splitting theorem stemming from [La 60a].

Theorem 4.2. *Let X be a non-singular curve of genus $\geqq 2$ defined over a function field F of characteristic* 0, *over the constant field k. Suppose $X(F)$ has infinitely many points of bounded height. Then there exists a curve X_0 defined over k such that X_0 is isomorphic to X over F, and all but a finite number of points in $X(F)$ are the images under this isomorphism of points in $X_0(k)$.*

The problem was to prove that the set of points $X(F)$ has bounded height. We shall describe some of the various geometric methods giving such bounds in Chapter IV.

In the direction of bounding the height of points on curves, we have conjectures. Suppose X is a curve defined over a field k. Let P be a point of X in some field containing k, and let $(z_1, \ldots, z_n)$ be a set of affine coordinates for P. We define

$$k(P) = k(z_1, \ldots, z_n).$$

If $\mathcal{O}_P$ is the local ring of regular functions at P and $\mathcal{M}_P$ is its maximal ideal, then we have a natural isomorphism

$$\mathcal{O}_P/\mathcal{M}_P \approx k(P).$$

Suppose X is defined over a number field F. We define

$$d(P) = d\big(F(P)\big) = d_F(P),$$

where $d\big(F(P)\big)$ is the normalized logarithmic discriminant already discussed in §2.

Vojta's conjecture 4.3. *Let X be a curve defined over a number field F. Let K be the canonical class of X. Given $\varepsilon > 0$, for all algebraic points $P \in X(\mathbf{Q}^{\mathrm{a}})$ we have*

$$h_K(P) \leqq (1 + \varepsilon)\, d(P) + O_\varepsilon(1).$$

The term $O_\varepsilon(1)$ is a bounded function of P, with a bound depending only on X and ε. Observe that if P ranges only over $X(F)$, then $d(P)$ is constant, and therefore the right-hand side is bounded, so Vojta's conjecture implies Mordell's conjecture at once, since for genus $\geqq 2$ the canonical class is ample.

The conjecture is stated here with a strong uniformity for all algebraic points, and Vojta himself sometimes feels it is safer to make the conjecture uniform only with respect to points of bounded degree.

As of today, no conjecture exists describing how big $O_\varepsilon(1)$ is, as a function of ε. In particular, the conjecture does not give an effective bound for the heights of algebraic or rational points.

Although the constant $1 + \varepsilon$ is conjecturally best possible, it would already be a great result to prove the inequality

$$h_K(P) \leqq C_1\, d(P) + O(1)$$

with some constant C_1. No such result is known today, but Vojta has proved a similar result with an **arithmetic discriminant** [Vo 90b], which is however usually much larger than the discriminant (see Chapter VII, §2).

Actually, denoting by E an ample divisor, the conjecture should be written

$$h_K(P) \leqq d(P) + O(\log h_E(P)) + O(1)$$

or still better

$$h_K(P) \leqq d(P) + (1 + \varepsilon) \log h_E(P) + O_\varepsilon(1).$$

Evidence for this comes from the function field case (Chapter VI, §2) and the holomorphic case of Nevanlinna theory (Chapter VIII, §5, Theorem 5.6), for instance.

Vojta's conjecture (the weaker form for points of bounded degree) implies all the concrete diophantine problems mentioned in §1. We list some of these specifically, with additional comments.

Corollary 4.4. *There exists an integer n_1 such that the Fermat equation*

$$x^n + y^n = z^n$$

has only the trivial solutions in relatively prime integers for $n \geqq n_1$.

See [Vo 87], end of Chapter V, and Vojta's appendix [Vo 88].

The extent to which n_1 can be determined effectively depends on the effectivity of the constant in Vojta's conjecture. If it ever turns out that this constant can be determined, and is sufficiently small that the remaining cases can be given to a computer, then one would have a solution of Fermat's problem.

Part of the importance of Vojta's conjecture is that it sometimes allows one to reduce the study of rational points of a family of curves over a given field to the set of rational points of a single curve, but over algebraic extensions of this field, of bounded degree. Thus the difficulty of studying rational points is shifted. One family is that of all Fermat curves of degree n for $n \geqq 3$. Vojta's conjecture reduces the study of their rational points to the algebraic points on the single Fermat curve of degree 4, i.e.

$$x_0^4 + x_1^4 = x_2^4.$$

This reduction is carried out by associating to each solution of $x^n + y^n = z^n$ with relatively prime integers x, y, z the point on the Fermat curve of degree 4 given by

$$P = (x^{n/4}, y^{n/4}, z^{n/4}),$$

which is an algebraic point of degree $\leqq 64$. Then immediately from the definitions we have

$$h(P) = \frac{n}{4} h(x, y, z) = \frac{n}{4} \log \max (|x|, |y|, |z|).$$

It is easy to get an upper bound for the logarithmic discriminant, namely

$$d(P) = O(\log \max (|x|, |y|, |z|)),$$

whence we obtain a bound on n by Vojta's conjecture. This example shows how more sophisticated conjectures imply classical concrete questions concerning diophantine equations. Cf. [Vo 87]. We also see how a problem which appeared isolated until recently now finds its place in the context of extensive structural theories in algebraic geometry mixed with number theory.

The Vojta conjecture also implies the *abc* conjecture quite generally.

Corollary 4.5. *Let F be a number field and let $\varepsilon > 0$. There exists a constant $C(F, \varepsilon)$ such that for all $a, b, c \in \mathfrak{o}_F$ with $a + b = c$ and $abc \neq 0$ we have*

$$H_F(a, b, c) \leqq C(F, \varepsilon) \prod_{\mathfrak{p} | abc} \mathbf{N}\mathfrak{p}^{1+\varepsilon}.$$

The idea of the proof is again to fix $n = 5/\varepsilon$ (or whatever), and to consider the point

$$(a^{1/n}, b^{1/n}, c^{1/n})$$

on the Fermat curve of degree n. See [Vo 87].

In the function field case, the result is a theorem of Mason [Mas], who gives an explicit bound for the constant $C(F, \varepsilon)$, as follows.

Theorem 4.6. *Let F be a function field of one variable over an algebraically closed field k of characteristic 0. Let $x, y \in F$ but $\notin k$ be such that $x + y = 1$. Let g be the genus of F and let s be the number of distinct zeros and poles of x, y. Then*

$$h_F(x) \text{ and } h_F(y) \leqq s + 2g - 2.$$

The height here is normalized so that

$$h_F(x) = \sum_v \max (0, \operatorname{ord}_v(x))$$

where v ranges over all the discrete valuations of F over k.

Observe that Mason's theorem concerns the **unit equation** if we take the point of view that x, y lie in the finitely generated group of units in the subring of F consisting of those functions which have poles only in a finite set of places.

Mason's theorem was extended to higher dimensions by Voloch as follows.

Theorem 4.7 (Voloch [Vol 85]). *Let Y be a complete non-singular curve of genus g, defined over an algebraically closed field k of characteristic 0. Let $u_0, \ldots, u_N$ be rational functions on Y such that*

$$u_0 + \cdots + u_N = 1.$$

Assume that there is no proper nonempty subset of $u_0, \ldots, u_N$, 1 whose elements are linearly dependent over k. Let s be the number of distinct zeros and poles of $u_0, \ldots, u_N$. Let $P = (u_0, \ldots, u_N)$ be the corresponding point in projective space $\mathbf{P}^N$, and let

$$h(P) = \sum_v -\min_i (0, \operatorname{ord}_v(u_i)).$$

Then

$$h(P) \leqq \tfrac{1}{2}N(N+1)(2g-2+s).$$

Here we see the same quantity $2g - 2 + s$ which will reappear Chapter VI, §3. See also Brownawell–Masser [BM 86]. But the factor $N(N+1)/2$ is not the best possible one conjecturally. For a discussion in the context of Schmidt's theorem, see Chapter IX, §2 and [Vo 89], §8.

Next we come to Vojta's conjecture for higher dimensions.

Vojta's conjecture 4.8. *Let X be a projective non-singular variety defined over a number field. Let E be an ample or pseudo ample divisor. Given $\varepsilon > 0$, there exists a proper Zariski closed subset Z such that for all algebraic points $P \in X(\mathbf{Q}^a) - Z$ we have*

$$h_K(P) \leqq d(P) + \varepsilon h_E(P) + O_\varepsilon(1).$$

In his Lecture Notes [Vo 87] Vojta actually puts the factor dim X in front of $d(P)$, but for complementary reasons, both he and I now believe this factor is unnecessary. From my point of view, analogous cases of Nevanlinna theory have been verified to hold without this factor. From his point of view, whatever reasons he gives in [Vo 87] actually do not apply. Furthermore, according to conjectures which I made long ago in connection with diophantine approximations, the part of the error term involving the ample height h_E should also be improved, so that the full inequality should read

$$h_K(P) \leqq d(P) + \tfrac{1}{2}(1 + \varepsilon) \log h_E(P) + O_\varepsilon(1).$$

Vojta also formulates another conjecture for coverings of X. Furthermore, the conjectures as we have stated them in one and higher dimension may be called the **absolute conjectures**, or the conjectures in the **compact case**. Vojta states generalizations, which contain another positive term on the left-hand side, to deal with quasi-projective varieties, so with the non-compact case. To define such a term requires other definitions, so we preferred to state the simplest case first, and we postpone the most general formulation until we discuss Weil functions in Chapter IX.

It is an important problem to determine the **exceptional set** Z. Note that when the canonical class is negative, or $-K$ is effective or when $-K$ is ample, then the inequality is vacuous. The inequality has content as stated only when the canonical class is effective, or ample. The nature of the exceptional set Z has to do with more qualitative conjectures, which we shall discuss in Chapter VIII.

Suppose that K is ample or pseudo ample. Then we may take $E = K$. If we look only at rational points $P \in X(F)$, then the term $d(P)$ is bounded, and the Vojta conjecture implies that the set of rational points P over F which do not lie in the Zariski closed set Z has bounded height, whence is finite. Thus the only possibility to have infinitely many rational points when K is ample is when these points lie in a proper Zariski closed subset. Thus Vojta's conjecture gives a quantitative estimate for the heights of points in the case discussed qualitatively in Chapter I, §3.

Furthermore, when K is ample, then I conjectured that the exceptional set Z in Vojta's inequality is the same as the special set defined in Chapter I, namely the Zariski closure of the union of all images of non-constant rational maps of group varieties or abelian varieties into X.

CHAPTER III

Abelian Varieties

The presence of a group structure on a variety gives rise to numerous additional relations for the height, and in particular, gives rise to a quadratic function associated with every divisor class. Furthermore, the group of rational points can be analyzed as a group, with a description of generators, bounds for the heights of generators, a description of the torsion, all emphasizing the group structure. Thus we collect such results in a separate chapter.

III, §0. BASIC FACTS ABOUT ALGEBRAIC FAMILIES AND NÉRON MODELS

We shall be dealing with algebraic families of abelian varieties and of curves, so we start with a summary of basic facts and terminology about such families in a fairly general context.

Let F be a field with a discrete valuation v, valuation ring $\mathfrak{o}_v$ and maximal ideal $\mathfrak{m}_v$. Let X be a projective non-singular variety over F. We first discuss a naive notion of reduction of X at v. Let $k(v)$ be the residue class field. We describe the reduction somewhat non-invariantly by having assumed the projective imbedding. We let I be the ideal in the projective coordinate ring $F[T_0, \dots, T_n]$ defining X over F, and we let I_v be the ideal in $\mathfrak{o}_v[T_0, \dots, T_n]$ consisting of those polynomials which have coefficients in $\mathfrak{o}_v$. Then we can reduce the coefficients of polynomials in I_v mod $\mathfrak{m}_v$, to get an ideal $I_{k(v)}$ in $k(v)[T_0, \dots, T_n]$. If $I_{k(v)}$ is the prime ideal defining a non-singular variety $X_{k(v)}$ over $k(v)$, then we say that X has good reduction at v. Given any variety X' over F, we say that it has **good reduction** at v if there exists a projective variety X over F, isomorphic to X' over F, and having good reduction at v. If $X = A$ is an abelian variety, and has good reduction at v, then $A_{k(v)}$ is then also an abelian variety, and the graph of the group law on $A_{k(v)}$ is the reduction of the graph of the group law on A. Let X be an abelian variety or a curve of genus $\geqq 1$. An early theorem of Chow–Lang asserts that if X' is another projective variety over F isomorphic to X over F, and X, X' have good reduction at v, then the isomorphism between X and X' reduces to an isomorphism between $X_{k(v)}$ and $X'_{k(v)}$ over $k(v)$. Thus we may say that the good reduction is uniquely determined.

If $\mathfrak{o}$ is a Dedekind ring with infinitely many prime ideals, then it is a basic fact that for all but a finite number of the discrete valuation rings $\mathfrak{o}_v$ corresponding to these maximal ideals, a non-singular variety X over the quotient field F of $\mathfrak{o}$ has good reduction at v. We shall usually say **almost all** instead of all but a finite number.

Good reduction is expressed more invariantly in the language of schemes as follows. Let X_F again be a non-singular variety over F. Then X_F has **good reduction** at v if and only if there exists a scheme

$$X \to \operatorname{spec}(\mathfrak{o}_v)$$

which is smooth and proper, and such that the generic fiber of this scheme is the given variety X_F. We can take this latter property as the definition, quite independent of a projective imbedding. But often in practice a variety may be defined by actual equations, as when we represent a curve by an equation

$$y^2 = x^3 - \gamma_2 x - \gamma_3$$

in affine coordinates, so we do want to know what reduction means in terms of such equations. Indeed, if the characteristic of $k(v)$ is $\neq 2, 3$ and $\gamma_2, \gamma_3 \in \mathfrak{o}_v$ then the elliptic curve has good reduction if and only if Δ is a unit in $\mathfrak{o}_v$, where Δ is the discriminant,

$$\Delta = 16(4\gamma_2^3 - 27\gamma_3^2).$$

Let $S = \operatorname{spec}(\mathfrak{o})$ where $\mathfrak{o}$ is a Dedekind ring, with quotient field F. Let A_F be an abelian variety over F. By a **Néron model** of A_F over $\mathfrak{o}$ we mean a group scheme $\mathbf{A}$ satisfying the following properties.

NM 1. $\mathbf{A}$ is smooth over S.

NM 2. The generic fiber $\mathbf{A}_F$ is the given abelian variety A_F.

NM 3. For every smooth morphism $X \to S$, a morphism $X_F \to A_F$ extends uniquely to a morphism $X \to \mathbf{A}$ over S. In other words, the natural map

$$\operatorname{Mor}_S(X, \mathbf{A}) \to \operatorname{Mor}_F(X_F, A_F)$$

obtained by extending the base from S to F is a bijection, and hence an isomorphism of abelian groups.

Note that the smoothness assumption implies in particular that $\mathbf{A}$ is regular, that is, all the local rings of points on $\mathbf{A}$ are regular; and therefore any Weil divisor on $\mathbf{A}$ is locally principal. Néron [Ne 55] proved the existence of Néron models. For a more recent discussion, see [Art 86] and [BLR 90]. A group scheme over spec($\mathfrak{o}$) whose generic fiber is an abelian variety is proper over spec($\mathfrak{o}$) if and only if the abelian variety has good reduction at all primes of $\mathfrak{o}$. In particular, the Néron model is not necessarily proper over a given discrete valuation ring $\mathfrak{o}_v$. It is proper if and only if A_F has good reduction at v.

Fix the discrete valuation ring $\mathfrak{o}_v$ and let $k = k(v)$. We denote by $\mathbf{A}_k$ or A_k the special fiber over the residue class field k. Then A_k is an algebraic group over k, not necessarily connected.

By the **connected Néron model**, denoted by $\mathbf{A}^0$, we mean the open subgroup scheme of the Néron model whose fibers are the connected components of the Néron model. Thus $\mathbf{A}_k^0$ is a group variety over k. By the general structure theorem for group varieties, we know that $\mathbf{A}_k^0$ is an extension of an abelian variety by a linear group. If this linear group is a torus (i.e. a product of multiplicative groups over k^{a}) and so $\mathbf{A}_k^0$ is a semiabelian variety, then we say that A_F has **semistable reduction**. The following are basic facts about good and semistable reduction. (For good reduction see Chapter IV, Corollary 4.2.)

Let B_F be an abelian variety over F, isogenous to A_F over F. Then:
A_F has good reduction at v if and only if B_F has good reduction.
A_F has semistable reduction at v if and only if B_F has semistable reduction.
If A_F has semistable reduction then taking the connected Néron model commutes with base change.

Of course, it is trivial from the definition that taking Néron models commutes with smooth base change.

If $\mathfrak{o}$ is a Dedekind ring with quotient field F, then we say that A_F has **good** (resp. **semistable**) **reduction** over $\mathfrak{o}$ or over spec($\mathfrak{o}$) if A_F has good (resp. semistable) reduction at every local ring $\mathfrak{o}_v$ of $\mathfrak{o}$.

For proofs of facts concerning the Néron model going beyond Néron, and specifically involving semistable reduction, see Grothendieck [Grot 70]. In particular, Grothendieck also proved the **semistable reduction theorem**:

Given an abelian variety over F there is a finite extension over which it has semistable reduction.

We shall now discuss the general notion of conductor for an abelian variety over F. Let A_k be the fiber of the Néron model over $\mathfrak{o}_v$, and suppose for simplicity that k is perfect. Then as already mentioned, A_k^0 is an extension of an abelian variety by a linear group, and this linear group is an extension of a torus by what is called a **unipotent group** (a tower of extensions of additive groups). We let:

u = dimension of the unipotent part of A_k,

t = dimension of the torus in A_k.

Then one defines the **order of the conductor** at the valuation v as in Serre–Tate [SeT 68], to be

$$f(v) = 2u + t + \delta,$$

where δ is an integer $\geqq 0$, which can also be described explicitly. This description gets lengthy and technical, but it already gives insight to list the following properties satisfied by δ, which I got from Serre–Tate [SeT 68].

If A has good reduction, then $u = t = \delta = 0$.
If A acquires semistable reduction over a Galois extension of F of degree prime to p, then $\delta = 0$. This occurs if $p > 2d + 1$, where $d = \dim A$, or if $p = 0$.

In fact, δ is defined for each prime l, in terms of Artin and Swan conductors related to the wild ramification of l. Grothendieck proved the independence from l. See [Grot 72], SGA 7.

In particular for $d = 1$, we see that the condition $p > 2d + 1$ is precisely the condition that $p \neq 2, 3$ for elliptic curves. For a systematic treatment of δ see [Ogg 67] for elliptic curves, and Raynaud [Ra 64–65] in higher dimensions. For a discussion of the discriminant and conductor of curves in general, including genus bigger than 1, and applications to other theories (for instance, Arakelov theory) see Saito [Sai$_{\rm T}$ 88].

III, §1. THE HEIGHT AS A QUADRATIC FUNCTION

Let A, B be abelian varieties.

As we know, relations between divisor classes give rise to relations between their associated heights. The principal relation between divisor classes is that given $c \in \mathrm{Pic}(B)$, the association

$$\alpha \mapsto \alpha^* c \quad \text{for} \quad \alpha \in \mathrm{Hom}(A, B)$$

is quadratic in α. In other word, if we let

$$D_c(\alpha, \beta) = (\alpha + \beta)^*c - \alpha^*c - \beta^*c$$

then $D_c(\alpha, \beta)$ is bilinear in (α, β). From this fundamental relation, one obtains [Ne 55]:

Theorem 1.1 (Néron–Tate). *Let A be an abelian variety defined over a finite extension of a field $\mathbf{F}$ with a proper set of absolute values satisfying the product formula. Let $c \in \mathrm{CH}^1(A)$. There exists a unique quadratic form q_c and a linear form l_c such that*

$$h_c = q_c + l_c + O(1) \text{ as functions on } A(\mathbf{F}^{\mathrm{a}}).$$

If c is even, that is $(-1)^*c = c$, then $l_c = 0$.

The sum $q_c + l_c$ which is uniquely determined by c will be denoted by $\hat{h}_c$ and is called the **Néron–Tate height**. The bilinear form

$$(P, Q) \mapsto \hat{h}_c(P + Q) - \hat{h}_c(P) - \hat{h}_c(Q) = \langle P, Q \rangle_c$$

will be called the **height pairing** associated with the class c. Some authors normalize the bilinear form with an extra factor 1/2 in front. The quadratic form can be obtained directly from a choice of h_c (in its class mod $O(1)$) by the limit

$$q_c(P) = \lim_{m \to \infty} h_c(2^m P)/2^{2m}.$$

This was Tate's fast way of getting the form, replacing more elaborate arguments of Néron, who expressed the height as a sum of local intersection numbers (but also used the limit argument locally). Estimates for the difference between the Néron–Tate height and the naive height obtained from a projective imbedding have been given, for instance in [Dem 68] and [Zi 76] for elliptic curves.

As an immediate consequence of the positivity properties of the height, one obtains Néron's theorem:

Theorem 1.2. *If c is ample, then its associated height pairing is a semipositive bilinear form*

$$A(\mathbf{F}^{\mathrm{a}}) \times A(\mathbf{F}^{\mathrm{a}}) \to \mathbf{R}.$$

In particular, the associated quadratic form q_c is semipositive. Then we get a seminorm

$$|P|_c = q_c(P)^{1/2} \quad \text{for} \quad P \in A(\mathbf{F}^{\mathrm{a}})/A(\mathbf{F}^{\mathrm{a}})_{\mathrm{tor}},$$

which we call the **Néron–Tate seminorm** associated with c. We shall discuss the set of P for which $|P|_c = 0$ below, in the cases of number fields and function fields.

We start with the number field case.

Theorem 1.3. *Let A be an abelian variety defined over a number field.*

(i) *The map $c \mapsto \hat{h}_c$ is a homomorphism from* Pic(A) *into the group of real valued functions on $A(\mathbf{Q}^a)$ whose kernel is the group of torsion elements in* Pic(A).

(ii) *If c is ample, then the kernel of the bilinear form*

$$(P, Q) \mapsto \langle P, Q \rangle_c$$

is the torsion subgroup of $A(\mathbf{Q}^a)$. If A is defined over the number field F, then the Néron–Tate height associated with an ample even class induces a positive definite quadratic form on the finite dimensional vector space $\mathbf{R} \otimes A(F)$.

The group of rational points modulo torsion can then be viewed as a lattice in $\mathbf{R} \otimes A(F)$. Let c be an even ample class. Then we get a *norm*

$$|P|_c = \hat{h}_c(P)^{1/2} \qquad \text{for} \quad P \in A(F)/A(F)_{\text{tor}},$$

which we may call the **Néron–Tate norm** associated with c. This norm extends to a *norm* on $\mathbf{R} \otimes A(F)$. One may ask for abelian varieties the same question about the asymptotic behavior of the number of points of bounded height. Since the height here is logarithmic, and is a positive definite quadratic form, the asymptotic formula for the number of points with height $\hat{h}_c(P) \leqq \log B$ has the shape

$$N(B) = \gamma(\log B)^{r/2} + O\big((\log B)^{(r-1)/2}\big) \quad \text{for some constant } \gamma,$$

again by counting lattice points in a homogeneously expanding domain as in Proposition 2.3 of Chapter II. This kind of asymptotic behavior was already given by Néron [Ne 52].

A fundamental problem is to find a basis for the lattice consisting of points of minimal height. We shall discuss this much deeper aspect in §4, how to give an upper bound for the heights of points in such a basis. Here we mention a related question which also has independent interest, and which concerns the possible lower bounds for the Néron–Tate height on non-torsion points. Say on an elliptic curve, we have a conjecture of mine [La 78]:

Conjecture 1.4. *There exists an absolute constant $C_{\mathbf{Q}} > 0$ such that for all elliptic curves A over $\mathbf{Q}$, with minimal discriminant Δ_A, a non-torsion* *point $P \in A(\mathbf{Q})$, satisfies*

$$h(P) \geqq C_{\mathbf{Q}} \log |\Delta_A|.$$

Hindry–Silverman showed that this conjecture is implied by the *abc* conjecture [HiS 88]. For some results on lower bounds of heights on elliptic curves and abelian varieties, see Masser [Mass 81], [Mass 84], [Mass 85]. For a discussion of the relation of Conjecture 1.4 with the upper bound problem, see the end of §5.

Remark. The notion corresponding to $\log|\Delta_A|$ in higher dimension is the Faltings height, which will be defined in the next chapter, §5.

We now pass to the function field case.

We let

$$\pi: A \to Y$$

be a morphism of projective varieties over a field k, which we take to be algebraically closed for simplicity. We assume Y non-singular in codimension 1, *and we let $F = k(Y)$ be the function field.*

Then we have the height h_Y defined as in our standard Example 3 of §1. We suppose that the generic fiber

$$\pi^{-1}(\eta) = A_\eta$$

for the generic point η of Y is an abelian variety, which is thus defined over the function field $k(\eta) = k(Y) = F$. We view a rational point $P \in A_\eta(F)$ as a rational section

$$P: Y \to A.$$

This section is a morphism on a non-empty Zariski open subset of Y, and we let $P(Y)$ denote its Zariski closure in A. From [LaN 59] we get:

Proposition 1.5. *Let $\tau: B \to A_\eta$ be the F/k-trace of A_η. Let P, Q be rational sections of $\pi: A \to Y$. If $P(Y)$ and $Q(Y)$ are in the same algebraic family in A, then P, Q are congruent modulo $\tau B(k)$, that is $P - Q \in \tau B(k)$.*

The analogue of the structure Theorem 1.3 then reads:

Theorem 1.6. *Let c be a divisor class on the generic fiber A_η, containing an ample divisor and even. Then for any point $P \in A(F)$ we have*

$h_c(P) = 0$ *if and only if*

$$P \in A(F)_{\mathrm{tor}} + \tau B(k).$$

The Néron–Tate height h_c extends to a positive definite quadratic form on the finite dimensional vector space

$$\mathbf{R} \otimes (A(F)/\tau B(k)).$$

Remark. Let F' be a finite extension of F. Then it may happen that the F/k-trace is trivial but the F'/k-trace is non-trivial. If we want the theorem to be formulated for all points $P \in A(F^{\mathrm{a}})$, rational over the algebraic closure, then one must assume the stability of the trace, i.e. that we have picked F sufficiently large so that the trace remains the same for finite extensions.

Applications. Over a finite field k the group $A(F)$ is finitely generated and modulo torsion admits the Néron–Tate positive definite quadratic form. One may thus view $\mathbf{R} \otimes A(F)$ as a lattice in $\mathbf{R}^r$ (r = rank of $A(F)$), with such a form, which can be normalized so that it is $\mathbf{Z}$-valued. Shioda [Shio 89a, b] has investigated this form and the lattices which arise from it, and found that one gets certain classical lattices from the theory of linear algebraic groups.

Having described the quadratic and bilinear forms associated with a divisor class on an abelian variety, we conclude this section with a description of another form arising from the dual variety, or Picard variety, which we encountered in Chapter I, Theorem 5.3. So we let $\mathbf{F}$ again be a field with a proper set of absolute values satisfying the product formula, and we let A be an abelian variety defined over $\mathbf{F}^{\mathrm{a}}$. We let (A', δ) be the dual variety. It is a fact that δ is even, that is

$$[-1]^*(\delta) = \delta.$$

Since $\delta \in \mathrm{Pic}(A \times A')$, we have the associated height $\hat{h}_\delta$. The basic properties of $\hat{h}_\delta$ are summarized in the next theorem.

Theorem 1.7

(1) *The height $(x, y) \mapsto \hat{h}_\delta(x, y)$ is a bilinear form on $A(\mathbf{F}^{\mathrm{a}}) \times A'(\mathbf{F}^{\mathrm{a}})$.*

(2) *For $x \in A(\mathbf{F}^{\mathrm{a}})$ and $y \in A'(\mathbf{F}^{\mathrm{a}})$ we have*

$$\hat{h}_{t_{\delta(y)}}(x) = \hat{h}_\delta(x, y).$$

(3) *Let $c \in \mathrm{Pic}(A)$ and let $\varphi_c \colon A \to A'$ be the homomorphism such that $\varphi_c(a)$ = element $a' \in A'$ such that ${}^t\delta(a') = c_a - c$. Let L_δ be the*

bilinear form derived from $\hat{h}_\delta$, that is

$$L_\delta(u, v) = \hat{h}_\delta(u + v) - \hat{h}_\delta(u) - \hat{h}_\delta(v).$$

Then

$$L_c(x, y) = -\hat{h}_\delta(x, \varphi_c y).$$

(4) *If c is even, then*

$$\hat{h}_c(x) = -\tfrac{1}{2}\hat{h}_\delta(x, \varphi_c x).$$

We call $\hat{h}_\delta$ the **Néron height pairing** on $A \times A'$.

III, §2. ALGEBRAIC FAMILIES OF HEIGHTS

The theorems in this section give some description how the height can vary in an algebraic family of abelian varieties. We work under the following assumptions.

Let $\mathbf{k}$ *be a field of characteristic* 0 *with a proper set of absolute values satisfying the product formula. Let* $k = \mathbf{k}^{\mathrm{a}}$. *Let*

$$\pi\colon A \to Y$$

be a flat morphism of projective non-singular varieties defined over a finite extension of $\mathbf{k}$, *such that the generic fiber* $\pi^{-1}(\eta) = A_\eta$ *is an abelian variety. Let* Y_0 *be the non-empty Zariski open subset where* π *is smooth over* Y_0, *so for all* $y \in Y_0(k)$, *the fiber* $\pi^{-1}(y) = A_y$ *is an abelian variety.*

Let $c \in \operatorname{Pic}(A, k)$ be a divisor class. Then we have associated a height function

$$h_c\colon A(k) \to \mathbf{R}$$

well defined mod $O(1)$. On the other hand, c has a restriction to each fiber, which we denote by c_y. For each $y \in Y_0(k)$ we have the Néron–Tate height

$$\hat{h}_{c_y}\colon A_y(k) \to \mathbf{R},$$

which is a quadratic function, differing by a bounded function from the restriction of h_c to $A_y(k)$. We are interested in seeing how this bound varies for $y \in Y_0(k)$. We shall give a bound in terms of the height on Y, which is given with a projective imbedding. We let h_Y be a height of Y associated with a very ample divisor class on Y, corresponding to this imbedding.

Theorem 2.1 (Silverman–Tate [Sil 84]). *Let $c \in \mathrm{Pic}(A, k)$ be a divisor class, and h_c a choice of height function associated with c. There exist numbers*

$$\gamma_1 = \gamma_1(\pi, c) \qquad \text{and} \qquad \gamma_2 = \gamma_2(Y, h_c)$$

such that for all $y \in Y_0(k)$ and $P \in A_y(k)$ we have

$$|\hat{h}_{c_y}(P) - h_c(P)| \leqq \gamma_1 h_Y(y) + \gamma_2.$$

Example. Consider the 3-dimensional variety A defined on an affine open set in characteristic 0 by

$$y^2 = x^3 + ax + b,$$

viewing a, b as variables, and thus affine coordinates of $\mathbf{P}^2$. Let c be the divisor class of $3(O)$, where O is the zero section on an affine open subset of $\mathbf{P}^2$. Then

$$h_c(P) = h\big((x(P), y(P), 1)\big).$$

Theorem 2.1 says that for all $y = (a, b, 1)$ in $\mathbf{P}^2(k)$ with $4a^3 + 27b^2 \neq 0$ and all $P \in A_y(k)$ with $P \neq O$ we have

$$|\hat{h}_{c_y}(P) - h\big((x(P), y(P), 1)\big)| \leqq \gamma_1 h\big((a, b, 1)\big) + \gamma_2,$$

where γ_1, γ_2 are absolute constants.

For abelian varieties, Manin–Zarhin give an estimate of the same kind as in Theorem 2.1, with respect to "canonical coordinates" [MaZ 72].

The restriction to characteristic 0 in Theorem 2.1 is made only because so far the proof uses resolution of singularities. Otherwise the arguments are rather formal.

For the next theorems, we shall also assume that the Néron–Severi group of Y is cyclic. This is the case when Y is a curve, and is the most important case for applications. Under this hypothesis, the projective degree gives an imbedding

$$\deg\colon \mathrm{NS}(Y) \to \mathbf{Z}.$$

This imbedding allows us to normalize heights on Y a little more than previously. Let $b \in \mathrm{Pic}(Y)$ with $\deg b \neq 0$. We let

$$h_Y = \frac{1}{\deg b} h_b.$$

By Theorem 3.5 of Chapter II, the height associated with a class algebraically equivalent to 0 has lower order of magnitude than the height associated with an ample divisor. Therefore asymptotically,

$$\frac{1}{\deg b} h_b$$

is independent of the choice of b with $\deg b \neq 0$, because for any other class b' with $\deg b' \neq 0$, we have that $\deg(b')b$ is algebraically equivalent to $\deg(b)b'$.

For the next theorems, we shall deal with **sections**

$$P\colon Y \to A \quad \text{of } \pi,$$

by which we shall means *morphisms* such that $\pi \circ P = \mathrm{id}$. Thus we are making a more severe restriction than when dealing with rational sections earlier. If $\dim Y = 1$, that is, Y is a curve, then any rational map $P\colon Y \to A$ such that $\pi \circ P = \mathrm{id}$ generically is necessarily a section in our restricted sense (morphism), because any rational map of a non-singular curve into a projective variety is a morphism. In the higher dimensional case, this is a relatively rare occurrence, but when

$$A = A_0 \times Y$$

is a product, so the family splits, then again any rational map $P\colon Y \to A_0$ is a morphism since Y is assumed non-singular. This is a theorem of Weil.

To suggest families of points, we shall write P_y instead of $P(y)$ for $y \in Y_0(k)$.

Theorem 2.2 (Silverman [Sil 84]). *Assume that* $\mathrm{NS}(Y)$ *is cyclic. Let* $P\colon Y \to A$ *be a section such that for arbitrarily large integers* n, $[n]P$ *is also a section. (This condition is satisfied if* Y *is a curve.) Let* $c \in \mathrm{Pic}(A, k)$. *Then*

$$\lim_{\substack{h(y)\to\infty \\ y \in Y_0(k)}} \hat{h}_{c_y}(P_y)/h_Y(y) = \hat{h}_{c_\eta}(P_\eta).$$

We shall now give an application by Silverman of his height theorem. Let (B, τ) be the $k(\eta)/k$-trace of the generic fiber A_η. By the Lang–Néron theorem, we know that

$$A_\eta(k(\eta))/\tau B(k)$$

is finitely generated. For each $y \in Y_0(k)$ we have the **specialization homomorphism**

$$\sigma_y\colon A_\eta(k(\eta)) \to A_y(k), \qquad \text{denoted by} \quad P \mapsto P_y,$$

under the assumption that sections are morphisms. We want to know how often this specialization homomorphism is not injective.

Theorem 2.3 (Silverman [Sil 84]). *Assume that $NS(Y)$ is cyclic, and that rational sections are morphisms. Let $k = \mathbf{F}^a$. Let Γ be a finitely generated free subgroup of $A_\eta(k(\eta))$ which injects in the quotient*

$$A_\eta(k(\eta))/\tau B(k).$$

Then the set of $y \in Y_0(k)$ such that σ_y is not injective on Γ has bounded height in $Y_0(k)$. In particular, if $k = \mathbf{Q}^a$, there is only a finite number of points $y \in Y_0(k)$ of bounded degree over $\mathbf{Q}$ such that σ_y is not injective on Γ.

As already mentioned in Chapter I, §7 Néron had obtained a weaker specialization theorem using the Hilbert irreducibility theorem, to show that the set of points where the specialization homomorphism σ_y is not injective is thinly distributed. But Silverman actually proves that this set is finite in fields of bounded degree.

As a corollary of Theorem 2.3, one then obtains a result of Demjanenko–Manin [De 66], [Man 69]:

Corollary 2.4. *Let A_0 be an abelian variety over a number field F. Let Y be a projective non-singular variety over F, and let $y_0 \in Y(F)$ be an F-rational point. Let Γ be the group of morphisms $f\colon Y \to A_0$ such that $f(y_0) = 0$. If* NS(Y) *is cyclic, and if*

$$\operatorname{rank} \Gamma > \operatorname{rank} A_0(F)$$

then $Y(F)$ is finite.

The corollary is essentially the split case of Theorem 2.3, and follows by applying that theorem to $A = A_0 \times Y$, with $\pi\colon A \to Y$ being the projection.

We shall now obtain a refinement of Theorem 2.2.

Let Y be a non-singular curve over k. Let $F = k(Y)$, and let A_F be an abelian variety over F. We can cover Y by two (or whatever) affine open subsets which are specs of Dedekind rings. Since we defined a Néron model over a Dedeking ring, we can then define a **Néron model** of A_F **over** Y in light of the uniqueness of a Néron model on the intersection of the two affine open subsets. We let

$$\pi\colon \mathbf{A} \to Y$$

be the Néron model, and $\mathbf{A}^0$ the connected Néron model. In [La 83a] I transposed to varieties of dimension $\geqq 1$ a theorem of Tate [Ta 83] concerning heights on a family of elliptic curves. (For a treatment of related results depending on moduli spaces and Faltings techniques, see Green [Gr_W 89].) In the proofs I gave in [La 83a] I assumed the conjectured existence of a good completion for the Néron model. However, Chai has pointed out to me that one could argue in essentially the same way without using this still unproved existence, by making use of other technical means, namely cubical sheaves on the Néron model, thereby proving the results unconditionally. I shall therefore follow Chai's suggestion.

If $P\colon Y \to \mathbf{A}$ is a section, we let

$$\tau_P\colon \mathbf{A} \to \mathbf{A}$$

be the corresponding translation.

Let $O\colon Y \to \mathbf{A}$ be the zero section and $P\colon Y \to \mathbf{A}$ any section. We have trivially

$$\tau_0 \circ \tau_P = \tau_P \circ \tau_0 = \tau_P \qquad \text{and} \qquad P = \tau_P \circ O.$$

The first relation is between automorphisms of $\mathbf{A}$, and the second between maps of Y into $\mathbf{A}$.

To simplify the notation, let us abbreviate formal linear combinations as follows:

$$[\tau_P, \tau_Q] = \tau_{P+Q} - \tau_P - \tau_Q + \tau_O,$$
$$[P, Q] = (P + Q) - (P) - (Q) + (O).$$

Then for instance if $c \in \text{Pic}(\mathbf{A})$, we have by definition

$$[P, Q]^*(c) = (P + Q)^*(c) - P^*c - Q^*c + O^*c,$$

and similarly for $[\tau_P, \tau_Q]^*c$ (inverse image). These two inverse images are related in a simple way, if we look only at the part in the connected Néron model.

Proposition 2.5. *Let $c_F \in \text{Pic}(A_F)$. Then there exists an extension c of c_F to an element of $\text{Pic}(\mathbf{A}^0)$ which satisfies the following property. For $P, Q \in \text{Sec}(T, \mathbf{A}^0)$ we have*

$$[\tau_P, \tau_Q]^*c = \pi^*[P, Q]^*c.$$

This is a variation of Proposition 5.1 of [La 83a] due to Chai. As he pointed out, this reformulation of the proposition follows as a direct consequence of Breen [Br 83], §3, pp. 30–31, or also of Moret–Bailly [MorB 85], Chapter II, Theorem 1.1, p. 40. Indeed, by those references, a line sheaf $\mathscr{L}_F$ on A_F extends uniquely to a cubical sheaf $\mathscr{L}$ on $\mathbf{A}^0$. The essential part of the definition of cubical sheaf is that $\mathscr{L}$ satisfies the theorem of the cube, i.e. the line sheaf

$$\bigotimes_I (\mathbf{m}_I^* \mathscr{L})^{(-1)^{|I|}} \quad \text{on} \quad \mathbf{A}^0 \times_T \mathbf{A}^0 \times_T \mathbf{A}^0$$

is trivial, where the tensor product is taken over all subsets I of $\{1, 2, 3\}$, and $\mathbf{m}_I$: $\mathbf{A}^0 \times_T \mathbf{A}^0 \times_T \mathbf{A}^0 \to \mathbf{A}^0$ is the addition taken over the factors in I. Pulling back to $\mathbf{A}^0 \times_T T \times_T T = \mathbf{A}^0$ via $\mathrm{id}_{\mathbf{A}^0} \times P \times Q$ we get Proposition 2.5.

Thus we call a class $c \in \mathrm{Pic}(\mathbf{A}^0)$ **cubical** if it satisfies the formula of Proposition 2.5. The existence of such a cubical class is all that is needed to deduce the following result, as in [La 83a], which is the higher dimensional version of Tate's theorem for elliptic curves.

Theorem 2.6. *Let $\pi: \mathbf{A} \to Y$ be the Néron model of A_F. Let $P \in \mathrm{Sec}(Y, \mathbf{A}^0)$. Let $c \in \mathrm{Pic}(\mathbf{A}^0)$ be cubical, and let $c_y = c|\mathbf{A}_y$ be the restriction of c to the fiber $\mathbf{A}_y$ for $y \in Y_0(k)$, where Y_0 is the open subset of points y such that $\mathbf{A}$ is proper over y. Let*:

$$\hat{h}_y = \hat{h}_{c_y} = \text{Néron–Tate height on } \mathbf{A}_y(k);$$

$$q_y = \text{homogeneous quadratic part of } \hat{h}_y.$$

Then for $y \in Y_0(k)$ we have

$$q_y(P(y)) = \tfrac{1}{2} h_{[P,P]^*c}(y) + O_P(1).$$

Remark 1. Note that $[P, P]^*c \in \mathrm{Pic}(Y)$, whence the height $h_{[P,P]^*c}$ makes sense since Y is a *complete* non-singular curve.

Remark 2. Since $[-1]_{\mathbf{A}}$ is an automorphism of the Néron model, one can write $2c$ as a sum of an even class and an odd class, and one can therefore catch the linear part of the height similarly. In the version of [La 83a] this posed a difficulty since one could not necessarily extend $[-1]$ as an automorphism of the good completion.

From the theorem, one gets a factorization of the height in algebraic families.

Corollary 2.7 (Silverman's conjecture). *Suppose $Y = \mathbf{P}^1$ is the projective line, and h_Y is the standard height on $\mathbf{P}^1$. Let η be the generic point of Y, and let q_η be the quadratic form of the Néron–Tate height $\hat{h}_{c_\eta}$ on the generic fiber. For $P \in \mathrm{Sec}(T, \mathbf{A})$ (and not necessarily in* $\mathrm{Sec}(T, \mathbf{A}^0)$*), P fixed and $y \in Y_0(k)$ variable, we have*

$$q_y(P(y)) = q_\eta(P) h_Y(y) + O_P(1).$$

Corollary 2.8. *Let Y be an arbitrary non-singular curve of genus $\geqq 1$. Let $a \in \operatorname{Pic}(Y, k)$ have degree 1. Then*

$$q_y(P(y)) = q_\eta(P)h_a(y) + O_P(h_a(y)^{1/2}) + O_P(1).$$

Both corollaries follow as in Tate and [La 83a], using Chai's versions of the preceding results, given here as Proposition 2.5 and Theorem 2.6. Hence whereas the corollaries were proved before in dimension > 1 only conditionally by the use of good completions for the Néron model, they are now proved unconditionally.

III, §3. TORSION POINTS AND THE l-ADIC REPRESENTATIONS

The group of torsion points on an abelian variety comes in the theory in several ways. First, one wants to determine the group of torsion points rational over a given field, for its own sake. But in addition, the torsion points give rise to representation spaces for the ring of endomorphisms and for the Galois group, thus intervening in a much more extensive way in the general diophantine study of abelian varieties. We shall collect here some basic facts which will be used in this chapter and even more importantly in the next chapter. In light of this multiple use, the present section will provide a common reference for those facts.

Classically, if A is an abelian variety of dimension d, defined over the complex numbers, then A is a compact complex Lie group, isomorphic to a complex torus, and thus for each positive integer m, if we let $A[m]$ be the subgroup of points of order m, then

$$A[m] \approx (\mathbf{Z}/m\mathbf{Z})^{2d} \quad \text{as abelian groups.}$$

By a theorem of Hasse in dimension 1 and Weil in arbitrary dimension, the same structure holds for all abelian varieties, provided m is prime to the characteristic of a field of definition.

We assume throughout that A is defined over a field k and that l is a prime number prime to the characteristic of k.

We define the **Tate module** $T_l(A)$ to consist of all sequences

$$(a_1, a_2, \ldots)$$

of points $a_i \in A(k^a)$ such that $la_1 = 0$ and $la_{i+1} = a_i$. Then $T_l(A)$ is a module over $\mathbf{Z}_l$, since for every p-adic number $c \in \mathbf{Z}_l$ we define its action

$$c(a_1, a_2, \ldots) = (ca_1, ca_2, \ldots).$$

A basic theorem of Weil (Hasse–Deuring in dimension 1) asserts:

Let $d = \dim A$. Then $T_l(A)$ is a free module of rank $2d$ over $\mathbf{Z}_l$.

We can identify $T_l/l^n T_l = A(k^a)[l^n] = A[l^n]$. We can also write

$$T_l(A) = \varprojlim A[l^n],$$

where the limit is the projective limit. We shall also consider the $\mathbf{Q}_l$-vector space

$$V_l(A) = \mathbf{Q}_l \otimes T_l(A) = \mathbf{Q}_l T_l(A),$$

which is then a $2d$-dimensional vector space over $\mathbf{Q}_l$.

Let G_k be the Galois group of k^a over k. It is a basic fact that the extensions $k(A[l^n])$ of k are separable and so Galois. Then $T_l(A)$ is a representation module for G_k, by defining for each $\sigma \in G_k$ the action componentwise:

$$\sigma(a_1, a_2, \ldots) = (\sigma a_1, \sigma a_2, \ldots).$$

Similarly, if $f\colon A \to B$ is a homomorphism over k, then $T_l(A)$ gives the representation

$$T_l(f)\colon T_l(A) \to T_l(B) \qquad \text{such that} \qquad f(a_1, a_2, \ldots) = \big(f(a_1), f(a_2), \ldots\big).$$

We let

$$V_l(f)\colon V_l(A) \to V_l(B)$$

be the natural extension to the vector spaces over $\mathbf{Q}_l$. Since $\dim V_l(A) = 2d$, after picking a basis, we get corresponding representations in $\mathrm{GL}_{2d}(\mathbf{Z}_l)$ and $\mathrm{GL}_{2d}(\mathbf{Q}_l)$. All these representations are called the **l-adic representations associated with** A over k.

Suppose next that $k = F$ is the quotient field of a discrete valuation ring $\mathfrak{o}_v$. Let w be a valuation of F^a extending v. By the **decomposition group** G_w of w we mean the subgroup of G_F consisting of all elements such that $\sigma w = w$. Then G_w can be identified with the Galois group of the completion, that is

$$G_w = \mathrm{Gal}(F_w^a/F_v).$$

If w, w' are two absolute values of F^a extending v, then w, w' are conjugate (i.e. there is an element of G_F sending w to w') so that the decomposition groups G_w and $G_{w'}$ are conjugate subgroups of G_F.

Let $k(v)$ and $k(w)$ be the residue class fields of $\mathfrak{o}_v$ and the corresponding valuation ring (not discrete) $\mathfrak{o}_w$ in F^a. Then there is a natural homomorphism

$$G_w \to \mathrm{Gal}\big(k(w)/k(v)\big)$$

which is easily shown to be surjective. The kernel of this homomorphism is defined to be the **inertia group**, denoted by I_w. Thus the inertia group is the subgroup of G_w inducing the identity on the residue class field extension. Again, any two inertia groups I_w, $I_{w'}$ are conjugate.

Suppose that $k(v)$ is a finite field with q elements. Then there exists a unique element $\sigma \in \mathrm{Gal}(k(w)/k(v))$ such that

$$\sigma x = x^q \qquad \text{for} \quad x \in k(w).$$

This element is called the **Frobenius automorphism**. Then there exists a unique element, denoted by Fr_w, in G_w/I_w such that Fr_w induces the Frobenius automorphism on $k(w)$. This element Fr_w is also called the **Frobenius element**. In general, it is well defined only modulo the inertia group. Once more, Frobenius elements Fr_w and $\mathrm{Fr}_{w'}$ are conjugate.

In light of these conjugacies, we frequently denote the decomposition group and the Frobenius element by G_v and Fr_v respectively, because we are interested in their properties only up to conjugacy. In particular, the characteristic polynomial of an element in a finite dimensional representation of the Galois group depends only on the conjugacy class.

Next, suppose that A_F is defined over the field F, quotient field of the discrete valuation ring $\mathfrak{o}_v$. Suppose that A_F has good reduction. We let A be the proper smooth model over $\mathrm{spec}(\mathfrak{o}_v)$ whose generic fiber is A_F. Let w be an extension of v to F^{a} as above. Then we have a natural homomorphism

$$A(F^{\mathrm{a}}) \to A(k(w)),$$

and therefore a natural homomorphism on the l-adic spaces

$$T_l(A) \to T_l(A_{k(v)})$$

commuting with the action of the decomposition group G_v. It is a basic theorem that:

Assuming l not equal to the characteristic of $k(v)$, then the extensions $F(A[l^n])$ are unramified for all n, and the natural reduction homomorphism gives G_v-isomorphisms

$$T_l(A) \xrightarrow{\approx} T_l(A_{k(v)}) \qquad \text{and} \qquad V_l(A) \xrightarrow{\approx} V_l(A_{k(v)}).$$

Let k be a finite field with q elements, and let A be an abelian variety defined over k and of dimension d. For each positive integer m there is a unique extension k_m of k of degree m. The set of points $A(k_m)$ is finite. The Frobenius automorphism of k^{a} which sends $x \mapsto x^q$ for $x \in k^{\mathrm{a}}$ induces an isogeny $\mathrm{Fr}_A \colon A \to A$, the **Frobenius isogeny**, whence an automorphism

$$\mathrm{Fr}_A \colon A(k^{\mathrm{a}}) \to A(k^{\mathrm{a}})$$

called the **Frobenius automorphism**. In fact, if $\mathrm{Fr} \in G_k$ is the Frobenius element in the Galois group, then the representation $V_l(\mathrm{Fr})$ on $V_l(A)$ is the same element as the representation of the Frobenius isogeny on $V_l(A)$. Let $\alpha_1, \ldots, \alpha_{2d}$ be the roots of the characteristic polynomial of $V_l(\mathrm{Fr})$. Then

$$\#A(k_m) = \prod_{i=1}^{2d} (1 - \alpha_i^m) = \text{degree of } (\mathrm{id} - \mathrm{Fr}_A^m)$$

because $A(k)$ is characterized as the set of points in $A(k^{\mathrm{a}})$ which are fixed under the Frobenius automorphism, so $A(k_m)$ is the kernel of $\mathrm{id} - \mathrm{Fr}_A^m$. We have:

The numbers α_i are algebraic and have absolute value $q^{1/2}$. The characteristic polynomial $\prod (T - \alpha_i)$ has coefficients in $\mathbf{Z}$, and is independent of l.

These properties are due to Hasse for $d = 1$ and to Weil in general.

III, §4. PRINCIPAL HOMOGENEOUS SPACES AND INFINITE DESCENTS

By an infinite descent one means a procedure by which, supposing given a rational point with some height, one finds another point with smaller height. The iteration of this procedure either shows that there was no rational point to start with, or ends up providing only a finite number, or a finite number of generators for the group of rational points if a group structure is involved.

Let A be an abelian variety over a number field F.

The infinite descent is most classically applied as follows to prove the Mordell–Weil theorem of Chapter I, Theorem 4.1. Let m be an integer $\geqq 2$. One proves first that $A(F)/mA(F)$ is finite. The proof effectively bounds the number of generators, but gives no indication as to their possible heights. To obtain a finite number of generators, one then has the following result whose proof is one version of descent.

Proposition 4.1. *Let Γ be an abelian group such that $\Gamma/m\Gamma$ is finite. Suppose given a real valued norm $|\ |$ on Γ. Let $a_1, \ldots, a_r$ be coset representatives of $\Gamma/m\Gamma$ in Γ. There exists a number c_1 and a subset B of Γ such that:*

(1) *$|P| \leqq c_1$ for all $P \in B$, i.e. B is a bounded.*

(2) *For any $P_0 \in \Gamma$, there exist integers $n_0, n_1, \ldots, n_r$ and a point $P \in B$*

such that

$$P_0 = n_0 P + n_1 a_1 + \cdots + n_r a_r.$$

Proof. We construct a sequence of points $(P_0, P_1, \ldots)$ in Γ by starting with our point P_0, and such that

$$mP_{\nu+1} = P_\nu - a_{i_\nu}.$$

By hypothesis,

$$|P_{\nu+1}| \leqq \frac{1}{m}|P_\nu| + \frac{c}{m}$$

where c is a bound for the norms of the elements $a_1, \ldots, a_r$. Iterating this estimate, we get

$$|P| \leqq \frac{1}{m^\nu}|P_0| + c\left(1 + \frac{1}{m} + \frac{1}{m^2} + \cdots\right),$$

which concludes the proof.

We apply the proposition when $\Gamma = A(F)$, with the seminorm equal to the square root of a Néron–Tate height (quadratic form) associated with an ample divisor class. Then the set B in the proposition is a set of bounded height, and in the number field case, B is therefore a finite set, thus showing that $A(F)$ is finitely generated. But the method does not bound the heights of generators.

We shall now describe another type of descent, more sophisticated. We start with a discussion of non-singular varieties over a field k which become isomorphic to an abelian variety over a finite extension of k, but which may not have any rational point over k. Let X be such a variety. Then X is a **principal homogeneous space** of an abelian variety A over k. By this we mean that there exists a morphism

$$A \times X \to X$$

defined over k, which is an action of A on X, and such that for each $x \in X$ the map $a \mapsto ax$ is an isomorphism A with X. Thus A is the Albanese variety of X. If X has a rational point, then X is isomorphic to A over k, but not otherwise. We denote by $\mathrm{PHS}(A, k)$ the isomorphism classes of principal homogeneous spaces of A over k.

We must now assume that the reader is acquainted with basic facts about the cohomology of groups. We then have the following cohomological description of $\mathrm{PHS}(A, k)$. We remind the reader that k^s denotes the separable closure of k.

Theorem 4.2. *Let X be a principal homogeneous space of A over k. For each point $x \in X(k^s)$ and $\sigma \in G_k$ there is a unique $a_\sigma \in A(k^s)$ such* *that*

$$\sigma x - x = a_\sigma,$$

and the function $\sigma \mapsto a_\sigma$ *is a* 1*-cocycle. The association* $X \mapsto \{a_\sigma\}$ *induces a bijection*:

$$\mathrm{PHS}(A, k) \to H^1(G_k, A) = H^1(G_k, A(k^s)).$$

In particular, the principal homogeneous spaces form an abelian group. We define the **Châtelet–Weil** group to be

$$\mathrm{WC}(A) = H^1(G_k, A(k^s)) = H^1(G_k, A).$$

Châtelet originated these constructions in special cases. For a comprehensive account, I refer to Colliot-Thélène's survey of Châtelet's works [CoT 88]. Weil defined the group law in geometric terms [We 55], but this is irrelevant for us here. The determination of this group is one of the standard diophantine problems. Actually, for every group variety (not even necessarily commutative) one can define a principal homogeneous space and a first cohomology *set* as we have done above. One has:

Theorem 4.3 ([La 56c]). *If* k *is a finite field and* X *is a group variety, then* $H^1(G_k, X)$ *is trivial. In particular, if* $X = A$ *is an abelian variety, then* $H^1(G_k, A) = 0$. *A principal homogeneous space of a group variety over a finite field always has a rational point.*

Perhaps the first result of this kind was due to F. K. Schmidt, who showed that a curve of genus 1 over a finite field has a rational point, and so is an elliptic curve in our sense.

Let m be a positive integer $\geqq 2$, and prime to the characteristic of k. The short exact sequence

$$0 \to A[m] \to A(k^s) \to A(k^s) \to 0$$

gives rise to the cohomology sequence, whose beginning can be rewritten in the form

$$0 \to A(k)/mA(k) \to H^1(G_k, A[m]) \to H^1(G_k, A(k^s))[m] \to 0.$$

See [LaT 58] for this as well as a general discussion of principal homogeneous spaces over abelian varieties. Tate [Ta 58] gave precise duality theorems over p-adic fields in this connection.

We now come to the study of descent in a cohomological context. Convenient references for proofs are [Ta 74] and [Sil 86], which although written for elliptic curves, apply without change to abelian varieties.

For simplicity, we assume that $k = F$ is a number field, because we are going to refine these exact sequences by using the absolute values. We also generalize them by considering an arbitrary isogeny

$$\varphi\colon A \to B \quad \text{defined over } F.$$

Then we obtain the two exact sequences:

ES 1. $$0 \to A[\varphi] \to A(F^{\mathrm{a}}) \to B(F^{\mathrm{a}}) \to 0$$

and

ES 2. $$0 \to B(F)/\varphi A(F) \to H^1(G_F, A[\varphi]) \to H^1(G_F, A)[\varphi] \to 0.$$

For each absolute value v on F denote by the same letter v an extension to the algebraic closure F^{a}. Then v on F^{a} is induced by an imbedding

$$F^{\mathrm{a}} \hookrightarrow F_v^{\mathrm{a}}.$$

We defined the **decomposition group** to be the subgroup G_v of G_F consisting of those elements $\sigma \in G_F$ such that $\sigma v = v$. Then G_v will also be viewed as a group of automorphisms of $A(F_v^{\mathrm{a}})$ and $B(F_v^{\mathrm{a}})$. In particular, we obtain the local exact sequence:

ES$_v$ 2. $$0 \to B(F_v)/\varphi A(F_v) \to H^1(G_v, A[\varphi]) \to H^1(G_v, A)[\varphi] \to 0.$$

The natural inclusions $G_v \subset G_F$ and $A(F^{\mathrm{a}}) \subset A(F_v^{\mathrm{a}})$ give rise to restriction homomorphisms on the cohomology groups, and hence to an exact and commutative diagram:

ES 3.
$$\begin{array}{ccccccc}
 & & & & 0 & & 0 \\
 & & & & \downarrow & & \downarrow \\
 & & & & S^{(\varphi)}(A_F) & \to & Ш(A_F)[\varphi] \\
 & & & & \downarrow & & \downarrow \\
0 \to & B(F)/\varphi A(F) & \to & & H^1(G_F, A[\varphi]) & \to & H^1(G_F, A)[\varphi] \to 0 \\
 & \downarrow & & & \downarrow & & \downarrow \\
0 \to & 0 & \to & & \prod_v H^1(G_v, A)[\varphi] & \to & \prod_v H^1(G_v, A)[\varphi] \to 0
\end{array}$$

The two groups $S^{(\varphi)}(A_F)$ and $Ш(A_F)$ are by definition the kernels of the central and right vertical maps, and are called the **Selmer group** and **Shafarevich–Tate groups** respectively. It is easy to see that they do not depend on the choice of extension of v to the algebraic closure, but depend only on A and F. By definition, in terms of the homogeneous spaces, we can describe Ш by saying:

The group Ш *is the subgroup of principal homogeneous spaces of* A *over* F *which have a rational point in* F_v *for all* v.

Example (Selmer [Sel 51]). The curve $3x^3 + 4y^3 + 5z^3 = 0$ has a rational point in $\mathbf{Q}_v$ for all v, but does not have a rational point in $\mathbf{Q}$.

Conjecture 4.4. *The Shafarevich–Tate group* Ш *is finite.*

Rubin [Ru 87] gave the first examples when it was proved that a Shafarevich–Tate group is finite. Further insight was given by Kolyvagin [Koly]. The direction of Kolyvagin and Rubin gives one connection between diophantine problems and the theory of cyclotomic fields.

Let A' be the dual (Picard) variety of A. Cassels [Cas 62] for elliptic curves followed by Tate for abelian varieties [Ta 62] defined a natural pairing

$$Ш(A) \times Ш(A') \to \mathbf{Q}/\mathbf{Z}.$$

Theorem 4.5 (Cassels–Tate). *Under the finiteness of* Ш (*Conjecture* 4.4) *the two group* $Ш(A)$ *and* $Ш(A')$ *are dual to each other.*

See also [Cas 65]. When A has dimension 1, so A is an elliptic curve, then A is self-dual, $Ш(A)$ and $Ш(A')$ can be identified, and because of a natural skew-symmetric non-degenerate form on $Ш(A)$, it follows that $\#Ш(A)$ is a perfect square (if finite). Extensive computations have shown this to be true experimentally.

Next we describe the role that Ш plays in making infinite descents.

Theorem 4.6. *Let* $\varphi\colon A \to B$ *be an isogeny of abelian varieties over* F. *Then we have an exact sequence*

$$0 \to B(F)/\varphi(A(F)) \to S^{(\varphi)}(A_F) \to Ш(A_F)[\varphi] \to 0.$$

Furthermore, the Selmer group $S^{(\varphi)}(A_F)$ *is finite.*

Observe that Theorem 4.6 gives a generalization of the weak Mordell–Weil theorem. Indeed, take for φ the isogeny m_A (multiplication by m on A). We write $S^{(m)}$ for the Selmer group in that case. Then the finiteness of $S^{(m)}(A_F)$ implies the finiteness of $A(F)/mA(F)$. Theorem 4.6 follows immediately from **ES 3** and the snake lemma. The injection

$$\delta\colon B(F)/\varphi(A(F)) \to S^{(\varphi)}(A_F)$$

is given by the coboundary operator from the long cohomology sequence.

We rewrite the exact sequence with m for the applications, namely

$$0 \to A(F)/mA(F) \to S^{(m)}(A_F) \to \text{Ш}(A_F)[m] \to 0.$$

The finite group $S^{(m)}(A_F)$ is effectively computable, but the problem lies with $\text{Ш}(A_F)[m]$. The infinite descent takes place by using powers of m. The exact sequences with m and with m^n fit into a commutative exact diagram

$$\begin{array}{ccccccc}
A(F) & \longrightarrow & S^{(m^n)}(A_F) & \longrightarrow & \text{Ш}(A_F)[m^n] & \longrightarrow & 0 \\
\downarrow{\scriptstyle \text{id}} & & \downarrow & & \downarrow{\scriptstyle m^{n-1}\cdot\text{id}} & & \\
A(F) & \longrightarrow & S^{(m)}(A_F) & \longrightarrow & \text{Ш}(A_F)[m] & \longrightarrow & 0.
\end{array}$$

It is now convenient to define:

$$S^{(m,n)}(A_F) = \text{image of } S^{(m^n)}(A_F) \text{ in } S^{(m)}(A_F) \text{ in the above diagram.}$$

Then directly from the definition, we obtain the exact sequence

ES 4. $\quad 0 \to A(F)/mA(F) \to S^{(m,n)}(A_F) \to m^{n-1}\text{Ш}(A_F)[m^n] \to 0.$

We reach the fundamental problem of finding generators of $A(F)$. We have on the one hand the decreasing Selmer groups:

$$S^{(m)}(A_F) = S^{(m,1)}(A_F) \supset S^{(m,2)}(A_F) \supset S^{(m,3)}(A_F) \supset \cdots$$

and on the other hand, we have the increasing groups

$$R_{(m,1)}(A_F) \subset R_{(m,2)}(A_F) \subset R_{(m,3)}(A_F) \subset \cdots$$

where

$R_{(m,s)}(A_F)$ is the subgroup of $S^{(m)}(A_F)$ generated by all points $P \in A(F)$ with height

$$h(P) \leqq s.$$

The height is the Néron–Tate height associated with an ample even divisor class, and so a positive definite quadratic form on the group of rational points.

Assume Conjecture 4.4 that Ш is finite. Then for some n the last term in **ES 4** is 0, so we get an isomorphism

$$A(F)/mA(F) \xrightarrow{\approx} S^{(m,n)}(A_F).$$

Since $S^{(m,n)}(A_F)$ is effectively computable, this allows us to find generators for $A(F)$. Alternatively, one could take for m the order of $Ш(A_F)$ and then get the easier isomorphism

$$A(F)/mA(F) \xrightarrow{\approx} S^{(m,2)}(A_F).$$

In any case, the finiteness of $Ш(A_F)$ guarantees that after a finite number of steps one obtains an equality

$$S^{(m,n)}(A_F) = R_{(m,s)}(A_F).$$

In particular, the determination of the bound s for the height of points generating $S^{(m,n)}(A_F)$ also would give a bound for the height of generators of $A(F)$. However, the procedure does not give a closed form for estimates of such heights. To give such estimates one has to dig deeper into the Birch–Swinnerton-Dyer conjecture, the zeta function, and their associated structures. It is not even clear from the literature, to my knowledge, if the finiteness of Ш would give an effective bound (to be determined in each case) for the heights of generators of $A(F)$.

III, §5. THE BIRCH–SWINNERTON-DYER CONJECTURE

Next let A be an abelian variety defined over a number field F. We let $\mathfrak{o} = \mathfrak{o}_F$ be the ring of algebraic integers, $\mathfrak{o}_v$ the local ring at one of the discrete valuations, and $k(v)$ the residue class field. *Suppose that A has good reduction at v.* We let G_v be a decomposition group, and we let:

$\mathrm{N}v = \#k(v)$;

Fr_v = Frobenius element in G_v, acting on $A(k(v)^{\mathrm{a}})$;

$\alpha_{i,v}$ $(i = 1,\ldots,2d)$ = characteristic roots of Fr_v;

$$P_v(T) = \prod_{i=1}^{2d} (1 - \alpha_{i,v}T).$$

Let S be a finite set of absolute values on F containing all archimedean ones, and all v where A does not have good reduction. The **Euler product**

$$L_{A,S}(s) = \prod_{v \notin S} \frac{1}{P_v(\mathrm{N}v^{-s})} = \prod_{v \notin S} \prod_{i=1}^{2d} \frac{1}{(1 - \alpha_{i,v}\mathrm{N}v^{-s})}$$

converges for $\mathrm{Re}(s) > 3/2$ because it is dominated by the product for

$$(\zeta_F(s - \tfrac{1}{2}))^{2d},$$

where ζ_F is the Dedekind zeta function of F. One can also define polynomials P_v for $v \in S_{\text{fin}}$, see Serre [Ser 69–70] and the subsequent article by Deligne [De 70], to form a complete **L-function**

$$L_A(s) = \prod_{v \notin S_\infty} \frac{1}{P_v(\mathbf{N}v^{-s})}.$$

We shall give the factors for bad reduction below. Then conjecturally, L_A has an analytic continuation to the whole plane, and satisfies a simple functional equation. Thus $L_{A,S}$ would also have such an analytic continuation.

The first **Birch–Swinnerton-Dyer conjecture** is:

Conjecture 5.1. *Let r be the rank of $A(F)$. Then $L_{A,S}$ has a zero of order r at $s = 1$.*

In particular, $L_{A,S}$ has the Taylor expansion

$$L_{A,S}(s) = C_A(S)(s-1)^r + \text{higher terms},$$

with some constant $C_A(S) \neq 0$. The second Birch–Swinnerton-Dyer conjecture concerns this constant. We must introduce other factors to get some constant independent of S. At first we follow Tate's presentation in [Ta 66b] to avoid having to define the bad factors for $v \in S$ by using another device as follows. One of the advantages of Tate's formulation is that it allowed him to transpose the whole set up of the conjecture to the function field case, and to formulate a coherent set of conjectures for arbitrary complete non-singular surfaces over finite fields. Although these conjectures are not proved today, they hang together more tightly than in the number field case, and the finiteness of the Brauer group which plays an analogous role to the Shafarevich–Tate group would imply the rest of the conjectures in the function field case. We shall not go any further into this case, however, except for indicating the following application. Ulmer [Ul 90] has constructed elliptic curves over a rational function field with finite constant field, such that the L-function of the elliptic curve vanishes of arbitrarily high order. The Birch–Swinnerton-Dyer conjecture would imply that these curves have arbitrarily high rank, but no construction is known today exhibiting explicitly independent points having this rank. Compare with [ShT 67].

At each absolute value v the completion F_v is a locally compact field. For each v we choose a Haar measure μ_v on F_v such that for all but a finite number of v the ring of local integers $\mathcal{O}_v$ has measure $\mu_v(\mathcal{O}_v) = 1$. Then for $x \in F_v$ and every open set U of K_v we have

$$\mu_v(xU) = \|x\|_v \mu_v(U),$$

where $\|x\|_v$ is the normalization of the absolute value described in Chapter II, §2.

The group $A(F_v)$ is a compact analytic group over F_v. Choose an invariant differential form ω of degree d on A over F. Then ω and μ_v determine a Haar measure $|\omega|_v \mu_v^d$ on $A(F_v)$. Define the **v-adic period**

$$\pi_v = \int_{A(F_v)} |\omega|_v \mu_v^d.$$

Let $\mathbf{A}_F$ be the additive group of **adeles**. By definition $\mathbf{A}_F$ consists of vectors

$$(\ldots, x_v, \ldots)$$

indexed by the absolute values of F, such that for all but a finite number of v, the v-component x_v lies in the local ring $\mathcal{O}_v$, and $x_v \in F_v$ for all v. Then the field F is imbedded discretely on the diagonal in $\mathbf{A}_F$, and it is a basic theorem that $\mathbf{A}_F/F$ is compact. The measures μ_v define a measure

$$\mu = \prod_v \mu_v \quad \text{on } \mathbf{A}_F,$$

and hence on the factor group $\mathbf{A}_F/F$. We define the **norm**

$$\|\mu\| = \mu(A_F/F).$$

We now define

$$L^*_{A,S}(s) = \frac{\|\mu\|^d}{\prod_{v \in S} \pi_v} \prod_{v \notin S} P_v(\mathbf{N}v^{-s})^{-1}.$$

It is easy to show that there exists a set S_0 (depending on A, ω, μ) such that for all finite sets S containing S_0 the function L^*_S is independent of the choice of ω and μ. Furthermore, for $S \supset S_0$, we have

$$\pi_v = P_v(\mathbf{N}v^{-1}) \qquad \text{for} \quad v \notin S.$$

Consequently the asymptotic behavior of $L^*_S(s)$ as $s \to 1$ is independent of S, and using Conjecture 4.1 we get

$$L^*_{A,S}(s) = C^*_A(s-1)^r + \text{higher order terms},$$

where C^*_A is independent of $S \supset S_0$.

We define the **regulator**

$$R_A = |\det\langle P_i, P'_j\rangle|,$$

where $\{P_1, \ldots, P_r\}$ is a basis of $A(F)$ modulo torsion, $\{P'_1, \ldots, P'_r\}$ is a basis of $A'(F)$ modulo torsion, and

$$\langle P, P' \rangle = \hat{h}_\delta(P, P')$$

is the Néron pairing of Theorem 1.5. We then have:

Conjecture 5.2. *The constant C_A^* has the value*

$$C_A^* = \frac{\#(\text{Ш})R_A}{\#\big(A(F)_{\text{tor}}\big)\#\big(A'(F)_{\text{tor}}\big)}$$

where Ш is the Shafarevich–Tate group discussed in §4.

The above form of the conjecture is convenient because it did not force a consideration of the factors in the L-function corresponding to the bad places v. On the other hand, it does *not* give something else we want, namely an explicit value for

$$\frac{1}{r!} L_A^{(r)}(1),$$

which is used to estimate the heights for the points in a basis of the Mordell–Weil group. Hence I shall also describe the more precise version worked out by Gross [Gross 82] for abelian varieties over any number field.

Let $\mathbf{A}$ be the Néron model of A_F over $\mathfrak{o}_F$, with connected Néron model $\mathbf{A}^0$. For v finite, we define

$$c_v(A) = c_v = \big(\mathbf{A}_{k(v)}(k(v)) : \mathbf{A}^0_{k(v)}(k(v))\big).$$

Thus c_v is an integer, equal to the index of the subgroup of points in the residue class field on the connected component of the special fiber, in the full group of $k(v)$-rational points on the whole fiber of the Néron model. We then define

$$C_{A,\text{fin}} = \prod_{v \in S_{\text{fin}}} c_v.$$

Let G_F be the Galois group of $\mathbf{Q}^{\text{a}}$ over F and let I_v be the inertia subgroup of G_v, inducing the identity on the residue class field extension at v. As before, let Fr_v be the Frobenius element in G_v/I_v. Let l be a rational prime unequal to the characteristic of $k(v)$, and as before let

$$T_l(A) = \varprojlim A(\mathbf{Q}^{\text{a}})[l^n],$$

where $A(\mathbf{Q}^{\mathrm{a}})[l^n]$ is the kernel of multiplication by l^n on the group of algebraic points of A, so $T_l(A)$ is a free $\mathbf{Z}_l$-module of rank $2d$. We define the **local** L**-factor of** A **at** v by the formula

$$L_{A,v}(s) = \det\bigl(\mathrm{id} - \mathbf{N}v^{-s}\mathrm{Fr}_v^{-1} \mid \mathrm{Hom}_{\mathbf{Z}_l}(T_l(A), \mathbf{Z}_l)^{I_v}\bigr)^{-1}.$$

The superscript I_v indicates the submodule of elements fixed by all elements of the inertia group I_v in G_v.

Thus we have defined the local factor for all v, including those with bad reduction, and

$$L_A(s) = \prod_v L_{A,v}(s).$$

Let $W_{\mathbf{A}}$ denote the projective $\mathfrak{o}_F$-module of invariant differentials on the Néron model $\mathbf{A}$. Then $\mathrm{rank}(W_{\mathbf{A}}) = d$ and $\bigwedge^{\max} W_A$ is a submodule of rank 1 of $H^0(A_F, \Omega^d)$. Let $\{\omega_1, \ldots, \omega_d\}$ be an F-basis of $H^0(A_F, \Omega^d)$ and let

$$\eta = \bigwedge_j \omega_j.$$

Then

$$\bigwedge^d W_{\mathbf{A}} = \eta \mathfrak{a}_\eta \quad \text{in } H^0(A_F, \Omega^d),$$

where $\mathfrak{a}_\eta$ is a fractional ideal in F.

Let v be a complex place of F and let $\sigma\colon F \to \mathbf{C}$ be an imbedding inducing v on F. Let $H = H_1(A_\sigma(\mathbf{C}), \mathbf{Z})$ be the integral homology of A_σ. Then H is a free $\mathbf{Z}$-module of rank $2d$. Let $\{\gamma_1, \ldots, \gamma_{2d}\}$ be a basis of H, and define

$$c_v(A, \eta) = \left|\det\left(\int_{\gamma_i} \omega_j, \int_{\gamma_i} \bar{\omega}_j\right)\right| = \int_{A(\mathbf{C})} \sqrt{-1}\eta \wedge \bar{\eta}.$$

The determinant of the $2d \times 2d$ matrix is non-zero and depends only on η, v.

Let v be a real place of F corresponding to the imbedding $\sigma\colon F \to \mathbf{R}$. Let H^+ denote the submodule of $H_1(A_\sigma(\mathbf{C}), \mathbf{Z})$ which is fixed by complex conjugation. Then H^+ is free of rank d. Let $\{\alpha_1, \ldots, \alpha_d\}$ be a basis and define

$$c_v(A, \eta) = \bigl(A_\sigma(\mathbf{R}) : A_\sigma(\mathbf{R})^0\bigr)\left|\det\left(\int_{\alpha_i} \omega_j\right)\right| = \int_{A_\sigma(\mathbf{R})} |\eta|.$$

Again the determinant is non-zero and depends only on η, v, while $A_\sigma(\mathbf{R})^0$ is the real connected component.

Let $\mathbf{D}_F$ be the absolute value of the discriminant of F over $\mathbf{Q}$. The product

$$C_{A,\infty} = \prod_{v \in S_\infty} c_v(A, \eta)\mathbf{N}\mathfrak{a}_\eta / \mathbf{D}_F^{d/2}$$

is independent of the choice of η. Finally we define

$$C_A = C_{A,\infty} C_{A,\mathrm{fin}}.$$

Then C_A is a positive real number. We now have all the local factors defined, and we can formulate the precise value of the constant in Birch–Swinnerton-Dyer:

Conjecture 5.3. *We have*

$$\frac{1}{r!} L_A^{(r)}(1) = \frac{\#Ш_A R_A C_A}{\#A(F)_{\mathrm{tor}} \#A'(F)_{\mathrm{tor}}}.$$

III, §6. THE CASE OF ELLIPTIC CURVES OVER Q

Both the fact that we deal with an abelian variety of dimension 1 and that we are over the rationals is significant for the structure of the L-function. Under these circumstances, there exists a "minimal" differential form $\omega_A = \omega$, which we shall now describe in almost complete generality. We suppose that the elliptic curve is defined by the homogeneous polynomial in $\mathbf{P}^2$:

$$y^2 z = x^3 - \gamma_2 x z^2 - \gamma_3 z^3 \qquad \text{with} \quad \gamma_2, \gamma_3 \in \mathbf{Q}.$$

The origin of the group law is at infinity in $\mathbf{P}^2$. The **discriminant** is given by

$$\Delta = 16(4\gamma_2^3 - 27\gamma_3^2).$$

An isomorphism over $\mathbf{Q}$ for elliptic curves in that form is given by a rational number $c \neq 0$ with the following effects on affine coordinates ($z = 1$):

$$x \mapsto c^{-2}x, \qquad y \mapsto c^{-3}y, \qquad \gamma_2 \mapsto c^{-4}\gamma_2, \qquad \gamma_3 \mapsto c^{-6}\gamma_3.$$

By such an isomorphism, we can always achieve that $\gamma_2, \gamma_3 \in \mathbf{Z}$. Suppose p is a prime such that p^4 divides γ_2 and p^6 divides γ_3. Then we can change the elliptic curve by an isomorphism using $p = c$, letting

$$\gamma_2 \mapsto p^{-4}\gamma_2 \qquad \text{and} \qquad \gamma_3 \mapsto p^{-6}\gamma_3.$$

After having done so repeatedly until no further possible, we obtain what is a **minimal model** for the curve over **Z**, and then Δ is called a **minimal discriminant**, denoted by Δ_A. This minimal discriminant is defined up to a factor of ± 1, and is an invariant of an isomorphism class of elliptic curves over **Q**. We let

$$\omega_A = \frac{dx}{2y}.$$

For $p \neq 2, 3$ we can then reduce the equation of the minimal model mod p. The curve has good reduction at p if and only if $p \nmid \Delta_A$.

To describe the fully general situation including the primes 2 and 3 requires longer formulas which the reader will find in [Ta 74], following Deuring [Deu 41]. Letting A be this minimal model, one then has (see [Art 86]):

The Néron model **A** *is the open subscheme of* A *consisting of the regular points.*

With respect to the minimal model, we define the **period at infinity**

$$c_\infty = \int_{A(\mathbf{R})} |\omega_A|.$$

The factors for the L-function can be described explicitly in the present case. For each prime p let $A_{k(p)}$ be the reduction of the minimal model for A mod p. Let

$$t_p = 1 + p - (\text{number of points of } A_{k(p)} \text{ in } k(p))$$

If A has good reduction at p, then $t_p = \alpha_1 + \alpha_2$ is the sum of the characteristic roots. On the other hand, if $p|\Delta$, then $t_p = 1, -1$, or 0. The L**-function** is then defined by

$$L_A(s) = \prod_{p|\Delta} \frac{1}{(1 - t_p p^{-s})} \prod_{p\nmid\Delta} \frac{1}{1 - t_p p^{-s} + p^{1-2s}}.$$

We can define the **conductor** N_A in various ways. It is a measure of bad reduction, and is a positive integer

$$N_A = \prod_{p|\Delta} p^{f(p)},$$

where the exponent $f(p)$ is 0 if $p \nmid \Delta$ and $\geqq 1$ if $p|\Delta$. If $n(p)$ denotes the number of irreducible components of $A_{k(p)}$ over the algebraic closure of

$k(p)$, then according to [Ogg 67],

$$f(p) = \operatorname{ord}_p(\Delta) + 1 - n(p) = \eta + \delta,$$

where $\eta = 0$, 1 or 2 according as A has good reduction, multiplicative reduction, or additive reduction; and δ is the integer defined by means of representation theoretic formulas like Artin and Swan conductors, already mentioned at the end of Chapter III, §1. If $p|\Delta$, but $p \neq 2, 3$ and p is not a common factor of γ_2, γ_3 in the minimal model, then $f(p) = 1$. So $f(p)$ has a tendency to be equal to 1 for $p|\Delta$.

Let

$$\xi_A(s) = N_A^{s/2}(2\pi)^{-s}\Gamma(s)L_A(s).$$

Conjecture 6.1. *The function $\xi_A(s)$ is holomorphic in the whole s-plane, and satisfies a functional equation*

$$\xi_A(s) = \varepsilon\xi_A(2-s) \qquad \textit{with} \quad \varepsilon = \pm 1.$$

We call ε the **sign of the functional equation**. The conjecture is a precise version of a conjecture of Hasse.

As before we let R_A be the regulator. Since $\dim A = 1$, the elliptic curve is self-dual, and if $\{P_1, \ldots, P_r\}$ is a basis for $A(\mathbf{Q})$ mod torsion, then

$$R_A = |\det\langle P_i, P_j\rangle|.$$

We let $\mathbf{A}^0$ be the connected Néron model, and as before,

$$c_p = (\mathbf{A}(\mathbf{F}_p) : \mathbf{A}^0(\mathbf{F}_p)),$$

so c_p is an integer for each $p|\Delta_A$. Then the Birch–Swinnerton-Dyer conjecture as formulated in [Ta 74] is:

Conjecture 6.2.

$$\frac{1}{r!}L_A^{(r)}(1) = \frac{\#(\text{Ш})R_A}{\#(A(Q)_{\text{tor}})^2}c_\infty \prod_{p|\Delta} c_p.$$

We want to estimate

$$\#(\text{Ш})R_A$$

from above, and for this purpose, having c_p integers helps us, so we don't need to know more about them. However, the period c_∞ can be small, and a lower bound must be determined. All in all, the product $\#(\text{Ш}_A)R_A$ occurs here much as the product of the class number and regulator occur together in the residue of the Dedekind zeta function of a number field. Cf. Theorem 2.1 of Chapter II.

By analogy with estimates for the residue of the Dedekind zeta function of number fields due to Landau, a conjecture of Montgomery for the leading coefficient of Dirichlet L-series, and other considerations described in [La 83b], I conjectured that one can estimate $\#(\text{Ш}_A)R_A$ as follows.

Conjecture 6.3. *Suppose we have a minimal equation* $y^2 = x^3 - \gamma_2 x - \gamma_3$ *for* A *over* $\mathbf{Z}$. *Let* $H(A) = \max(|\gamma_2|^3, |\gamma_3|^2)$, *and* $N = N_A$. *Then*

$$\#(\text{Ш}_A)R_A \leqq b_1 H(A)^{1/12} N^{\varepsilon(N)} b_2^r (\log N)^r$$

where b_1, b_2 *are universal constants independent of* A, *and* $\varepsilon(N) \to 0$ *as* $N \to \infty$. *In fact,* $\varepsilon(N)$ *may have the explicit form*

$$\varepsilon(N) = b_3(\log N \log\log N)^{-1/2}.$$

Since $\#(\text{Ш}_A)$ is an integer, the conjectured bound applies just to the regulator R_A. We actually want bounds for the heights $\hat{h}(P_1), \ldots, \hat{h}(P_r)$ of points in a basis $\{P_1, \ldots, P_r\}$ for $A(\mathbf{Q})$ modulo torsion. Here we set

$$\hat{h}(P) = \langle P, P \rangle.$$

Since $A(\mathbf{Q})/A(\mathbf{Q})_{\text{tor}}$ is a lattice in the finite dimensional vector space $\mathbf{R} \otimes A(\mathbf{Q})$, we cannot quite find an orthogonalized basis for this lattice, but we can find an almost orthogonalized basis, by a general **theorem of Hermite**. (See [La 83a], Chapter 5, Theorem 7.7 and Corollary 7.8 for precise statements and proofs.) As a result, I arrived at the conjecture [La 83b]:

Conjecture 6.4. *There exists a basis* $\{P_1, \ldots, P_r\}$ *for* $A(\mathbf{Q})$ *modulo torsion, ordered by ascending height, such that:*

$$\hat{h}(P_1) \ll H(A)^{1/12r} N^{\varepsilon(N)} (\log N) \left(\frac{1}{\sqrt{2}}\right)^{(r-1)/2},$$

$$\hat{h}(P_r) \ll H(A)^{1/12} N^{\varepsilon(N)} (\log N) c^{r(r-1)/2},$$

where the constant implicit in $\ll$, *and* c, *are absolute, independent of* A.

The point is that the regulator is a determinant, and a bound for R_A gives a bound for the product of the heights

$$\hat{h}(P_1) \ldots \hat{h}(P_r)$$

if $\{P_1, \ldots, P_r\}$ is almost orthogonalized. The bound for the smallest height $\hat{h}(P_1)$ comes from the bound for the product, which in turn came from the bound for the regulator by using only linear algebra.

On the other hand, the upper bound for $\hat{h}(P_r)$ comes from dividing by $\hat{h}(P_i)$ for $i < r$, and so we need a *lower bound* for these heights. Such a lower bound is given by my Conjecture 1.4.

CHAPTER IV

Faltings' Finiteness Theorems on Abelian Varieties and Curves

This chapter gives an account of Faltings' finiteness theorems, and structure theorems for l-adic representations. These theorems were outstanding conjectures regarded as having independent interest. Faltings proved them all simultaneously with the Mordell conjecture. In retrospect, it is hard to remember, for instance, that the isogeny theorem for elliptic curves was not known before Faltings, and that a proof of this theorem would have been regarded as a major result by itself, just in this special case.

In the late sixties, Parshin reduced the Mordell conjecture to a conjectured property of abelian varieties, the Shafarevich conjecture, which we shall describe in §2, after a preliminary recall of Torelli's theorem in §1. The Shafarevich conjecture actually splits into two parts, which we shall call Finiteness I and Finiteness II. Finiteness I concerns finiteness of isomorphism classes within an isogeny class, while Finiteness II is a statement about finiteness of the number of certain isogeny classes. Faltings gave a proof for Finiteness I based on an extensive theory using the moduli spaces for abelian varieties and curves, and using a new notion of height, his modular height. He also used results of Raynaud on p-power group schemes. It was not entirely made clear in several expositions of Faltings' work, that the first part implies the second part in a more elementary way, following ideas of Tate and Faltings. I shall therefore give details for this implication. After that I shall briefly discuss Faltings' method used to prove the first part.

The proof (due to Zarhin) that Finiteness I implies the semisimplicity of certain l-adic representations and a conjecture of Tate will be given in §3. Faltings' proof that it implies Finiteness II will be given in §4. The discussion of the Faltings height and some of its properties leading to the proof of Finiteness I will be given in §5.

In a last section I shall give results of Masser–Wustholz showing how to bound the degree of isogenies of abelian varieties in terms of the Faltings height of the varieties and polarization degrees. This gives an alternative approach to prove Finiteness I.

IV, §1. TORELLI'S THEOREM

We recall from Chapter I, §5 that a **polarized abelian variety** is a pair (A, c) consisting of an abelian variety A and a divisor class for algebraic equivalence (i.e. an element of $\mathrm{NS}(A)$) containing an ample divisor. We also recall the isogeny

$$\varphi_c\colon A \to A' \qquad \text{such that} \quad \varphi_c(a) = c_a - c.$$

We defined c to be a **principal polarization** if the degree of c (i.e. the degree of φ_c) is equal to 1, so φ_c is an isomorphism. We shall now give an example of a principal polarization.

Let C be a curve (complete, nonsingular as always), and let

$$f\colon C \to J$$

be a canonical morphism into its Jacobian. By taking the sum of $f(C)$ with itself $g - 1$ times (where g is the genus), we obtain a divisor on J, called the **theta divisor**, namely

$$\Theta = f(C) + \cdots + f(C) \qquad (g - 1 \text{ times}).$$

Let θ denote the algebraic equivalence class of Θ. It is a basic fact that:

Theorem 1.1. *The class θ is a principal polarization.*

We call (J, θ) **the principally polarized Jacobian of** C. An isomorphism of curves induces an isomorphism of their Jacobians, but even more induces an isomorphism as principally polarized abelian varieties, where the polarization is defined by the classes of the theta divisors. **Torelli's theorem** asserts the converse. A proof of the following version can be found in Weil [We 57], see also [Mil 86a].

Theorem 1.2. *Let C_1, C_2 be curves defined over a field k, and imbedded in their Jacobians J_1, J_2 over k. Let*

$$\alpha\colon (J_1, \theta_1) \to (J_2, \theta_2)$$

be an isomorphism of principally polarized abelian varieties defined over k. Then α restricted to C_1 gives an isomorphism of C_1 with $\pm C_2 + a$, for some point $a \in J(k)$. In particular, C_1 is isomorphic (as a curve) with C_2 over k.

It is not true in general that the isomorphism class of J (without polarization) determines the isomorphism class of the curve. The first examples are due to Humbert [Hu 1899].

Using basic properties of abelian varieties having to do with the existence of a certain involution (the Rosati involution) and its positive definite trace, one can prove a finiteness statement about principally polarized abelian varieties due to Narasimhan–Nori [NaN 81].

Theorem 1.3. *Let A be an abelian variety defined over an algebraically closed field k of characteristic zero. Then there exists only a finite number of polarizations of given degree, and in particular, there is only a finite number of principal polarizations.*

For an investigation of the actual number of principal polarizations possible, see [Lange 87].

The standard books on abelian varieties develop the theory of the Rosati involution (e.g. [La 59] or [Mu 70]). For a reasonably detailed account of the general algebra concerning semisimple algebras with involutions implying the finiteness statement, following Narashiman–Nori, see Milne [Mil 86a], Proposition 18.2. What we really want is the combination of this finiteness theorem with Torelli's theorem, also as in [NaN 81], which yields:

Corollary 1.4. *Let A be an abelian variety over a field k. There is only a finite number of isomorphism classes of curves over k, imbeddable in their Jacobians over k, whose Jacobian is isomorphic to A over k.*

IV, §2. THE SHAFAREVICH CONJECTURE

Let R be a discrete valuation ring with maximal ideal M. Let F be the quotient field of R and let C be a complete non-singular curve over F. Unless otherwise specified, a **curve** will be assumed complete and non-singular. Let y be the closed point of spec(R). We say that C has **good reduction** at y if there exists a scheme X over spec(R) which is proper and smooth, and such that if X_F is the extension of the base from spec(R) to F then $X_F = C$. An early theorem of Chow–Lang asserts that if the genus of C is $\geqq 1$, then such a scheme X is unique up to isomorphism over spec(R). If k is the residue class field R/M, then the fiber X_k obtained from X by reduction modulo M is then also a non-singular curve, over k, and having the same genus as X_F.

Suppose R is a Dedekind ring, with quotient field F, and let $Y = \operatorname{spec}(R)$. Let C be a curve over F. We say that C has **good reduction** at a closed point $y \in Y$ if C has good reduction over the local ring $\mathcal{O}_{y,Y}$, which is a discrete valuation ring.

The **Shafarevich conjecture** [Sha 63], proved by Faltings [Fa 83], runs as follows.

Number field case. *Let R be the ring of integers of a number field F. Given $g \geqq 1$ and a finite set of points S of* spec(R), *there exists only a finite number of curves of genus g over F (up to isomorphism) having good reduction outside S.*

Now let us pass to the function field case. Let k be a field and let $F = k(Y)$ be the function field of a complete non-singular curve Y over k. Then Y is covered by two open sets spec(R) and spec(R') where R, R' are Dedekind rings. For instance, if $t \in k(Y)$ is a non-constant function, we let R be the integral closure of $k[t]$ in F and we let R' be the integral closure of $k[1/t]$. Let $y \in Y$ be a closed point. Let X_F be a curve over F. We say that X_F has **good reduction** at y if X has good reduction at the local ring $\mathcal{O}_{y,Y}$.

The function field analogue of the Shafarevich conjecture was proved by Parshin and Arakelov [Pa 68], [Ar 71] ten to fifteen years before Faltings, and is formulated as follows.

Function field case. *Let $F = k(Y)$ be the function field of a curve Y over an algebraically closed field k of characteristic* 0. *Let S be a finite set of closed points of Y. Let g be an integer $\geqq 1$. Then there is only a finite number of curves of genus g over F, up to isomorphism, which are not split, and which have good reduction at all points of Y outside S.*

The proof made extensive use of intersection theory on the surface X whose generic fiber is a curve X_F of genus g.

Parshin at the end of his 1968 paper observed that the Shafarevich conjecture implies the Mordell conjecture, and in particular, in the function field case implies Manin's theorem. The implication was given by constructing certain ramified coverings of the curve, and may be summarized as follows. For concreteness, we express the construction over number fields, but the same works in the function field case mutatis mutandis.

Lemma 2.1. *Let C be a complete non-singular curve of genus $g \geqq 1$, defined over a number field F, with good reduction outside a finite set of points S of* spec($\mathfrak{o}_F$). *Then there exists a finite extension F' of F and a finite set of points S' of* spec($\mathfrak{o}_{F'}$) *(containing all points lying over* 2) *having the following property. For every rational point $P \in C(F)$ there exists a covering W_P of $C' = C_{F'}$ defined over F', with good reduction outside S', with*

$$[W_P : C'] \leqq 2 \cdot 2^{2g} \qquad \textit{and ramification index } \leqq 2 \textit{ over } P,$$

unramified over all other points of C, and so of bounded genus.

By a **covering** we mean a possibly ramified covering, and $[W:C]$ denotes the degree of a covering W of C. The lemma can be proved as follows. We suppose the curve C imbedded in its Jacobian J. We first extend the base field F to a finite extension over which the points of order 2 on J are rational. The covering $J \to J$ obtained by multiplication by 2 restricts to a covering C' of the curve which is unramified. Let P_1, P_2 be two distinct points on C' lying above P. Let D be a divisor of degree 0 on C' such that

$$2D = (P_1) - (P_2) + (f)$$

where f is a rational function on C'. Dividing the divisor class of $(P_1) - (P_2)$ by 2 is unramified outside a fixed set of primes, independent of the choice of P, so after extending the ground field again to a finite extension F' we may assume that D is rational over F'.

Now let R be the localization of $\mathfrak{o}_{F'}$ at enough primes to include those where dividing $(P_1) - (P_2)$ might ramify, and also enough primes so that R is a principal ring, so has unique factorization. Then the function f above, which can be changed by an element of F, can be selected without loss of generality so that C' is the generic fiber of a scheme X proper and smooth over spec(R), and such that the divisor of f on X does not have any fibral components. Let X_P be the normalization of X in the function field $F'(f^{1/2})$. Then X_P is smooth over spec(R), and if we let W_P be the generic fiber of X_P then W_P satisfies the desired coditions. The fact that the genus of W_P is bounded independent of P follows from the Hurwitz formula for the genus of a covering. If W is a covering of C of degree n, then

$$2g(W) - 2 = n(2g(C) - 2) + \sum_Q (e_Q - 1)$$

where the sum is taken over all points of W and e_Q is the ramification index of such a point.

Remark. The idea of constructing all abelian coverings of a curve by pull back from isogenies of commutative algebraic groups was first used in connection with the class field theory in the geometric case [La 56a, b]. Unramified coverings come from the Jacobian, and ramified coverings come from the generalized Jacobians of Rosenlicht. Parshin's original construction used the generalized Jacobians, but several people noticed that the above Kummer construction can be used instead.

We may now give Parshin's application.

Theorem 2.2. *The Shafarevich conjecture implies Mordell's conjecture.*

Proof. Suppose for concreteness that we are in the number field case, and C is a curve over a number field F. Suppose that there are infinitely many rational points P of X in F. By Shafarevich, there are infinitely many W_P isomorphic to each other, or say to a fixed curve W, which therefore has infinitely many rational maps onto C', contradicting the theorem of de Franchis, and concluding the proof. The same argument applies mutatis mutandis in the function field case.

The proof of the Shafarevich conjecture is reduced to the analogous theorem about abelian varieties, which reads as follows.

Shafarevich conjecture for abelian varieties. *Let F be a number field and let S be a finite set of primes of $\mathfrak{o}_F$. Then there is only a finite number of isomorphism classes of abelian varieties A over F with good reduction outside S.*

To get from abelian varieties to curves, one must use mostly Torelli's theorem. By a basic elementary fact recalled in §1, a given abelian variety has only a finite number of polarizations of given degree, and in particular has only a finite number of principal polarizations. Suppose we have infinitely many isomorphism classes of curves over F with good reduction outside S. By Torelli's theorem, their Jacobians provide infinitely many isomorphism classes of principally polarized abelian varieties with good reduction outside S. Then the Shafarevich conjecture for abelian varieties yields a contradiction of the above elementary fact.

For the rest of this chapter, when referring to the **Shafarevich conjecture**, we shall mean the conjecture for abelian varieties unless otherwise specified.

We recall the basic fact that if two abelian varieties A, B are isogenous over F, then A has good reduction at v if and only if B has good reduction at v. Thus good reduction depends only on an isogeny class. See also Theorem 4.1 below.

To prove the Shafarevich conjecture, one goes through two parts:

Finiteness I. *Given a number field F and an abelian variety A over F, there is only a finite number of isomorphism classes of abelian varieties over F which are isogenous to A over F.*

Finiteness II. *Given a number field F and a finite set of primes S, there is only a finite number of isogeny classes of abelian varieties over F of given dimension and having good reduction outside S.*

The next sections describe the properties of abelian varieties used to deal with these steps, and other theorems concerning abelian varieties, which contain both geometric aspects and diophantine aspects. Among other things, we shall prove:

Finiteness I implies Finiteness II

IV, §3. THE l-ADIC REPRESENTATIONS AND SEMISIMPLICITY

In Chapter III, §3 we defined the l-adic representations of the Galois group and of the endomorphisms. We consider these representations now in greater detail. As before, we suppose that l is a prime number not equal to the characteristic of the field k. We let $G_k = \mathrm{Gal}(k^a/k)$ be the Galois group. For simplicity we omit the sign $\otimes$ when tensoring with $\mathbf{Q}_l$ or $\mathbf{Z}_l$. Consider the following properties of a field k and abelian varieties over k.

Property 3.1 (Semisimplicity). *For every abelian variety A over k, the representation of G_k on $V_l(A)$ is semisimple.*

Property 3.2 (Tate property). *For every abelian variety over k, the natural maps*

$$\mathbf{Z}_l\,\mathrm{End}_k(A) \to \mathrm{End}_{G_k}(T_l(A)), \tag{1}$$

$$\mathbf{Q}_l\,\mathrm{End}_k(A) \to \mathrm{End}_{G_k}(V_l(A)), \tag{2}$$

are isomorphisms.

Note that the first isomorphism (1) is equivalent with the second (2). Indeed, all modules involved are torsion free finitely generated, so tensoring an isomorphism (1) with $\mathbf{Q}_l$ yields an isomorphism (2). Conversely, the map in (1) is injective, and its cokernel is torsion free because a homomorphism which vanishes on the points of order l factors through multiplication by l. Hence if (2) is an isomorphism, it follows that (1) is an isomorphism.

As a consequence of Property 3.2, we obtain a seemingly more general statement which is an alternative form of the Tate property [Ta 66].

Corollary 3.3. *Let A, B be abelian varieties defined over k satisfying Properties* 3.1 *and* 3.2. *Then the natural maps*

$$\mathbf{Z}_l \operatorname{Hom}_k(A, B) \to \operatorname{Hom}_{G_k}(T_l(A), T_l(B)),$$
$$\mathbf{Q}_l \operatorname{Hom}_k(A, B) \to \operatorname{Hom}_{G_k}(V_l(A), V_l(B)),$$

are isomorphisms.

Proof. As in Property 3.2, it suffices to prove the second statement, for vector spaces over $\mathbf{Q}_l$. We then apply (2) of Property 3.2 to the product $A \times B$, and use the formula

$$\operatorname{End}_k(A \times B) = \operatorname{End}_k(A) \times \operatorname{Hom}_k(A, B) \times \operatorname{Hom}_k(B, A) \times \operatorname{End}_k(B)$$

as well as the analogous formula for the ring of G_k-endomorphisms of

$$V_l(A \times B) = V_l(A) \times V_l(B).$$

In addition, we also get the immediate consequence:

Corollary 3.4 (Isogeny theorem). *Let A, B be abelian varieties over k satisfying Properties* 3.1 *and* 3.2. *If $V_l(A)$ and $V_l(B)$ are G_k-isomorphic, then A, B are isogenous over k.*

Proof. Let $\varphi\colon V_l(A) \to V_l(B)$ be a G_k-isomorphism. Multiplying by a suitable l-adic integer, we may assume without loss of generality that we have a homomorphism

$$\varphi\colon T_l(A) \to T_l(B)$$

of finite cokernel. By Corollary 3.3, if $\alpha_1, \ldots, \alpha_r$ is a basis of $\operatorname{Hom}_k(A, B)$ over $\mathbf{Z}$, we can write

$$\varphi = \sum_{i=1}^{r} z_i \alpha_i \qquad \text{with} \quad z_i \in Z_l.$$

Let n_i be ordinary integers l-adically very close to z_i. Then $\Sigma\, n_i \alpha_i$ has finite kernel in A_{l^∞}, and so is an isogeny, thus proving the theorem.

Note that the argument actually proves that Corollary 3.3 implies Corollary 3.4.

The two finiteness properties may be formulated for other fields besides number fields over which they may be true. We denote them by

Finiteness I(k) and Finiteness II(k)

for a field k. Let us also define an **l-isogeny** to be an isogeny whose degree is a power of the prime l. Then we have a theorem of Zarhin:

Theorem 3.5 ([Zar 75]). *Finiteness $I(k)$ for l-isogenies implies Property 3.1 (semisimplicity) and Property 3.2 (Tate property) for the field k, and hence also the isogeny theorem over k, Corollary 3.4.*

Because the arguments used to prove properties 3.1 and 3.2 from Finiteness I are not very long, we shall give them in full. Part of them stem from [Ta 66]. These arguments are contained in two lemmas. The first gives us projection operators on the way to semisimplicity.

Lemma 3.6. *If the l-isogeny class of A over k contains only a finite number of k-isomorphism classes, then for all $\mathbf{Q}_l[G_k]$-submodules W of $V_l(A)$ these exists $u \in \mathbf{Q}_l \operatorname{End}_k(A)$ such that*

$$u(V_l(A)) = W.$$

Proof. Abbreviate $T_l = T_l(A)$ and $V_l = V_l(A)$. Let:

$$W^0 = W \cap T_l,$$

$$W_n^0 = W^0 + l^n T_l \subset T_l \qquad \text{for} \quad n \geqq 1,$$

$$W_n = \gamma_n(W_n^0) \qquad \text{where} \quad \gamma_n\colon T_l \to T_l/l^n T_l \text{ is the canonical map.}$$

We can identify W_n as a subgroup of $A[l^n]$. We consider the isogeny

$$\alpha_n\colon A \to A/W_n = B_n$$

and the corresponding

$$\beta_n\colon B_n \to A \qquad \text{such that} \quad \beta_n \alpha_n = l_A^n.$$

Both α_n and β_n are defined over k.

We claim that $\beta_n(T_l(B_n)) = W_n^0$.

It suffices to prove that both the right-hand side and left-hand side contain $l^n T_l$ and have the same canonical image mod $l^n T_l$, in $A[l^n]$. But first

$$\beta_n T_l(B_n) \supset \beta_n \alpha_n T_l(A) = l^n T_l(A).$$

And second, by definition $W_n = \ker \alpha_n$. Then trivially

$$\beta_n(B_n[l^n]) \subset \ker \alpha_n.$$

Conversely, let $\alpha_n(x) = 0$. There exists $y \in B$ such that $x = \beta_n y$ since β_n is surjective. But then

$$0 = \alpha_n(x) = \alpha_n \beta_n y = l^n y$$

so $y \in B[l^n]$ whence $\ker \alpha_n \subset \beta_n(B_n[l^n])$ thus proving the claim.

By the hypothesis that there is only a finite number of isomorphism classes in the isogeny class of A, we can find a sequence I of integers with smallest element n, and isomorphisms

$$v_i\colon B_n \to B_i \qquad \text{for} \quad i \in I.$$

Let

$$u_i = \beta_i v_i \beta_n^{-1} \in \mathbf{Q}\ \mathrm{End}_k(A).$$

Then

$$\begin{aligned} u_i(W_n^0) &= \beta_i v_i \beta_n^{-1}(W_n^0) \\ &= \beta_i v_i \beta_n^{-1} \beta_n T_l(B_n) \quad \text{by the claim} \\ &= \beta_i v_i T_l(B_n) \\ &= \beta_i T_l(B_i) = W_i^0 \subset W_n^0. \end{aligned}$$

Therefore u_i is actually an endomorphism of W_n^0. Since $\mathrm{End}(W_n^0)$ is compact, after taking a subsequence of I we may assume without loss of generality that the sequence $\{u_i\}$ converges to an endomorphism u of W_n^0. But $\mathbf{Q}_l W_n^0 = V_l(A)$. Hence the sequence $\{u_i\}$ also converges in $\mathbf{Q}_l\ \mathrm{End}_k(A)$, so we may view u as the restriction to W_n^0 of an element of $\mathbf{Q}_l\ \mathrm{End}_k(A)$, denoted by the same letter. Since W_n^0 is compact, every element $x \in u(W_n^0)$ is a limit

$$x = \lim x_i \qquad \text{with} \quad x_i \in u_i(W_n^0) = W_i^0.$$

Hence

$$u(W_n^0) = \bigcap_{i \in I} W_i^0 = W \cap T_l.$$

Thus finally $u(V_l(A)) = W$, and the lemma is proved.

The next lemma shows how to apply the conclusion of Lemma 3.6.

Lemma 3.7. *Let again A be an abelian variety defined over a field k. Suppose that for each $\mathbf{Q}_l[G_k]$-submodule W of $V_l(A)$ there exists an element $u \in \mathbf{Q}_l\ \mathrm{End}_k(A)$ such that $u(V_l(A)) = W$. Then:*

(1) *$V_l(A)$ is G_k-semisimple.*

(2) *Let $E =$ image of $\mathbf{Q}_l[G_k]$ in $\mathrm{End}\ V_l(A)$. Then E is the commutant of $\mathbf{Q}_l\ \mathrm{End}_k(A)$ in $\mathrm{End}\ V_l(A)$.*

(3) $\mathbf{Q}_l\ \mathrm{End}_k(A) = \mathrm{End}_{G_k}(V_l)$.

Proof. Let $V_l = V_l(A)$. It is a basic fact from the theory of abelian varieties that $\mathbf{Q}_l \operatorname{End}_k(A)$ is a semisimple $\mathbf{Q}_l$-algebra. To prove (1), let W be a G_k-subspace of V_l. By hypothesis, there exists $u \in \mathbf{Q}_l \operatorname{End}_k(A)$ such that $u(V_l) = W$. The right ideal $u(\mathbf{Q}_l \operatorname{End}_k(A))$ is generated by an idempotent e and $eV_l = W$. Since elements of G_k commute with elements of $\operatorname{End}_k(A)$, the G_k-semisimplicity of V_l follows.

To prove (2), we follow Tate. Let C be the commutant of $\mathbf{Q}_l \operatorname{End}_k(A)$ in $\operatorname{End}(V_l)$. Then $E \subset C$. Conversely, let $c \in C$. Then for every G_k-submodule W of V_l we have $cW \subset W$. Indeed, there exists $u \in \mathbf{Q}_l \operatorname{End}_k(A)$ such that $uV = W$, whence

$$cW = cuV = ucV \subset W$$

so W is stable under c. Take for W a simple factor of V_l, corresponding to a simple factor S of E. Given $x \in W$ there exists $s \in S$ such that $cx = sx$, and therefore c_W commutes with elements of $\operatorname{End}_S(W)$. Since W is S-simple, it follows that c_W is the same as multiplication by an element of $\mathbf{Q}_l$. It follows that $c \in E$, thus proving (2).

As for (3), since $\mathbf{Q}_l \operatorname{End}_k(A)$ is semisimple, it follows that V_l is semisimple over $\mathbf{Q}_l \operatorname{End}_k(A)$, so (3) follows directly from (2) and the basic theory of semisimple algebras. [Use for instance Jacobson's bicommutant theorem, as in my *Algebra*, Chapter XVII, Theorem 3.2.]

Combining the two lemmas proves Properties 3.1 and 3.2, and therefore also their consequences Corollaries 3.3 and 3.4. Thus we have proved Theorem 3.5.

Theorem 3.8 (Faltings). *Let $k = F$ be a number field. Then abelian varieties over F satisfy semisimplicity and the Tate property (i.e. Properties* 3.1 *and* 3.2), *and therefore their corollaries, e.g. the isogeny theorem.*

To prove Theorem 3.8, it will now suffice to prove Finiteness I for a number field. This will be described in §5. The next section, which is more elementary, will prove that Finiteness I implies Finiteness II, and so implies the Shafarevich conjecture.

Remark 1. When $\dim A = 1$, the semisimplicity of the representation of G_k on $V_l(A)$ was proved by Serre [Ser 72], who proved that either A has complex multiplication, that is $\operatorname{End}_{\mathbf{C}}(A)$ has rank 2 over $\mathbf{Z}$, or the image of G_k in $\operatorname{Aut} T_l(A)$ is open in $\mathrm{GL}_2(\mathbf{Z}_l)$.

Remark 2. All of Faltings' theorems which we are stating for number fields have also been proved by Faltings for finitely generated extensions of $\mathbf{Q}$, specifically: the conjecture of Shafarevich, the semisimplicity of

l-adic representations, the Tate conjecture, the isogeny theorem. See Faltings' Chapter VI in [FaW 84]. As we already pointed out in Chapter I, §7, the Mordell conjecture followed from the number field case in the more general case by specialization, e.g. via the Hilbert irreducibility theorem.

In addition, the function field case over finite fields was known by work of Tate [Ta 66] and the sequence of papers of Zarhin [Zar 74–76].

IV, §4. THE FINITENESS OF CERTAIN l-ADIC REPRESENTATIONS FINITENESS I IMPLIES FINITENESS II

The previous section consisted of theorems over an arbitrary field. Next:

We let R be a discrete valuation ring and F its quotient field. We let v be the associated valuation, $k(v)$ the residue class field, and l a prime not equal to the characteristic of $k(v)$.

Let V be a finite dimensional vector space over the l-adic field $\mathbf{Q}_l$, and let

$$\rho\colon G_k \to \mathrm{Aut}(V)$$

be a representation (continuous homomorphism). Such a representation is called an **l-adic representation**. The kernel of ρ is a closed normal subgroup, whose fixed field is a possibly infinite Galois extension K of F.

Let w be a valuation of K extending v. The **decomposition group** G_w is the subgroup of $G_{K/F}$ which leaves w fixed, i.e. it is the stabilizer of w. An element of G_w induces an automorphism of the residue class field $k(w)$ over the residue class field $k(v)$, and we thus get a homomorphism

$$G_w \to \bar{G}_w \subset G\big(k(w)/k(v)\big).$$

The kernel of this homomorphism is called the **inertia group**. If the inertia group is trivial, then w is said to be **unramified over** v. Since all the decomposition groups G_w for $w|v$ are conjugate in $G_{K/F}$, it follows that if some w is unramified, then all $w|v$ are unramified, so we simply say that v itself is **unramified** in K. Equivalently, the representation ρ is said to be **unramified** at v if its kernel contains the inertia groups of G_w for $w|v$.

Let A be an abelian variety defined over F. If A has good reduction at v, then it is part of the basic general theory that the reduction map

$$T_l(A) \to T_l(A_{k(v)})$$

is an isomorphism, and that the representation of G_F on $T_l(A)$ is unramified. Much deeper is the converse due to Ogg for elliptic curves, and mostly Néron in general, but see Serre–Tate [SeT 68].

Theorem 4.1. *Let A be an abelian variety defined over the field F which is the quotient field of a discrete valuation ring. Let l be a prime unequal to the characteristic of the residue class field. Then A has good reduction if and only if the representation of G_F on $V_l(A)$ is unramified.*

Corollary 4.2 (Koizumi–Shimura). *If B is isogenous to A over F and A has good reduction, then B has good reduction also.*

Proof. An isogeny induces a G_F-isomorphism $V_l(A) \to V_l(B)$, and we can apply Theorem 4.1 to conclude the proof.

From now on, let F be a number field and let v be a finite absolute value. If ρ is unramified at v, then there exists a unique element $\mathrm{Fr}_w \in G_w$ such that Fr_w induces the Frobenius automorphism on the residue class field of w. This **Frobenius automorphism** is the unique element such that, if $k(v)$ has q elements, then

$$\overline{\mathrm{Fr}}_v(x) = x^q \qquad \text{for all} \quad x \in k(w).$$

All such elements Fr_w are conjugate in $G_{K/F}$. Hence the trace

$$\mathrm{tr}\big(\rho(\mathrm{Fr}_w)\big)$$

depends only on v and not on w. Thus we shall write it $\mathrm{tr}\big(\rho(\mathrm{Fr}_v)\big)$. We now have a basic result.

Theorem 4.3 (Faltings). *Let F be a number field and let d be a positive integer. Let S be a finite set of finite places of F, and for each $v \notin S$ let Z_v be a finite set of elements of $\mathbf{Q}_l$. Up to isomorphism, there exists only a finite number of semisimple representations of G_F of dimension d over $\mathbf{Q}_l$, unramified outside S, and such that for $v \notin S$ the traces of Frobenius elements Fr_v lie in Z_v.*

The proof of Theorem 4.3 is based on the following lemma.

Lemma 4.4. *Given a finite set S of finite places of F, there exists a finite set S' of places disjoint from l and S, such that if ρ, ρ' are two semisimple l-adic representations of dimension d, of G_F, unramified outside S, and if*

$$\mathrm{tr}\, \rho(\mathrm{Fr}_v) = \mathrm{tr}\, \rho'(\mathrm{Fr}_v) \qquad \textit{for all} \quad v \in S',$$

then ρ is G_K-isomorphic to ρ'.

Proof. Since G_F is compact, in the l-adic spaces V and V' of ρ and ρ' respectively, there exist l-adic lattices T in V and T' in V' which are G_F-stable. (An l-adic **lattice** is a free $\mathbf{Z}_l$-module of rank d.) Let R be the $\mathbf{Z}_l$-algebra generated by the image of G_F in

$$\mathrm{End}_{\mathbf{Z}_l}(T) \times \mathrm{End}_{\mathbf{Z}_l}(T').$$

Then $\dim_{\mathbf{Z}_l}(R) \leqq 8d^2$. We have a natural homomorphism

$$G_F \to (R/lR)^*,$$

and $\#(R/lR) \leqq l^{\dim R}$. Hence the cardinality of R/lR is bounded. Furthermore, by Nakayama's lemma, representatives of R/lR generate R over $\mathbf{Z}_l$, and R/lR is itself generated over Z_l/lZ_l by the images of the elements of G_F in $(R/lR)^*$. The representation of G_F in the finite group $(R/lR)^*$ is unramified outside S. By a theorem of Hermite, there exists only a finite number of extensions of a number field of bounded degree, unramified outside S. Thus in fact, there is a subgroup H which is closed and of finite index in G_F such that for all representations ρ, ρ' as in the lemma, the kernel contains H, and we are actually representing the finite group G/H in $(R/lR)^*$. Now by Tchebotarev's theorem, there exists a finite set S' of absolute values such that the Frobenius elements Fr_v for $v \in S'$ have images which cover the image of G_F in $(R/lR)^*$. If

$$\mathrm{tr}\,\rho(\mathrm{Fr}_v) = \mathrm{tr}\,\rho'(\mathrm{Fr}_v) \qquad \text{for} \quad v \in S',$$

then we obtain $\mathrm{tr}\,\rho(\alpha) = \mathrm{tr}\,\rho'(\alpha)$ for all $\alpha \in R$, whence ρ is G_F-isomorphic to ρ' since the representations are assumed semisimple. This proves the lemma.

If we now apply the hypothesis that the traces of Frobenius are bounded, in the sense that they can take only a finite number of values for each v, then Theorem 4.3 follows directly from Lemma 4.4.

Example. Let ρ_l be the representation on $V_l(A)$ for some abelian variety A of dimension d over F. By Weil's theorem (Hasse in dimension 1) the trace of Frobenius $\mathrm{tr}\,\rho_l(\mathrm{Fr}_v)$ is an ordinary integer, satisfying

$$|\mathrm{tr}\,\rho_l(\mathrm{Fr}_v)| \leqq 2dq_v^{1/2},$$

where q_v is the number of elements in the residue class field. Hence Theorem 4.3 applies to the l-adic representations with which we are concerned.

Remark. Faltings' lemma is remarkably simple. The difficulty of the lemma lay in discovering it!

Corollary 4.5. *Finiteness I implies Finiteness II. Actually, we have the following implications*:

l-Finiteness I $\Rightarrow$ *Semisimplicity and Tate's property for l*

$\Rightarrow$ *Semisimplicity and Isogeny theorem for l*

$\Rightarrow$ *Finiteness II.*

Proof. Recall that Finiteness II states:

Given a positive integer d, a number field F and a finite set of primes S, there exists only a finite number of isogeny classes of abelian varieties over F which have dimension d and good reduction outside S.

To prove this, we fix a prime l. We let S_l consist of S together with all the primes of F dividing l. If A is an abelian variety of dimension d and good reduction outside S_l then $V_l(A)$ is a semisimple representation of G_F by Theorem 3.5, and we can apply Theorem 4.3 to the family of such representations. We conclude that there is only a finite number up to isomorphism, and by Tate's property in the form of the Isogeny Theorem 3.4, we conclude that there is only a finite number of isogeny classes over F with good reduction outside S_l. This proves Corollary 4.5.

IV, §5. THE FALTINGS HEIGHT AND ISOGENIES: FINITENESS I

Isomorphism classes of abelian varieties are parametrized by what are called moduli varieties. Faltings defined the height of an abelian variety directly, and showed how it was related to the height of the associated point on the moduli variety with respect to some divisor class. He proved that the heights of abelian varieties isogenous to a given one are bounded, whence follows Finiteness I. Unless otherwise specified, the results of this section are also due to Faltings.

As in Chapter III, §4 we let $\mathfrak{o}$ be a Dedekind ring with quotient field F and we let $S = \mathrm{spec}(\mathfrak{o})$. We let A_F be an abelian variety over F, and we let $A_\mathfrak{o}$ be its Néron model over S.

Let $\zeta\colon \mathrm{spec}(\mathfrak{o}) \to A_\mathfrak{o}$ be the zero section. Let Ω^1 denote the usual sheaf of differentials. Since $A_\mathfrak{o}$ is smooth over spec($\mathfrak{o}$), it follows that $\Omega^1(A_\mathfrak{o}/S)$ is locally free of rank equal to the relative dimension of $A_\mathfrak{o}$ over S. We abbreviate $\Omega^1(A_\mathfrak{o}/S)$ by $\Omega^1(A_\mathfrak{o})$. We then have the determinant

$$\det \Omega^1(A_\mathfrak{o}) = \bigwedge^{\max} \Omega^1(A_\mathfrak{o}),$$

the dual of the Lie algebra $\mathrm{Lie}(A_\mathfrak{o})^\vee = \zeta^*\Omega^1(A_\mathfrak{o})$, and what we call the

co-Lie determinant

$$\mathscr{L}(A_{\mathfrak{o}}) = \zeta^* \det \Omega^1(A_{\mathfrak{o}}) = \det \operatorname{Lie}(A_{\mathfrak{o}})^{\vee},$$

which is the determinant of the dual Lie algebra $\zeta^*\Omega^1(A_{\mathfrak{o}})$. The co-Lie determinant is a line sheaf over spec($\mathfrak{o}$).

We now define an Arakelov metric and degree for a line sheaf over spec($\mathfrak{o}$) in the case of a number field F whose ring of integers is $\mathfrak{o} = \mathfrak{o}_F$. First let L be a vector space of dimension 1 over F. Suppose given for each absolute value v on F a v-adic absolute value $|\ |_v$ on L, satisfying

$$|as|_v = |a|_v |s|_v \qquad \text{for all} \quad a \in F \quad \text{and} \quad s \in L.$$

We also assume that for $s \in L$, $s \neq 0$ we have $|s|_v = 1$ for all but a finite number of v. Let $N_v = [F_v : \mathbf{Q}_v]$, and let

$$\|s\|_v = |s|_v^{N_v}.$$

We define the **degree** of L, and the **degree** of the line sheaf $\mathscr{L} = L^{\sim}$ associated to L, to be

$$\deg \mathscr{L} = \deg L = -\sum_v \log \|s\|_v.$$

The right-hand side is independent of the choice of s by the product formula. The elements $s \in L$ are just the sections of $\mathscr{L}$.

By an $\mathfrak{o}$-**form** of L we mean a module $L_{\mathfrak{o}}$ locally free of rank 1 over $\mathfrak{o}$, so that $F \otimes_{\mathfrak{o}} L_{\mathfrak{o}} \approx L$. Let v be a finite absolute value on F. Each isomorphism of F_v-vector spaces $F_v \to L_v$ transports the absolute value of F_v to L_v. We pick such an isomorphism sending the local ring $\mathfrak{o}_v$ to $L_{\mathfrak{o},v}$. Then given $s \in L$, $s \neq 0$ we have $\|s\|_v = 1$ for almost all v. For $s \in L_{\mathfrak{o}}$,

$$(L_{\mathfrak{o}} : \mathfrak{o}_s) = \prod_{v \text{ fin}} \|s\|_v^{-1}.$$

For v at infinity, let $\sigma: F \to \mathbf{C}$ be an imbedding inducing v on F. Suppose given a hermitian structure, i.e. a positive definite hermitian form $\langle\ ,\ \rangle_\sigma$ on the $\mathbf{C}$-vector space $L_\sigma = \mathbf{C} \otimes_{F,\sigma} L$. Two complex conjugate imbeddings give rise to complex conjugate spaces L_σ, and we assume that for $s \in L$ we have

$$\langle s, s \rangle_\sigma = \langle \bar{s}, \bar{s} \rangle_{\bar{\sigma}}.$$

Then we let

$$\|s\|_v = \langle s, s \rangle_\sigma^{1/2} \qquad \text{for } v \text{ real}$$

$$\|s\|_v = \langle s, s \rangle_\sigma \qquad \text{for } v \text{ complex}.$$

The above constructions give rise to a set of absolute values on L of the type which then allows us to define $\deg L$.

We apply this degree to the Lie determinant. Let $A_{\mathbf{C}}$ be an abelian variety over the complex numbers. Let $H^0\Omega^1(A_{\mathbf{C}})$ be the space of holomorphic 1-forms. Then

$$\mathrm{Lie}(A_{\mathbf{C}})^{\vee} = H^0\Omega^1(A_{\mathbf{C}}) \qquad \text{and} \qquad \mathscr{L}(A_{\mathbf{C}}) = \textstyle\bigwedge^{\max} H^0\Omega^1(A_{\mathbf{C}}).$$

We define the hermitian structure such that for $\omega \in \mathscr{L}(A_{\mathbf{C}})$ we have

$$\langle \omega, \omega \rangle = \frac{1}{(2\pi)^d} \int_{A(\mathbf{C})} |\omega \wedge \bar{\omega}|, \qquad \text{where} \quad d = \dim A.$$

Globally, starting with our abelian variety A over F, for each imbedding σ of F into $\mathbf{C}$ we obtain an abelian variety A_σ over $\mathbf{C}$, and the above formula defines the Faltings hermitian structure on $\mathscr{L}(A_o)$ at the infinite absolute values. As a result, we can define the **Faltings height**

$$h_{\mathrm{Fal}}(A) = \frac{1}{[F:\mathbf{Q}]} \deg \mathscr{L}(A_o).$$

Theorem 5.1. *Let F be a number field, n, d, h_0 positive integers. Then there is only a finite number of isomorphism classes of polarized abelian varieties (A, c) over F, of dimension d, and polarization degree n such that $h_{\mathrm{Fal}}(A) \leq h_0$.*

Note that for the application to the proof of Finiteness I, it suffices to prove the finiteness of Theorem 5.1 for all abelian varieties in a given isogeny class.

Theorem 5.1 is of course analogous to the theorem that in projective space, there is only a finite number of points of $\mathbf{P}^N(F)$ of bounded height. To prove Theorem 5.1, Faltings compares his height with an ordinary height associated with an ample divisor class on a variety which parametrizes isomorphism classes of certain abelian varieties.

At the beginning of Chapter III we defined Néron models and semistability. We also mentioned the basic fact that taking connected components of Néron models commutes with base change under the hypothesis of semistability. We now have:

Proposition 5.2. *Let F be a number field. If A_F has semistable reduction, and if F' is a finite extension of F, then the Faltings heights of A_F and $A_{F'}$ are equal.*

Consequently, we may define the **stable Faltings height**

$$h_{\mathrm{Fal}}^{\mathrm{st}}(A_F) = h_{\mathrm{Fal}}(A_{F'})$$

for any finite extension F' such that $A_{F'}$ is semistable.

Let N be a positive integer. By a **level N structure** for an abelian variety A of dimension d we mean an isomorphism

$$\varepsilon\colon (\mathbf{Z}/N\mathbf{Z})^{2d} \to A[N].$$

Let A be defined over the field k of characteristic 0 which is all we want here. The isomorphism is to be a Galois isomorphism, and consequently this means that all the points of order N are rational over k. We consider **triples** (A, c, ε) consisting of an abelian variety of dimension d, a principal polarization c, and a level N structure with a positive integer N divisible by at least two primes $\geqq 3$. It follows that such a triple has no automorphisms other than the identity. Then there exists a **moduli space** $M_{d,N}$ for such triples, in the following sense.

Theorem 5.3. *Given a positive integer d and a positive integer N divisible by at least two primes $\geqq 3$, there exist:*

A finite disjoint union of projective varieties $\overline{M}_{d,N}$, defined over $\mathbf{Q}$ in the sense that this union is stable under the action of the Galois group $G_{\mathbf{Q}}$, such that all components have the same dimension $d(d+1)/2$;

an open Zariski dense subset $M_{d,N}$ which is non-singular, defined over $\mathbf{Q}$ as above, and whose complement in $\overline{M}_{d,N}$ has codimension d;

a proper smooth morphism $f\colon \mathbf{A} \to M_{d,N}$ over $\mathbf{Q}$ which is an abelian scheme of relative dimension d;

a level N structure $\boldsymbol{\varepsilon}\colon (\mathbf{Z}/N\mathbf{Z})^{2d} \to \mathbf{A}[N]$;

an algebraic equivalence class $\mathbf{c}$ on $\mathbf{A}$;

such that for every extension field k of $\mathbf{Q}$ the association

$$x \mapsto (\mathbf{A}_x, \mathbf{c}_x, \boldsymbol{\varepsilon}_x) = f^{-1}(x) \qquad \text{for} \quad x \in M_{d,N}(k)$$

is a bijection between the set of rational points $M_{d,N}(k)$ and the isomorphism classes of triples over k.

The algebraic set $\overline{M}_{d,N}$ is called the **Baily–Borel compactification** of $M_{d,N}$. In addition, since we are dealing with a smooth family of abelian varieties over $M_{d,N}$, we can define the **co-Lie determinant** as before as a line sheaf $\mathscr{L}$ on $M_{d,N}$. This line sheaf has an extension to a line sheaf over $\overline{M}_{d,N}$ (for a proof, see for instance [ChF 90], V, Theorem 2.5(1)); and when $d \geqq 2$ this extension is unique because the complement of $M_{d,N}$ in $\overline{M}_{d,N}$ has codimension $\geqq 2$. This co-Lie determinant corresponds to a divisor class in $\mathrm{Pic}(\overline{M}_{d,N})$, which we shall call the **co-Lie determinant class**

λ. As such, we can associate to λ a height function (up to $O(1)$)

$$h_\lambda: \overline{M}_{d,N} \to \mathbf{R},$$

which we call the **Lie height** on $\overline{M}_{d,N}$. Finally when $d \geqq 2$ it is a fact that the canonical sheaf $\bigwedge^{\max} \Omega^1_{M/\mathbf{Q}}$ has a unique extension to a line sheaf on $\overline{M}_{d,N}$, corresponding to a class in $\operatorname{Pic}(\overline{M}_{d,N})$, also called the **canonical class**, and denoted by K. Furthermore, we have the relation

$$(d + 1)\lambda = K,$$

and both λ, K are ample. These properties summarize the relevant facts about the moduli variety for the moment. Proofs can be found in Chai–Faltings [CF 90], and I am much indebted to Chai for his guidance on moduli facts.

If (A, c, ε) is a triple, we denote by

$$\mathbf{m}(A, c, \varepsilon)$$

the associated point in the moduli space $M_{d,N}$. Faltings then established a connection between his height and the Lie height as follows.

Theorem 5.4. *For every triple (A, c, ε) over a number field we have*

$$h^{\mathrm{st}}_{\mathrm{Fal}}(A) = h_\lambda(\mathbf{m}(A, c, \varepsilon)) + O(\log h_\lambda(\mathbf{m}(A, c, \varepsilon))).$$

Since λ is ample, the error term with a big O on the right makes sense, because we can take h_λ to be positive. The elementary fact that there is only a finite number of points in $\overline{M}_{d,N}(F)$ of bounded height, for every number field F, because λ is ample, now implies immediately the finiteness statement for triples with bounded Faltings height over F. This is the analogue of Theorem 5.1 for triples. It is a technical matter to reduce Theorem 5.1 to Theorem 5.4. It is a fact that in a given triple (A, c, ε) over a number field, the abelian variety A is semistable, so in Theorem 5.4, we are actually dealing with the semistable height on the left-hand side. To reduce the general case of Theorem 5.1 to Theorem 5.4, one has to adjoin points of finite order to create semistability; and one can use a theorem of Zarhin ([Zar 74], [Zar 77]) to deal only with principal polarizations. Indeed, if A' is the dual variety of A, then Raynaud has proved in general, even in the non-semistable case, that

$$h_{\mathrm{Fal}}(A') = h_{\mathrm{Fal}}(A).$$

and Zarhin pointed out that $(A \times A')^4$ is principally polarizable. These are the essential ingredients which go into a full proof.

For a more detailed exposition of Faltings' theorem, the reader can consult Faltings' own write up on heights and moduli spaces [Fa 84a, b], as well as Deligne [De 83]. Faltings stated explicitly that it would require an entire book to justify properly the use he was making of the moduli spaces, and Deligne pointed out certain properties (a), (b), (c) which, if available, would simplify the exposition considerably. These properties have since been proved, and will be found in the book by Chai–Faltings [ChF 90].

We now come to the second ingredient which enters into Faltings' proof of Finiteness I. This finiteness comes by putting together Theorem 5.1 for a given isogeny class, together with the next result from [Fa 83], [Fa 84b].

Theorem 5.5. *Let F be a number field and A_F an abelian variety over F with semistable reduction. Then the set of Faltings heights of abelian varieties B_F which are F-isogenous to A_F is a finite set.*

Raynaud ([Ra 85], Theorem 4.4.9) has given an effective bound for the height of abelian varieties isogeneous to A_F over F, and for the differences

$$|h_{\mathrm{Fal}}(A_F) - h_{\mathrm{Fal}}(B_F)|,$$

by a modification and extension of Faltings' methods. If $u_F\colon A_F \to B_F$ is an isogeny over F, and $u\colon A_{\mathfrak{o}} \to B_{\mathfrak{o}}$ is its extension to the Néron models, then he gives an explicit formula for the difference of the heights, extending Faltings' results to the non-semistable case. Indeed, define

$$\mathscr{L}_u = \mathscr{L}(A_{\mathfrak{o}}) \otimes \mathscr{L}(B_{\mathfrak{o}})^{-1}.$$

The canonical homomorphism $\mathscr{L}(B_{\mathfrak{o}}) \to \mathscr{L}(A_{\mathfrak{o}})$ is injective. Hence we get a canonical injection

$$\mathcal{O} \hookrightarrow \mathscr{L}_u,$$

where $\mathcal{O} = \mathfrak{o}^{\sim}$ is the sheaf associated with the ring of integers $\mathfrak{o}$. There exists an ideal $\mathfrak{d}_u$ ($\neq 0$ as usual) such that $\mathscr{L}_u = \mathcal{O}(\mathfrak{d}_u)$. We call $\mathfrak{d}_u$ the **different** of u. The Faltings metrics on $\mathscr{L}(A_{\mathfrak{o}})$ and $\mathscr{L}(B_{\mathfrak{o}})$ give rise to the tensor product metric on $\mathscr{L}_u$. Then **Raynaud's formula** is:

$$\deg \mathscr{L}(A_{\mathfrak{o}}) - \deg \mathscr{L}(B_{\mathfrak{o}}) = \deg(\mathscr{L}_u) = -\tfrac{1}{2}[F:\mathbf{Q}] \deg(u) + \log(\mathfrak{o}:\mathfrak{d}_u).$$

On the other hand, if A_F has semistable reduction, then instead of $\log(\mathfrak{o}:\mathfrak{d}_u)$, Faltings used another expression. Indeed, let ζ be the zero section and let $G = \ker u$. If A_F is semistable, then

$$(\mathfrak{o}:\mathfrak{d}_u) = \operatorname{rank} \zeta^*\Omega^1_{G/\mathfrak{o}},$$

which is what Faltings used in his original formula.

The finiteness of heights in Theorem 5.5 comes partly from the following result.

Theorem 5.6. *Let A_F have semistable reduction. There exists a finite set S of primes, such that for any isogeny $u_F: A_F \to B_F$ of degree not divisible by any primes in S, we have*

$$h_{\mathrm{Fal}}(A) = h_{\mathrm{Fal}}(B).$$

This result reduces Theorem 5.5 to the study of p-isogenies for a finite number of primes p, and to refined properties of ramification. It uses other results of Raynaud [Ra 74], among other things. See also the expositions of Faltings' work in Deligne [De 83], Schappacher [Sch 84], Szpiro [Szp 85b], and Wustholz [Wu 84].

Finally we note the technical point that Faltings in [Fa 83] proved the Shafarevich conjecture for abelian varieties for a given polarization degree. Making explicit that one can extend the result without mentioning the polarization was done by Zarhin [Zar 85], using his product trick.

Different types of finiteness bounds occur here. One of them is a bound on the number of solutions of a given diophantine problem, in this case the number of isomorphism classes, but another requires more, namely a bound for the height of the solutions, or in this case a bound for the degree of the isogeny. We shall deal with this from another point of view in the next section.

IV, §6. THE MASSER–WUSTHOLZ APPROACH TO FINITENESS I

Let A be an abelian variety over a number field. We want to bound in some sense isomorphism classes of abelian varieties isogenous to A. One approach runs as follows. Let

$$\alpha: A \to B$$

be an isogeny. One **basic isogeny problem** then is: *find another isogeny $\beta: A \to B$ whose degree is bounded only in terms of A.* This is the approach taken by Masser–Wustholz [MaW 91] who solve part of this problem effectively as in the next theorem.

Theorem 6.1. *Let d, m, n be positive integers. Then there exist effective constants $c = c(m, n, d)$ and $k = k(n)$ having the following property. If A, B are abelian varieties of dimension n over a number field F of degree d, having polarizations of degrees at most m, and if they are isogenous over F, then there exists an isogeny over F of degree at most*

$$c \cdot \max\left(1, h_{\mathrm{Fal}}^{\mathrm{st}}(A)\right)^k.$$

As in Faltings' proof the above result implies Finiteness I by using the Zarhin remark that $(A \times A')^4$ and $(B \times B')^4$ are principally polarized. However although Theorem 6.1 is effective, at the moment the reduction of the basic problem to the case of principally polarized varieties is not, for various reasons. For instance, if we are given a polarization on A with known degree, we do not know a priori a polarization on B whose degree is bounded in terms of $h_{\mathrm{Fal}}^{\mathrm{st}}(A)$ and the degree of the polarization on A. As Masser–Wustholz have observed, such a reduction requires ideal class estimates for endomorphism rings of abelian varieties which are not known to be effective today.

In particular, the above approach leads to the following questions raised in their forthcoming papers:

Bound the degree of some polarization on A in terms of its Faltings height.

Bound the degree of an isogeny as stated at the beginning of the section.

Bound the discriminant of the ring of endomorphisms of an abelian variety in terms of the Faltings height.

Of course one can require more than just bounds, namely bounds having roughly a form similar to the estimate in Theorem 6.1. Since current results concerning these questions are very partial at the moment, I shall not go any further into them.

We shall return to a discussion of the proof of Theorem 6.1 in Chapter IX because this proof in some ways follows a pattern whose origins lie in the Baker method of diophantine approximation. We note here already that the proof in the manuscript I have seen, like Faltings' proof also uses a good dose of moduli theory as in Chai–Faltings, and also uses an explicit version of Theorem 5.4 which makes effective the constants implicit in that theorem. However, the Masser–Wustholz proof replaces Raynaud's p-adic theory by arguments at infinity.

CHAPTER V

Modular Curves Over **Q**

Among all curves, there is a particularly interesting family consisting of the modular curves, which we shall describe in §1. These parametrize elliptic curves with points of order N, or cyclic subgroups of order N. They form the prototype of higher dimensional versions, modular varieties, which parametrize abelian varieties with other structures involving points of finite order. We have already seen the use of such varieties in Faltings' proof of the Mordell conjecture, and more specifically of the Shafarevich conjecture, in Chapter IV, §5.

Here we concentrate on elliptic curves and points of finite order, as parametrized by modular curves. In this case, Mazur was able to describe completely the rational points in the most classical sense, i.e. over **Q**, lying on the modular curves, and we shall describe Mazur's results in §2. These results also involve the determination of certain Mordell–Weil groups for the Jacobian, or quotients of the Jacobian, of modular curves.

A famous conjecture of Taniyama–Shimura asserts that every elliptic curve over **Q** is modular, in the sense that it is a rational image of a modular curve. Frey had the idea of associating an elliptic curve to every solution of the Fermat equation in such a way that the curve would exhibit remarkable properties which would contradict the Taniyama–Shimura conjecture. It turned out that there were serious difficulties in carrying out this idea, having to do with the Galois representation on points of finite order of the curve. Ribet succeeded in proving a result on such representations which was strong enough to show that Taniyama–Shimura implies Fermat's last theorem. We shall indicate the main idea in §3.

In §4 we give an application of modular theory to a classical problem concerning Pythagorean triples, stemming from Tunnell.

Finally, in §5, we show in the case of rank 1 how one can construct a rational point of infinite order, following work of Gross–Zagier, with additional information due to Kolyvagin. At the present time, it is not known what to do when the rank is greater than 1, or what to do in higher dimension. The modular construction and the insight via the Birch–Swinnerton-Dyer conjecture provide an explicit solution of a diophantine problem, and also establish a connection with the theory of cyclotomic and modular units. The general pattern behind these special phenomena is also unknown at present.

Thus the present chapter can be seen as giving concrete instances of the more general theory of previous chapters, and also showing how more specific results have been proved in the case of curves defined over the ultimate base field, the rational numbers themselves.

V, §1. BASIC DEFINITIONS

A general pattern of algebraic geometry is that isomorphism classes of certain geometric objects are parametrized by varieties. We have already seen an example of this situation with the Picard variety, which parametrizes divisor classes, or isomorphism classes of line sheaves. Here we shall be concerned with isomorphism classes of elliptic curves and some additional structure arising from points of finite order. We start with a complex analytic description of the situation.

Let $\mathfrak{H}$ be the upper half plane, that is the set of complex numbers

$$\tau = x + iy \qquad \text{with} \quad y > 0.$$

Let $\Gamma(1) = \mathrm{SL}_2(\mathbf{Z})$ be the modular group, that is the group of matrices

$$\gamma = \begin{pmatrix} a & b \\ c & d \end{pmatrix}$$

with integer coefficients and determinant 1. Then $\Gamma(1)$ operates on $\mathfrak{H}$ by

$$\tau \mapsto \frac{a\tau + b}{c\tau + d},$$

and ± 1 operates trivially, so we get a faithful representation of $\Gamma(1)/\pm 1$ in $\mathrm{Aut}_{\mathbf{C}}(\mathfrak{H})$. The coset space has a representative fundamental domain with a well-known shape pictured below.

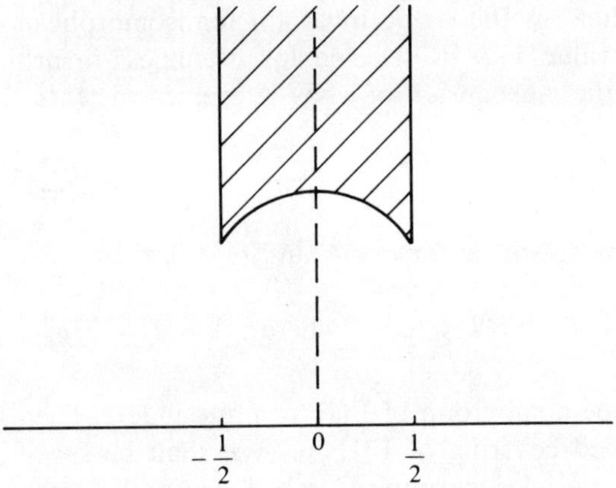

There is a classical function, holomorphic on $\mathfrak{H}$ and invariant under $\Gamma(1)$ called the j-**function**, which gives a holomorphic isomorphism

$$j: \Gamma(1)\backslash\mathfrak{H} \to \mathbf{P}^1(\mathbf{C}) - \{\infty\}$$

with the affine line (projective 1-space from which infinity is deleted). If one takes

$$q = e^{2\pi i\tau}$$

as a local parameter at infinity, then one can compactify $\Gamma(1)\backslash\mathfrak{H}$ by adjoining the point at infinity, thus obtaining a compact Riemann surface isomorphic to $\mathbf{P}^1(\mathbf{C})$. In terms of q, the function j has a Laurent expansion

$$j = \frac{1}{q} + 744 + 196884q + \text{higher terms.}$$

One can characterize j analytically by stating that $j(i) = 1728$ and $j(e^{2\pi i/3}) = 0$, while $j(\infty) = \infty$. This is rather ad hoc. A better way to conceive of j is in terms of isomorphism classes of complex toruses as follows.

Let $\Lambda = [\omega_1, \omega_2]$ be a lattice in $\mathbf{C}$, with basis ω_1, ω_2 over the integers. This means that Λ is an abelian group generated by ω_1 and ω_2, and that these two elements are linearly independent over the real numbers. In addition, we shall always suppose that ω_1/ω_2 lies in the upper half plane, so we put $\omega_1/\omega_2 = \tau$. Then the invariance of j under $SL_2(\mathbf{Z})$ shows that the value $j(\tau)$ is independent of the choice of basis as above, and in addition is the same if we replace $[\omega_1, \omega_2]$ by $[c\omega_1, c\omega_2]$ for any complex number $c \neq 0$. Thus we may define $j(\Lambda) = j(\tau)$, and we have $j(c\Lambda) = j(\Lambda)$. But $\mathbf{C}/\Lambda$ is a complex torus of dimension 1, and the above arguments show that j is the single invariant for isomorphism classes of such toruses. The value 1728 is selected for usefulness in arithmetic applications, so that the q-expansion of j has integer coefficients. If we put

$$g_2(\Lambda) = 60 \sum \omega^{-4} \qquad \text{and} \qquad g_3(\Lambda) = 140 \sum \omega^{-6},$$

where the sums are taken for $\omega \in \Lambda$ and $\omega \neq 0$, then

$$j = 1728 g_2^3/\Delta, \qquad \text{where} \quad \Delta = g_2^3 - 27 g_3^2.$$

Now let Γ be a subgroup of $\Gamma(1)$, of finite index. Then $\Gamma\backslash\mathfrak{H}$ is a finite possibly ramified covering of $\Gamma(1)\backslash\mathfrak{H}$. We shall be specifically interested in some very special subgroups Γ, which we now describe. Let N be a positive integer. We define:

$\Gamma(N)$ = subgroup of elements $\gamma \equiv \begin{pmatrix} 1 & 0 \\ 0 & 1 \end{pmatrix}$ mod N;

$\Gamma_1(N)$ = subgroup of elements $\gamma \equiv \begin{pmatrix} 1 & b \\ 0 & 1 \end{pmatrix}$ mod N with arbitrary b;

$\Gamma_0(N)$ = subgroup of elements $\gamma \equiv \begin{pmatrix} a & b \\ 0 & d \end{pmatrix}$ mod N with arbitrary a, b.

We may view $1/N$ as a point of order N on the torus $\mathbf{C}/[\tau, 1]$. Let Z_N be the cyclic group generated by $1/N$. Then we may consider the pair $(\mathbf{C}/[\tau, 1], Z_N)$ as consisting of a torus and a cyclic subgroup of order N. We have the following parametrizations:

The association

$$\tau \mapsto (\mathbf{C}/[\tau, 1], 1/N)$$

gives a bijection between $\Gamma_1(N)\backslash\mathfrak{H}$ and isomorphism classes of toruses together with a point of order N.

The association

$$\tau \mapsto (\mathbf{C}/[\tau, 1], Z_N)$$

gives a bijection between $\Gamma_0(N)\backslash\mathfrak{H}$ and isomorphism classes of toruses together with a cyclic subgroup of order N.

Furthermore, there exist affine curves $Y_1(N)$ and $Y_0(N)$, defined over $\mathbf{Q}$, such that

$$Y_1(N)(\mathbf{C}) \approx \Gamma_1(N)\backslash\mathfrak{H} \qquad \text{and} \qquad Y_0(N)(\mathbf{C}) \approx \Gamma_0(N)\backslash\mathfrak{H},$$

and such that $Y_1(N)$ parametrizes isomorphism classes of pairs (E, P) algebraically, where E is an elliptic curve and P is a point of order N, in the following sense. If k is a field containing $\mathbf{Q}$, then a point of $Y_1(N)(k)$ corresponds to such a pair (E, P) with E defined over k and P rational over k. Similarly, for $Y_0(N)$ parametrizing pairs (E, Z), where E is defined over k and Z is invariant under the Galois group G_k.

The affine curve $Y_1(N)$ can be compactified by adjointing the points which lie above $j = \infty$. Its completion is denoted by $X_1(N)$, and the points at infinity are called the **cusps**. The set of cusps on $X_1(N)$ is denoted by $X_1^\infty(N)$.

Similarly, we have the completion $X_0(N)$ of $Y_0(N)$ with its **cusps** $X_0^\infty(N)$.

The ramified covering $Y_1(N) \to Y_0(N)$ extends to a ramified covering

$$X_1(N) \to X_0(N)$$

defined over $\mathbf{Q}$. This covering is Galois, with Galois group isomorphic to $(\mathbf{Z}/N\mathbf{Z})^*/\pm 1$, under the association

$$a \mapsto \sigma_a \equiv \begin{pmatrix} a & 0 \\ 0 & a^{-1} \end{pmatrix} \bmod N \qquad \text{for} \quad a \in (\mathbf{Z}/N\mathbf{Z})^*.$$

V, §2. MAZUR'S THEOREMS

We consider the modular curves $X_0(N)$ and $X_1(N)$ over $\mathbf{Q}$ and ask for a description of the sets of rational points. If p is a prime dividing N, then we have natural finite morphisms

$$X_0(N) \to X_0(p) \qquad \text{and} \qquad X_1(N) \to X_1(p).$$

The essential problem is to describe the rational points when N is prime. Mazur gives a complete description, in the following theorems [Maz 77] and [Maz 78].

Theorem 2.1. *Let N be a prime number such that the genus of $X_0(N)$ is > 0, that is, $N = 11$ or $N \geqq 17$. Then the only rational points of $X_0(N)$ in $\mathbf{Q}$ are at the cusps, except when $N = 11, 17, 19, 37, 43, 67$, or 163.*

For the exceptional cases of N listed above, Mazur gives a complete table describing $X_0(N)(\mathbf{Q})$ in [Maz 78].

In light of the representation of pairs (E, P) of an elliptic curve and a rational point of order N by the points on $X_0(N)$, one gets an interpretation in terms of elliptic curves as follows.

Theorem 2.2. *Let N be a prime number such that some elliptic curve over $\mathbf{Q}$ admits a cyclic subgroup of order N stable under the Galois group $G_{\mathbf{Q}}$. Then:*

$$N = 2, 3, 5, 7, 13 \textit{ when } X_0(N) \textit{ has genus } 0;$$

$$N = 11, 17, 19, 37, 43, 67, \textit{ or } 163 \textit{ otherwise.}$$

As to rational torsion points on elliptic curves, Mazur gets:

Theorem 2.3. *Let E be an elliptic curve over $\mathbf{Q}$. Then the torsion subgroup $E(\mathbf{Q})_{\text{tor}}$ is isomorphic to one of the following groups:*

$$\mathbf{Z}/m\mathbf{Z} \qquad \textit{with} \quad 1 \leqq m \leqq 10 \quad \textit{or} \quad m = 12;$$

$$\mathbf{Z}/2\mathbf{Z} \times \mathbf{Z}/2d\mathbf{Z} \qquad \textit{with} \quad 1 \leqq d \leqq 4.$$

Part of the investigation of these sets of rational points lies in an analysis of the rational points on the Jacobians of $X_0(N)$ and $X_1(N)$, which are denoted by $J_0(N)$ and $J_1(N)$. We map these curves into their Jacobians over $\mathbf{Q}$, and send the cusp at infinity to the origin of the Jacobian.

Now take N prime $\geqq 3$. It turns out that there are precisely two cusps on $X_0(N)$ lying above $j = \infty$. The degree of the above covering is given by

$$[X_1(N) : X_0(N)] = \frac{N-1}{2},$$

and it is easy to see that the covering is unramified over the cusps. If we let Q_0, Q_∞ be the two cusps (lying above 0 and ∞ respectively in the complex representation of the upper half plane), then

$$(Q_0) - (Q_\infty)$$

is a 0-cycle of degree 0 on $X_0(N)$, which corresponds to a point in the Jacobian. We define the **cuspidal (divisor class) group** to be the group generated by the associated point in the Jacobian, and denote this group by C. It is a fact that C is finite, and that

$$\text{order of } C = n = \text{numerator of } \frac{N-1}{12}.$$

We suppose from now on that the genus of $X_0(N)$ is > 0, so $X_0(N)$ gets imbedded in its Jacobian.

We shall describe the algebra of endomorphisms of $J_0(N)$ by giving generators, in terms of the modular representation via elliptic curves and cyclic subgroups of order N. Such endomorphisms are induced by correspondences on $X_0(N)$, and a correspondence is described by associating a divisor to a point. We now describe these correspondences.

First there is an **involution** (automorphism of order 2) on $X_0(N)$ which consists of the association

$$(E, Z) \mapsto (E/Z, E[N]/Z).$$

Recall that $E[N]$ is the kernel of multiplication by N on E. In the representation from the upper half plane, this involution is induced by the map

$$\tau \mapsto -1/N\tau.$$

Second, for a prime number $l \nmid N$, we define the **Hecke correspondence** by

$$(E, Z) \mapsto \sum_H (E/H, (Z + H)/H)$$

where the sum is taken over all cyclic subgroups H of $E(\mathbf{Q}^a)$, of order l.

These correspondences induce endomorphisms of $J_0(N)$, denoted respectively by

$$w_{N,J} = w_J \quad \text{and} \quad T_l.$$

We let $\mathbf{T}$ be the subalgebra of End $J_0(N)$ generated by w_J and all T_l with $l \nmid N$. We call $\mathbf{T}$ the **Hecke algebra**. It is a fact that:

$$\mathbf{T} = \mathrm{End}_{\mathbf{C}}\, J_0(N) \qquad \text{(proof in [Maz 77], Chapter II, 9.5).}$$

Mazur defines the **Eisenstein ideal** I to be the ideal of $\mathbf{T}$ generated by

$$1 + w_J \quad \text{and} \quad 1 + l - T_l \quad (\text{for } l \nmid N).$$

On the other hand, define the **cuspidal ideal** of $\mathbf{T}$ to be the ideal annihilating the cuspidal group C. Then Mazur proved that *the Eisenstein ideal is equal to the cuspidal ideal.* However, as Mazur points out, it is possible to define Eisenstein ideals (related to Eisenstein series) in other contexts when there are no cusps, so the Eisenstein terminology had precedence over cuspidal terminology.

Let

$$I^\infty = \bigcap_{m=1}^{\infty} I^m.$$

Then I^∞ is an ideal of $\mathbf{T}$, and $I^\infty J_0(N)$ is an abelian subvariety of $J_0(N)$ defined over $\mathbf{Q}$. We let the **Eisenstein quotient** be

$$B = J_0(N)/I^\infty J_0(N).$$

Theorem 2.4. *Let C be the cuspidal group on $J_0(N)$. Then*

$$J_0(N)(\mathbf{Q})_{\text{tor}} = C.$$

The natural map of C into $B(\mathbf{Q})$ (as quotient of $J_0(N)$) gives an isomorphism

$$C \approx B(\mathbf{Q}),$$

so in particular, $B(\mathbf{Q})$ is finite.

For the proof see [Maz 77], Chapter III, Theorem 1.2 (Conjecture of Ogg) and Theorem 3.1. Mazur's proof engages in a very jazzed up form of descent.

After proving that $B(\mathbf{Q})$ is finite, one sees as an immediate corollary that the set of rational points $X_0(N)(\mathbf{Q})$ is finite, but Mazur still had to do substantial work to prove the full strength of Theorem 2.1 above, showing that the only rational points on $X_0(N)$ are at the cusps.

V, §3. MODULAR ELLIPTIC CURVES AND FERMAT'S LAST THEOREM

We consider again the curve $X_0(N)$ defined over $\mathbf{Q}$. Let $S(N)$ be the complex vector space of differentials of first kind on $X_0(N)$. We call N the **level**. In terms of the complex variable on the upper half plane covering $Y_0(N)(\mathbf{C})$ by the map

$$\Gamma_0(N)\backslash\mathfrak{H} \to Y_0(N)(\mathbf{C}),$$

such a differential can be expressed in the form

$$\omega = f(\tau)\, d\tau,$$

where f is holomorphic on $\mathfrak{H}$. In terms of the parameter $q = e^{2\pi i\tau}$, we can write this differential as

$$\omega = \sum_{n=1}^{\infty} a_n q^n \frac{dq}{q} = f_\infty(q)\frac{dq}{q}.$$

The coefficients a_n are called the **Fourier coefficients** of the form. We redefine the **Hecke operators** in a manner suitable for the current application. For each prime $p \nmid N$ there is an operator T_p on $S(N)$ whose effect on the Fourier coefficients of a form as above is:

$$T_p\colon f_\infty = \sum a_n q^n \mapsto \sum a_{pn} q^n + p \sum a_n q^{pn}.$$

If $p|N$, there is also an operator T_p such that

$$T_p: f_\infty = \sum a_n q^n \mapsto \sum a_{pn} q^n.$$

An **eigenform** in $S(N)$ is a non-zero form which is an eigenvector for each of the operators T_p. For $p \nmid N$, the effect of T_p is the same as the effect of the correspondences defined in §2, applied to differential forms.

One way to express the Taniyama–Shimura conjecture is to say that every elliptic curve over **Q** is modular. Taniyama expressed this conjecture roughly in problems at a conference in Kyoto in 1955, when he wrote:

> 12. Let C be an elliptic curve defined over an algebraic number field k, and $L_C(s)$ denote the L-function of C over k. Namely,
>
> $$\zeta_C(s) = \frac{\zeta_k(s)\zeta_k(s-1)}{L_C(s)}$$
>
> is the zeta function of C over k. If a conjecture of Hasse is true for $\zeta_C(s)$, then the Fourier series obtained from $L_C(s)$ by the inverse Mellin transformation must be an automorphic form of dimension -2, of some special type (cf. Hecke). If so, it is very plausible that this form is an elliptic differential of the field of that automorphic function. The problem is to ask if it is possible to prove Hasse's conjecture for C, by going back this considerations, and by finding a suitable automorphic form from which $L_C(s)$ may be obtained.
>
> 13. Concerning the above problem, our new problem is to characterize the field of elliptic modular functions of "Stufe" N, and especially to decompose the jacobian variety J of this function field into simple factors, in the sense of isogeneity. It is well known, that, in case $N = q$ is a prime number, satisfying $q \equiv 3 \pmod 4$, J contains elliptic curves with complex multiplication. Is this true for general N?

In his reference to Hecke, Taniyama thought that functions other than the ordinary modular functions parametrizing $X_0(N)$ might be necessary. See [Shi 89]. Shimura, around 1962, in conversations with Serre and Weil, among others, made the conjecture more precise. He explained the role of the rational numbers **Q** as a ground field (as distinguished from an arbitrary "algebraic number field k" as in Taniyama's Problem 12), and said that he expected that the ordinary modular curves sufficed. These ideas, astonishing at the time, have now been adopted. A precise statement runs as follows.

Taniyama–Shimura conjecture. *Let E be an elliptic curve over* **Q**, *and let N be its conductor. Then there is an eigenform in $S(N)$ such that for* *each prime $p \nmid N$ the eigenvalue a_p of T_p for this form has the property that*

$$\#E(\mathbf{F}_p) = 1 + p - a_p.$$

Thus a_p is also the trace of Frobenius in the l-adic representations. In addition, there is a rational map

$$\pi\colon X_0(N) \to E$$

defined over $\mathbf{Q}$, such that if $\omega_{E,\pi}$ is a suitably normalized differential of first kind on E, then $\pi^*\omega_{E,\pi}$ is the above eigenform for the Hecke operators, and

$$\pi^*\omega_{E,\pi} = \sum_{n=1}^{\infty} a_n q^n \frac{dq}{q},$$

where $a_1 = 1$ and a_p is the eigenvalue as above.

The special role played by the conductor arose in Weil [We 67], from the point of view of the functional equation of the zeta function, but as Weil writes: "Nach einer Mitteilung von G. Shimura" the differential of first kind on E corresponds to the Hecke eigenform on $X_0(N)$.

When there exists an eigenform as in the statement of the conjecture, one says that E is **modular**. We note that the coefficients a_p are integers, since they are expressible as

$$a_p = p + 1 - \#E(\mathbf{F}_p).$$

It also follows that all coefficients a_n are integers.

Let $\mathbf{T} = \mathbf{T}_N$ now be the subring of $\mathrm{End}_{\mathbf{C}}(S(N))$ generated by the operators T_p over $\mathbf{Z}$. Then $\mathbf{T}$ is a free $\mathbf{Z}$-module of rank equal to the genus $g(N)$ of $X_0(N)$, which is equal to the dimension of $S(N)$. See for instance [Shi 71b]. I shall follow Ribet [Ri 90b] in describing how one obtains certain representations of the Galois group $G_{\mathbf{Q}}$ and how one can combine a theorem of Ribet with Frey's basic idea to deduce Fermat's last theorem from the Taniyama–Shimura conjecture.

Let $\mathfrak{m}$ be a maximal ideal of $\mathbf{T}$. Then the residue field $k_{\mathfrak{m}} = T/\mathfrak{m}$ is a finite field, say of residue characteristic l. One can show ([DeS 74] or [Ri 90a]) that there is a semisimple continuous homomorphism

$$\rho_{\mathfrak{m}}\colon G_{\mathbf{Q}} \to \mathrm{Gl}(2, k_{\mathfrak{m}})$$

having the properties:

1. $\det \rho_{\mathfrak{m}} = \chi_l$, the cyclotomic character (definition recalled below).
2. $\rho_{\mathfrak{m}}$ is unramified at all primes p not dividing N.
3. $\mathrm{tr}\, \rho_{\mathfrak{m}}(\mathrm{Fr}_v) = T_p \bmod \mathfrak{m}$ for all $p \nmid N$.

Recall that the **cyclotomic character**

$$\chi_l\colon G_{\mathbf{Q}} \to \mathbf{F}_l^* \subset k_{\mathfrak{m}}^*$$

is the character such that for every $\sigma \in G_{\mathbf{Q}}$ and an l-th root of unity ζ, we have $\sigma\zeta = \zeta^{\chi(\sigma)}$. The Frobenius element Fr_v was defined in Chapter IV, §4, for v lying above p.

The representation $\rho_{\mathfrak{m}}$ is unique up to isomorphism. This follows from the Čebotarev density theorem, which implies that all elements of the image of $\rho_{\mathfrak{m}}$ are conjugate to Frobenius elements $\rho_{\mathfrak{m}}(\mathrm{Fr}_v)$, together with the fact that a semisimple 2-dimensional representation is determined by its trace and determinant.

Example. Let E be a modular elliptic curve of conductor N, and let ω be an eigenform in $S(N)$ whose eigenvalue under T_p is a_p for each prime $p \nmid N$. The action of $\mathbf{T}$ on ω is given by the homomorphism $\varphi\colon \mathbf{T} \to \mathbf{C}$ which takes each $T \in \mathbf{T}$ to the eigenvalue of ω under T. This homomorphism is in fact $\mathbf{Z}$-valued. If l is a prime number, and

$$\mathfrak{m} = \varphi^{-1}((l)),$$

then $\rho_{\mathfrak{m}}\colon G_{\mathbf{Q}} \to \mathrm{GL}(2, \mathbf{F}_l)$ is the semisimplification ρ^{ss} of the representation

$$\rho\colon G_{\mathbf{Q}} \to \mathrm{Aut}(E[l])$$

on the points of order l in $E(\mathbf{Q}^{\mathrm{a}})$. This **semisimplification** is defined to be the direct sum of the Jordan–Hölder factors of ρ.

Next let $\mathbf{F}$ be a finite field and let

$$\gamma\colon G_{\mathbf{Q}} \to \mathrm{GL}(2, \mathbf{F})$$

be a continuous semisimple representation. We say that γ is **modular of level** N if there is a maximal ideal $\mathfrak{m}$ of $\mathbf{T}$ and an imbedding $\iota\colon \mathbf{T}/\mathfrak{m} \to \mathbf{F}^{\mathrm{a}}$ such that the $\mathbf{F}^{\mathrm{a}}$-representations

$$G_{\mathbf{Q}} \xrightarrow{\ \gamma\ } \mathrm{GL}(2, \mathbf{F}) \subset \mathrm{GL}(2, \mathbf{F}^{\mathrm{a}}),$$
$$G_{\mathbf{Q}} \xrightarrow{\ \rho_{\mathfrak{m}}\ } \mathrm{GL}(2, \mathbf{T}/\mathfrak{m}) \xhookrightarrow{\ \iota\ } \mathrm{GL}(2, \mathbf{F}^{\mathrm{a}}),$$

are isomorphic. Equivalently, one requires a homomorphism

$$\alpha\colon \mathbf{T} \to \mathbf{F}^{\mathrm{a}}$$

such that

$$\mathrm{tr}(\gamma(\mathrm{Fr}_p)) = \alpha(T_p) \qquad \text{and} \qquad \det \gamma(\mathrm{Fr}_p)) = \bar{p}$$

for all but a finite number of primes p. Here, Fr_p denotes Fr_v for some $v|p$, and $\bar{p}$ is the image of p in $\mathbf{F}$.

Finally, we need to define the notion of a representation γ being finite. Let p be a prime number, and choose $v|p$. The decomposition group D_v is then the Galois group $\mathrm{Gal}(\mathbf{Q}_p^a/\mathbf{Q}_p)$. By restriction, we view $\mathbf{F}^2$ as a $\mathrm{Gal}(\mathbf{Q}_p^a/\mathbf{Q}_p)$-module. We say that γ is **finite at** p if there is a finite flat group scheme $\mathscr{G}$ over $\mathbf{Z}_p$, with an action of $\mathbf{F}$ on $\mathscr{G}$ making $\mathscr{G}$ of rank 2 over $\mathbf{F}$, such that γ and the representations of $\mathrm{Gal}(\mathbf{Q}_p^a/\mathbf{Q}_p)$ on $\mathscr{G}(\mathbf{Q}_p^a)$ are isomorphic. If $p \neq l$, this means simply that γ is unramified at p. One needs a test for finiteness, given by the next proposition. Note that the proposition uses the minimal discriminant, and that the fully general minimal model must be used here, taking into account the primes 2 and 3.

Proposition 3.1. *Let E be a semistable elliptic curve over $\mathbf{Q}$, put in minimal model over $\mathbf{Z}$. Let ρ be the representation of $G_{\mathbf{Q}}$ on $E[l]$ for some prime l. Let p be a prime. Let Δ be the minimal discriminant. The representation ρ is finite at p if and only if*

$$\mathrm{ord}_p(\Delta) \equiv 0 \bmod l.$$

For a proof, see [Ser 87].

The next theorem is the principal result of [Ri 90b]. It shows how the level of a representation can be diminished. It proves a special case of conjectures of Serre [Ser 87], and is weaker than what is proved in [Ri 90a], but easier to understand and sufficient for the application.

Theorem 3.2 (Ribet [Ri 90b]). *Let γ be an irreducible 2-dimensional representation of $G_{\mathbf{Q}}$ over a finite field of characteristic $l > 2$. Assume that γ is modular of square free level N, and that there is a prime $q|N$, $q \neq l$ such that γ is not finite at q. Suppose further that p is a divisor of N at which γ is finite. Then γ is modular of level N/p.*

This theorem immediately gives:

Conjecture of Taniyama–Shimura $\Rightarrow$ *Fermat's last theorem.*

Indeed, suppose that conjecture true. It suffices to show that there is no triple (a, b, c) of relatively prime non-zero integers satisfying the equation

$$a^l + b^l + c^l = 0,$$

where $l \geqq 5$ is prime. Suppose there is such a triple. Without loss of generality, dividing out by any common factor, we may assume that a, b, c are relatively prime. After permuting these integers, we may suppose that b is even and that $a \equiv 1 \bmod 4$. Following Frey [Fr], we consider the elliptic curve E defined by the equation

$$y^2 = x(x - a^l)(x + b^l).$$

One can compute the conductor of E to be $N = abc$, and the minimal discriminant in a minimal equation to be

$$\Delta = (abc)^{2l}/2^8.$$

We then obtain the representation ρ of $G_{\mathbf{Q}}$ on $E[l]$, which is proved to be irreducible by results of Mazur [Maz 77] and Serre [Ser 87] Proposition 6, §4.1. Frey already had noted that if $p \neq l$ and p divides N but $p \neq 2$, then ρ is unramified at p, so finite at p. By using Proposition 3.1, one sees that this representation is not finite at 2. By Taniyama–Shimura, the representation is modular. Applying Theorem 3.2 inductively, we deduce that ρ is modular of level 2. This is impossible, because $S(2)$ has dimension $g(2) = 0$, so the Hecke algebra is 0, and we are done.

V, §4. APPLICATION TO PYTHAGOREAN TRIPLES

Classically, a non-zero rational number is called a **congruent number** if it occurs as the area of a right triangle with rational sides. Dickson in [Di 20] traces the question of whether a given number is congruent back to Arab manuscripts and the Greeks prior to that. Tunnell [Tu 83] showed that this old problem is intimately connected with coefficients of certain modular forms, and we shall give a brief summary of some of his results and the context in which they occur.

The problem whether a number is congruent or not can be reduced to a problem about elliptic curves and modular forms as follows. From the Pythagoras formula, it is clear that a rational number D is the area of a right triangle with rational sides and hypotenuse h if and only if $(h/2)^2 \pm D$ are both rational squares. Hence D is a congruent number if and only if the simultaneous equations

$$u^2 + Dv^2 = w^2,$$
$$u^2 - Dv^2 = z^2,$$

have a solution in integers (u, v, w, z) with $v \neq 0$. Geometrically, these two hypersurfaces in $\mathbf{P}^3$ intersect in a smooth quartic in $\mathbf{P}^3$ which contains the point $(1, 0, 1, 1)$. The intersection is thus an elliptic curve over $\mathbf{Q}$, and projection from $(1, 0, 1, 1)$ to the plane $z = 0$ gives a birational isomorphism with a plane cubic whose Weierstrass form is

$$E^D\colon y^2 = x^3 - D^2x \qquad \text{or also} \qquad Dy^2 = x^3 - x.$$

The points on the space curve with $v = 0$ correspond to the points where $y = 0$ and the point at infinity on E^D. It is easy to see by reducing modulo primes that these points on $E^D(\mathbf{Q})$ are precisely those of finite order. Thus we see that:

D is the area of a rational right triangle if and only if the group $E^D(\mathbf{Q})$ is infinite.

We recall that an elliptic curve E is said to have **complex multiplication** if $\mathrm{End}_{\mathbf{C}}(E)$ has rank 2 over $\mathbf{Z}$, and so is a subring of an imaginary quadratic field over $\mathbf{Q}$. A curve E^D defined by the above equation with $D \neq 0$ has complex multiplication by $\mathbf{Z}[\sqrt{-1}]$. A test whether $E^D(\mathbf{Q})$ is finite comes from applying a result of Coates–Wiles, along the lines of the Birch–Swinnerton-Dyer conjecture [CoW 77]:

Theorem 4.1. *Let E be an elliptic curve over $\mathbf{Q}$ with complex multiplication by the ring of integers of a quadratic field of class number 1. If $L_E(1) \neq 0$ then $E(\mathbf{Q})$ is finite.*

On the other hand, Shimura [Shi 71] proved that every elliptic curve over $\mathbf{Q}$ with complex multiplication is modular, and the modularity can be used to analyze the behavior of L_E at $s = 1$. We make the matter a bit more explicit.

Consider the power series $f_\infty(q)$ associated with a form as at the beginning of §3. We have the corresponding function f of τ in the upper half plane, and this function satisfies the equation

$$f\big((a\tau + b)/(c\tau + d)\big) = (c\tau + d)^2 f(\tau) \qquad \text{for} \quad \begin{pmatrix} a & b \\ c & d \end{pmatrix} \in \mathrm{SL}_2(\mathbf{Z}).$$

Thus one says that f is a form of **weight** 2. Suppose a function f on $\mathfrak{H}$ is defined by a power series

$$f(\tau) = f_\infty(q) = \sum_{n=1}^{\infty} a_n q^n$$

with $a_1 = 1$, and satisfies an equation as above, but with exponent 3/2 instead of 2, and a sign factor which is relatively complicated, see [Sh 73]. Then we say that f is a form of **weight** 3/2.

If E is a modular elliptic curve over $\mathbf{Q}$, and ω_E is a differential of first kind, normalized as in the statement of the Taniyama–Shimura conjecture in §3, then the L-function $L_E(s)$ has the form

$$L_E(s) = \sum \frac{a_n}{n^s}.$$

In particular, the behavior of $L_E(s)$ at $s = 1$ can be deduced from modular properties of E. Indeed, let

$$g = g_\infty(q) = q \prod_{n=1}^{\infty} (1 - q^{8n})(1 - q^{16n})$$

and for each positive integer t, let

$$\theta_t = \sum_{-\infty}^{\infty} q^{tn^2}.$$

Let

$$g\theta_2 = \sum_{n=1}^{\infty} a(n)q^n \quad \text{and} \quad g\theta_4 = \sum_{n=1}^{\infty} b(n)q^n.$$

Waldspurger [Walp 81] showed:

Theorem 4.2. *For d a square free odd positive integer, we have*

$$L(E^d, 1) = a(d)\beta d^{-1/2}/4,$$
$$L(E^{2d}, 1) = b(d)^2\beta(2d)^{-1/2}/2,$$

where

$$\beta = \int_1^\infty (x^3 - x)^{-1/2}\, dx = 2.62205\ldots$$

is the real period of E.

Thus the value of the L-function at 1 is a non-zero multiple of coefficients of $g\theta_2$ and $g\theta_4$. Furthermore, $g\theta_2$ and $g\theta_4$ are modular forms of weight 3/2. Putting everything together, we have Tunnell's theorem, the main result of [Tu 83]:

Theorem 4.3. *If $a(n) \neq 0$ then n is not the area of any right triangle with rational sides. If $b(n) \neq 0$ then $2n$ is not the area of any right triangle with rational sides.*

As Tunnell also remarks, the Birch–Swinnerton-Dyer conjecture implies the converse of the statements in Theorem 4.3 when n is a square free positive integer.

V, §5. MODULAR ELLIPTIC CURVES OF RANK 1

In this section, we consider conjectures and results related to the Birch–Swinnerton-Dyer conjecture for elliptic curves over the rationals, which describe the group of rational points modulo torsion when this group has rank 1. So we let E be a modular elliptic curve over $\mathbf{Q}$, with a parametrization

$$\pi: X_0(N) \to E$$

over $\mathbf{Q}$ as in the Taniyama–Shimura conjecture stated in §3. The differential form $\omega_{E,\pi}$ is normalized as in that statement, so

$$\pi^*\omega_{E,\pi} = f_\infty(q)\frac{dq}{q}.$$

We begin with a *construction of a certain rational point in* $E(\mathbf{Q})$. Let D be the discriminant of an imaginary quadratic field

$$K = \mathbf{Q}(\sqrt{D}).$$

We assume throughout that D (and so K) is chosen so that every prime $p|N$ splits completely in K. Then there exists an ideal $\mathfrak{n}$ of $\mathfrak{o}_K$ such that

$$\mathfrak{o}_K/\mathfrak{n} = \mathbf{Z}/N\mathbf{Z},$$

and in particular, $\mathfrak{o}_K/\mathfrak{n}$ is cyclic of order N. Let $\mathfrak{a}$ be an ideal of $\mathfrak{o}_K$, with ideal class a. We define the point $x_a = x(a, \mathfrak{n})$ (depending also on the choice of $\mathfrak{n}$) to be the point in $X_0(N)(\mathbf{C})$ represented over the complex numbers by the pair $(\mathbf{C}/\mathfrak{a}, \mathfrak{a}\mathfrak{n}^{-1}/\mathfrak{a})$. In the canonical model $X_0(N)$, it can be proved that

$$x_a \in X_0(N)(H),$$

where H is the Hilbert class field of K.

By the theory of complex multiplication and class field theory, if $\sigma_b \in G_{H/K}$ is the element of the Galois group corresponding to the ideal class b, it follows that

$$\sigma_b x_a = x_{ab^{-1}}.$$

We assume that the rational map $\pi: X_0(N) \to E$ is normalized so that the cusp (∞) goes to the origin in E, that is $\pi(\infty) = 0$. Then following Birch, we define the **Heegner point** on E to be the trace of πx_a to the rationals, that is

$$Q_D = \mathrm{Tr}_{H/\mathbf{Q}}(\pi x_a) = \sum_{\sigma \in G_{H/\mathbf{Q}}} \sigma(\pi x_a).$$

It is easily shown from the basic theory of modular curves and complex multiplication that in the group $E(\mathbf{Q})$ mod torsion, the point Q_D depends up to sign only on the parametrization $\pi: X_0(N) \to E$ and on the choice of D, but not on the ideal $\mathfrak{n}$ and the class a.

We shall now consider when Q_D is a torsion point. As before, we define the **twisted elliptic curve** E^D: If

$$y^2 = x^3 - \gamma_2 x - \gamma_3$$

is a Weierstrass equation for E, then E^D is defined by the Weierstrass equation

$$y^2 = x^3 - D^2\gamma_2 x - D^3\gamma_3 \quad \text{or} \quad Dy^2 = x^3 - \gamma_2 x - \gamma_3.$$

Let ε be the sign of the functional equation as in Conjecture 6.1 of Chapter III. In terms of the modular curve, one can describe ε also as follows. Let $w_N = w$ be the involution Hecke operator as described in §2. Then

$$w_N f = -\varepsilon f.$$

Theorem 5.1 (Gross–Zagier [GrZ 86]). *The point Q_D in $E(\mathbf{Q})$ has infinite order if and only if the following three conditions are satisfied*:

(a) $\varepsilon = -1$;
(b) $L'(E, 1) \neq 0$;
(c) $L(E^D, 1) \neq 0$.

Thus we get necessary and sufficient conditions under which Q_D has infinite order, and under these conditions, Gross–Zagier have constructed a non-torsion rational point. In particular, if $\varepsilon = 1$, then Q_D is a torsion point. The three conditions are natural if one thinks in terms of the factorization of the L-functions associated with the elliptic curves, namely one has the formula:

Theorem 5.2 ([GrZ 86]). *If $\varepsilon = -1$, then for $D \neq -3, -4$,*

$$L'(E, 1)L(E^D, 1) = L'(E_K, 1) = \frac{\hat{h}(Q_D)}{2\sqrt{|D|}} \int_{E(\mathbf{C})} \sqrt{-1}\,\omega_{E,\pi} \wedge \bar{\omega}_{E,\pi}.$$

The canonical height $\hat{h}(Q_D)$ is 0 if and only if Q_D is a torsion point, so Theorem 5.1 essentially comes from the factorization formula of Theorem 5.2.

Theorem 5.3 (Kolyvagin [Koly 88]). *If the Heegner point Q_D has infinite order, then $E^D(\mathbf{Q})$ is finite, Ш(E_K) is finite, and the rank of $E(\mathbf{Q})$ is* 1.

The theorem is predicted by the Birch–Swinnerton-Dyer conjecture and the condition in Theorem 5.1(c). It will play a role in the next conjecture where the order of the finite group $E^D(\mathbf{Q})$ enters in the formula. This conjecture describes more properties of the Heegner point, and in particular gives the index of the subgroup generated by this point in the group of all rational points. We recall some notation.

For each $p|N$ we let c_p as before be the integers in the Birch–Swinnerton-Dyer conjecture (Conjecture 6.2 of Chapter III).

Aside from the form $\omega_{E,\pi}$ there is also the form $\omega_{E,\min}$, the **minimal (Néron) differential** associated with the minimal model of E over **Z**. If this minimal model has the Weierstrass form, then

$$\omega_{E,\min} = dx/2y.$$

For the definition of this minimal form in general, see Tate [Ta 74]. It can be shown that there is an integer $c(\pi)$ such that

$$c(\pi)\omega_{E,\pi} = \omega_{E,\min}.$$

The integer $c(\pi)$ describes the extent to which the parametrization π is not minimal, in an appropriate sense. See Remark 1 below.

If one combines the formula for $L'(E_K, 1)$ with the Birch–Swinnerton-Dyer conjecture, one is led to the following conjecture on the index of the point

$$P_D = \mathrm{Tr}_{H/K}(\pi x_a).$$

Conjecture 5.4 (Gross–Zagier [GrZ 86], V, 2.2 and also [Gros 90]). *If P_D has infinite order, then*

$$(E(K) : \mathbf{Z}P_D) = \#\text{Ш}(E_K)^{1/2} c(\pi) \prod_{p|N} c_p.$$

Most of Conjecture 5.4 has been proved by Kolyvagin [Ko 88], see also [Gros 90]. A corollary would be that when the three conditions of Theorem 5.1 are satisfied, then

$$(E(\mathbf{Q}) : \mathbf{Z}Q_D)\,\#E^D(\mathbf{Q}) = \#\text{Ш}(E_K)^{1/2} c(\pi) \prod_{p|N} c_p \cdot 2^t$$

where t is an integer with $|t| \leqq 3$, depending on the action of Galois on $E(K)/2E(K)$.

To the extent diophantine geometry is concerned with rational points, we see that the construction of Heegner points provides an explicit way of getting part of the group of rational points when the rank is 1. To my knowledge, no construction is known or conjectured today to give a subgroup of finite index when the rank of $E(\mathbf{Q})$ is greater than 1 for modular elliptic curves.

Remark 1. The integer $c(\pi)$ is interesting independently of the context of elliptic curves of rank 1. It is sometimes called the **Manin constant**, because Manin [Ma 72] conjectured that in every isogeny class of modular elliptic curves over **Q**, there is a curve whose Manin constant is 1. Stevens [St 89] refines this conjecture to the case of parametrizations by $X_1(N)$ rather than $X_0(N)$, in which case he conjectures that every modular elliptic curve admits a parametrization by $X_1(N)$ with Manin constant equal to 1.

Remark 2. The index formula above follows the same formalism as the classical index formula for cyclotomic units in the group of all units, or the more recent formulas pertaining to the modular units. It is not clear today in general under which conditions such formulas should exist, but the reader should be aware of a very broad formalism concerning such index formulas in which the Mordell–Weil group plays the role of units and Ш plays the rôle of a class group. The explicit construction of the Heegner point corresponds to the explicit construction of cyclotomic units in the cyclotomic case, and of modular units. Taking the trace corresponds to a similar operation in the theory of modular units, which give rise to a subgroup of the cyclotomic units as in [KuL 79].

The Gross–Zagier construction depends on the choice of an auxiliary D. The question may be raised about more precise information on this dependence. I shall reproduce one simple statement due to Gross–Zagier which explains some of this dependence. For convenience, we make a definition. Let N be a prime number. Suppose we have a modular form of weight 3/2, with q-expansion

$$g_\infty(q) = \sum_{n=1}^{\infty} b_n q^n.$$

We say that g is **Shimura correspondent to** f if for all primes $l \nmid 4N$, g is an eigenfunction of T_{l^2} with eigenvalue a_l (the T_l-eigenvalue of f).

Theorem 5.5 ([GrZ 86]). *Assume N prime and* rank $E(\mathbf{Q}) = 1$, *so that*

$$E(\mathbf{Q})/\text{torsion} = \mathbf{Z}Q_0$$

with some generator Q_0. For each D such that N splits completely in $\mathbf{Q}(\sqrt{D})$ let $m_D \in \mathbf{Z}$ be the integer such that

$$Q_D = m_D Q_0.$$

Then there exists a modular form g of weight 3/2 invariant by $\Gamma_0(4N)$, which is Shimura correspondent to f, and such that the coefficient b_n for $n = |D|$ is given by

$$b_{|D|} = m_D.$$

Furthermore $g \neq 0$ if and only if $L'(E, 1) \neq 0$.

Although this section has been fairly specialized, I have included it because it gives prototypes for possible much more extensive results, some of which are not even conjectured today. I am much indebted to Gross for his advice and help in writing this section.

CHAPTER VI

The Geometric Case of Mordell's Conjecture

As we saw already in Chapter I, there is a geometric analogue to the theorem that a curve over a number field has only a finite number of rational points. We consider a projective non-singular surface X, and a morphism onto a curve

$$\pi: X \to Y$$

defined over an algebraically closed field of characteristic 0, so the generic fiber is a non-singular curve over the function field of Y. Rational points of this curve over finite extensions of $k(Y)$ amount to sections of this fibering over finite coverings of Y. The height of these sections has a geometric definition, and we want to give bounds for those heights. There have been several methods in the function field case to obtain such bounds, which are of independent interest since they exhibit the diophantine geometry in a context independent of more refined arithmetic invariants found in the number field case. The purpose of this chapter is to describe some of these methods. The original proof of finiteness (without explicit bounds on heights, conjectured in [La 60a]) is due to Manin [Man 63] and the ideas of this proof will be given in §4.

VI, §0. BASIC GEOMETRIC FACTS

In this chapter we assume that the reader is well acquainted with basic properties of intersection theory on surfaces. Let X be a complete variety. The group of Cartier divisor classes $\mathrm{Pic}(X)$ is isomorphic to the group of isomorphism classes of line sheaves on X. This isomorphism is given as follows. Let D be a (Cartier) divisor on X. Then $\mathcal{O}_X(D)$ is the sheaf such that, if D is represented by the pair (U, φ) on an open set U, then the section of $\mathcal{O}_X(D)$ over U are given by

$$\mathcal{O}_X(D)(U) = \mathcal{O}_X(U)\varphi^{-1}.$$

The association $D \mapsto \mathcal{O}_X(D)$ induces the isomorphism. If $\mathcal{L}$ is a line sheaf on X, then we define

$$\mathcal{L}(D) = \mathcal{L} \otimes \mathcal{O}_X(D).$$

The tensor product is taken over $\mathcal{O}_X$ itself.

Suppose X is a curve. If $\mathcal{L} \approx \mathcal{O}_X(D)$ we define the **degree**

$$\deg(\mathcal{L}) = \deg(D) = \deg(c),$$

where c is the class of $\mathcal{L}$ in $\mathrm{Pic}(X)$ or the class of D. The **degree on a singular curve** X is defined to be the degree on the pull back to the normalization X'. So if $f\colon X' \to X$ is the normalization of X, then by definition

$$\deg(c) = \deg f^*(c).$$

Suppose that X has arbitrary dimension, but Z is a curve on X. Let

$$f\colon Z' \to X$$

be the finite morphism, such that Z' is the normalization of Z, and f is the composed morphism

$$Z' \to Z \subset X.$$

Then the **intersection number** $(c\,.\,Z)$ is defined as

$$(c\,.\,Z) = \deg f^*(c).$$

If $\mathcal{L}$ is a line sheaf in the class c, we define $(\mathcal{L}\,.\,Z) = (c\,.\,Z)$.

The intersection symbol extends bilinearly to a pairing between $\mathrm{Pic}(X)$ and the group generated by the curves. In particular, if X has dimension 2, so X is what we call a **surface**, and is non-singular, then we get a pairing between line sheaves into the integers called the **intersection pairing**. The intersection number of two line sheaves $\mathcal{L}$, $\mathcal{M}$ is denoted by $(\mathcal{L}\,.\,\mathcal{M})$. We often write $(\mathcal{L}^2)$ instead of $(\mathcal{L}\,.\,\mathcal{L})$ on a surface.

By a **vector sheaf** $\mathcal{E}$ we mean a locally free sheaf of finite rank. If this rank is r, we let the **determinant** be

$$\det(\mathcal{E}) = \bigwedge^r \mathcal{E} = \bigwedge^{\max} \mathcal{E}.$$

Such determinants already occurred in the discussion of the Faltings height, but they will occur from now more frequently and systematically.

VI, §1. THE FUNCTION FIELD CASE AND ITS CANONICAL SHEAF

Let Y be a complete non-singular curve of genus q defined over an algebraically closed field k of characteristic 0. *Let*

$$\pi: X \to Y$$

be a flat proper morphism, such that X is a complete non-singular surface. We let g be the genus of the generic fiber, and we suppose that $g \geqq 2$. We denote this generic fiber by

$$C = \pi^{-1}(\eta) = X_\eta \qquad \textit{where } \eta \textit{ is a generic point of } Y.$$

We let $F = k(Y)$ be the function field of Y, and $L = k(X)$ function field of X.

The first thing to do is to describe a canonical line sheaf on X. If π were smooth (which it usually is not), then this line sheaf would just be $\Omega^1_{X/Y}$, the ordinary sheaf of differential forms. Since π is not necessarily smooth, we must go around this possibility. Since X is assumed non-singular, there exists an imbedding

$$j: X \to S$$

of X into a smooth scheme S over Y, and such that X is a **local complete intersection** in S. This means that the ideal sheaf $\mathcal{J}$ defining X in S is generated at every point by a regular sequence. The **conormal sheaf** is the vector sheaf

$$\mathscr{C}_{X/S} = \mathcal{J}/\mathcal{J}^2.$$

We then define the **canonical sheaf of the imbedding** j to be

$$\omega_{X/Y} = \det j^*\Omega^1_{S/Y} \otimes \det \mathscr{C}^\vee_{X/S},$$

where $\mathscr{C}^\vee_{X/S}$ denotes the dual sheaf. The determinant is the maximal exterior power. It is easily shown that the canonical sheaf is independent of the imbedding, up to a natural isomorphism. Furthermore, if U is an open subset of X such that the restriction of π to U is smooth, then

$$\omega_{U/Y} = \Omega^1_{U/Y}.$$

There is a canonical choice for $\omega_{X/Y}$ as a subsheaf of the constant sheaf $\Omega^1_{L/F}$. In fact, let z be the generic point of X. Then $\omega_{X/Y}$ is a subsheaf of the constant sheaf with fiber

$$\omega_{X/Y,z} = \det(j^*\Omega^1_{S/Y})_z \otimes \det(\mathscr{C}^\vee_{X/S})_z.$$

Since the canonical sequence

$$0 \to (\mathscr{C}_{X/S})_z \to (j^*\Omega^1_{S/Y})_z \to (\Omega^1_{X/Y})_z \to 0$$

is exact and $\Omega^1_{X/Y,z} = \Omega^1_{L/F}$, we obtain a natural isomorphism

$$\omega_{X/Y,z} \approx \Omega^1_{L/F} = \det \Omega^1_{L/F}$$

which gives us our imbedding. The image of $\omega_{X/Y}$ in $\Omega^1_{L/F}$ is independent of the factorization $X \subset S \to Y$.

The line sheaf $\omega_{X/Y}$ correspond to a divisor class which is called the **relative canonical class**, which we denote by

$$K = K_{X/Y} \qquad \text{so that} \qquad \omega_{X/Y} \approx \mathscr{O}_X(K).$$

A point P of C in a finite extension of $k(Y)$ corresponds to a morphism

$$s = s_P \colon Y' \to X$$

where Y' is a (possibly ramified) non-singular covering of Y, making the diagram commutative:

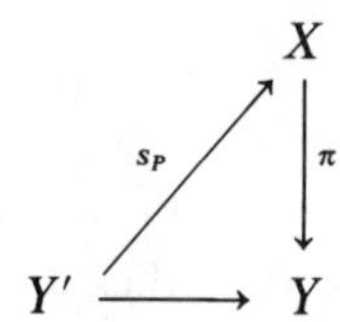

and such that s_P is generically an injection. For such a covering Y', we let q' be its genus. We define the **geometric canonical height** by the formula

$$h_k(P) = \frac{1}{[Y' : Y]} \deg s_P^* \omega_{X/Y}.$$

Finally, we define the **geometric logarithmic discriminant** by

$$d(P) = \frac{1}{[Y' : Y]}(2q' - 2).$$

The above notation and definitions will be in force throughout §2 and §3, which will describe two approaches to prove the boundedness of heights of sections of the family $\pi\colon X \to Y$.

VI, §2. GRAUERT'S CONSTRUCTION AND VOJTA'S INEQUALITY

In the function field case, Vojta was able to give a remarkably good estimate for the height of algebraic points as follows. We continue with the notation of §1.

Theorem 2.1 (Vojta [Vo 90c]). *Let C be the generic curve of a family defined by our morphism π. Given $\varepsilon > 0$, for all algebraic points P of C over $k(Y)$, we have*

$$h_K(P) \leqq (2 + \varepsilon)d(P) + O_\varepsilon(1).$$

The bound on the right-hand side is remarkable in that the factor of $d(P)$ does not depend on the genus of the generic fiber. Vojta's conjecture is that this factor can be replaced by $1 + \varepsilon$. See also [Vo 90b].

Before we go into the ideas of the proof, we recall a basic fact from elementary algebraic geometry [Ha 77], Chapter II, Proposition 7.12. Given a vector sheaf $\mathscr{E}$ over a base scheme X, one can form the **projective bundle**

$$p\colon \mathbf{P}(\mathscr{E}) \to X \qquad \text{defined by} \qquad \mathbf{P}(\mathscr{E}) = \operatorname{proj}(\operatorname{Sym}(\mathscr{E})),$$

where $\operatorname{Sym} \mathscr{E}$ is the symmetric algebra, which is a graded algebra. To give a morphism $Z \to P(\mathscr{E})$ of a scheme Z into $\mathbf{P}(\mathscr{E})$ over X, is equivalent to giving a morphism $s\colon Z \to X$, a line sheaf $\mathscr{L}$ on Z, and a surjective homomorphism of sheaves $s^*\mathscr{E} \to \mathscr{L}$ on Z. We apply this to the case of our surface X.

Vojta's proof is then based on the following construction of Grauert. Let Z be a non-singular curve, and let $s\colon Z \to X$ be a finite morphism. We get a homomorphism

$$s^*\Omega^1_X \to \mathscr{L} \subset \Omega^1_Z,$$

where $\mathscr{L}$ is the image of $s^*\Omega^1_X$ in Ω^1_Z. We have abbreviated $\Omega^1_X = \Omega^1_{X/k}$ as usual in the theory of surfaces, and similarly for Ω^1_Z. Then we obtain the corresponding mapping $t_s\colon Z \to \mathbf{P}(\Omega^1_X)$ making the following diagram commutative.

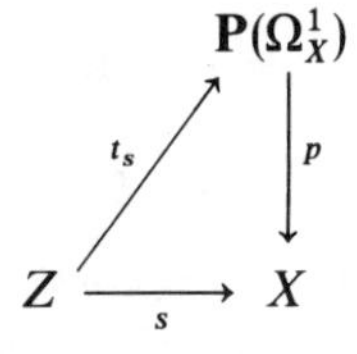

If Z is a singular curve with a finite morphism into X, then the above construction is meant to be applied to its desingularization. The map t_s

is essentially the differential of s, taken in the projective bundle. We let $\mathbf{P}\mathscr{E} = \mathbf{P}(\Omega_X^1)$. The sheaf $\mathscr{L}$ is the inverse image

$$\mathscr{L} = t_s^* \mathcal{O}_{\mathbf{P}\mathscr{E}}(1),$$

where $\mathcal{O}_{\mathbf{P}\mathscr{E}}(1)$ is the hyperplane line sheaf on $\mathbf{P}\mathscr{E}$. Note that $\mathbf{P}(\Omega_X^1)$ is a variety of dimension 3.

The above is Grauert's construction. To prove Theorem 2.1, Vojta then proceeds in two steps. In the first place, one has the inequality

$$(*) \qquad \deg t_s^* \mathcal{O}_{\mathbf{P}\mathscr{E}}(1) \leqq \deg \Omega_Z^1 = 2g(Z) - 2.$$

Let $\deg_{\mathbf{P}} Z$ = intersection number of Z and the generic fiber of $\pi \circ p$. Then:

Proposition 2.2. *Fix a rational number $\varepsilon > 0$. There exists a constant c and an effective divisor D on $\mathbf{P}(\Omega_X^1)$ such that for all irreducible curves Z on $\mathbf{P}(\Omega_X^1)$ not contained in D, letting $\mathscr{E} = \Omega_X^1$ we have*

$$(Z \,.\, p^*\omega_{X/Y}) \leqq (2 + \varepsilon)\big(Z \,.\, \mathcal{O}_{\mathbf{P}\mathscr{E}}(1)\big) + c \deg_{\mathbf{P}} Z.$$

Theorem 2.1 follows from $(*)$ and Proposition 2.2 for those points P and corresponding section $s = s_P$ such that $t_s(Y')$ is not contained in D.

For those points P such that $t_s(Y')$ is contained in D, one must dig deeper, in a way which leads to solutions of certain algebraic differential equations. We describe the main step.

Let for the moment X be an arbitrary projective non-singular variety in characteristic 0. Let D be a divisor on X, and let $\omega \in H^0(X, \Omega_X^1(D))$, where H^0 denotes the global sections, and as usual,

$$\Omega_X^1(D) = \Omega_X^1 \otimes \mathcal{O}_X(D).$$

We may then think of ω as a rational differential 1-form on X. We define a **pfaffian divisor** W with respect to ω to be a divisor, which when represented by a pair (U, f_U) on an open subset U of X has the property that

$$\omega \wedge \frac{df_U}{f_U} \in H^0\big(U, \Omega_X^2(D)\big).$$

The next theorem provides a key finiteness result.

Theorem 2.3 (Jouanolou [Jo 78]). *Let X, ω and D be as above. Then:*

(1) *There are infinitely many irreducible Pfaffian divisors with respect to ω if and only if ω is of the form $\omega = \varphi\, d\psi$, for some rational functions φ and ψ.*

(2) *If there are finitely many such divisors, then their number is bounded by*

$$\dim[H^0(X, \Omega^2_X(D))/\omega \wedge H^0(X, \Omega^1_X)] + \rho + 1,$$

where ρ is the rank of the Néron–Severi group NS(X).

Then Theorem 2.1 is an immediate consequence of the following corollary.

Corollary 2.4. *Let X be a non-singular projective surface. Let D be an effective divisor on $\mathbf{P}(\Omega^1_X)$. Then the set of irreducible curves Z on X which lift via the Grauert construction to curves contained in D is a union of finitely many algebraic families.*

Indeed, an algebraic family of curves in X lifts to an algebraic family of curves in $\mathbf{P}(\Omega^1_X)$. Intersection numbers and degrees for elements of a family are constant in the family, so Corollary 2.4 and Proposition 2.2 conclude the proof of Theorem 2.1.

Although the proof bounds the number of solutions (to a differential equation), like other proofs in the subject at the moment, it does not give a bound for the degrees (or heights) of solutions.

One limitation of the above proof, using the Grauert construction, is that it involves horizontal differentiation, for which no equivalent, even conjecturally, is known today in number fields.

VI, §3. PARSHIN'S METHOD WITH $(\omega^2_{X/Y})$

This method depends on bounding the self intersection $(\omega^2_{X/Y})$. Neat bounds are obtained under an additional condition besides the basic situation described in §1, namely semistability, which we must now define. For this general definition, we temporarily use X, Y in more generality than the basic situation described at the beginning of §1. Let $\mathfrak{o}$ be a discrete valuation ring with quotient field F and residue class field k, which we assume perfect. Let $Y = \mathrm{spec}(\mathfrak{o})$, and let $X \to Y$ be a flat proper morphism, of relative dimension 1. We say that X is **regular semistable** over Y if the following conditions are satisfied:

SS 1. *The scheme X is integral, regular, and the generic fiber X_F is geometrically irreducible (i.e. remains irreducible under base change from F).*

By a theorem of Zariski, it follows that the special fiber is geometrically connected ([Ha 77] Chapter III, Corollary 11.5). We define the **geometric fiber** above the closed point $y \in Y$ to be the base extension of the fiber to the algebraic closure of the residue class field $k(y)$. Then an irreducible component of the fiber X_y splits into a finite number of conjugate geometric components in the geometric fiber.

SS 2. *The geometric fiber above the closed point is reduced and has only ordinary double points.*

SS 3. *If Z is a non-singular irreducible component of the geometric fiber, and Z has genus 0, then Z meets the other components of the geometric fiber in at least two points.*

Semistability is preserved by making a base change from Y to $\mathrm{spec}(\hat{\mathcal{O}}_y)$ where $\hat{\mathcal{O}}_y$ is the completion, and it is also preserved under unramified extensions of $\mathcal{O}_y$.

If $Y = \mathrm{spec}(R)$ where R is a Dedekind ring, or Y is a curve over a field k, then we say that $X \to Y$ is **semistable** if it is semistable over the local ring of every point of $\mathrm{spec}(R)$.

Grothendieck defined the notion of semistability for abelian varieties, and he proved that given an abelian variety over a Dedekind ring, there always exists a finite extension over which the abelian variety is semistable. Using Grothendieck's theorem, Deligne–Mumford [DelM 69] proved the corresponding result for curves. See also Artin–Winters [ArW 71]. Let C be a curve over F. We say that C is **semistable** if there is a regular semistable $X \to Y$ such that $X_F = C$. Note that C is semistable if and only if its Jacobian over F is semistable.

In addition to semistability, we also have the notion of **stability**. The definition is the same as for semistability, except for the following two modifications:

In **SS 1**, we do not assume that X is regular, so X may have singularities.

In **SS 3**, a non-singular rational component of a geometric fiber must meet the other components of the geometric fiber in at least three points.

There is a bijection between stable and semistable models. The singularities of a stable model can be resolved be a sequence of blow ups in a canonical fashion, resulting in a semistable model. Conversely, rational curves with self intersection -2 can be blown down, resulting in a stable model. If X is semistable, we denote by $X^{\#}$ the corresponding stable model.

Suppose now that Y is either $\mathrm{spec}(\mathfrak{o})$ for a discrete valuation ring or Y is a complete non-singular curve over an algebraically closed field of characteristic 0.

If X/Y is a semistable family of curves of genus $g \geqq 2$ and y is a closed point of Y, then we let

δ_y = number of double points on the geometric fiber over y.

$\delta = \delta_{X/Y} = \sum_y \delta_y$.

If v is the valuation corresponding to the closed point y of Y, we also write $\delta_v = \delta_y$.

If $X^\#/Y$ is a stable family and y is a closed point of Y, then we let

$\delta^\#_y$ = number of double points on the geometric fiber over y.

One has the upper bound of Arakelov [Ara 71]:

Proposition 3.1. $\delta^\#_y \leqq 3g - 3$.

We now return to the basic assumptions of §2, but with the added hypothesis of semistability, in order to get neat bounds for various objects. For the rest of this section, unless otherwise specified:

We let $X \to Y$ be a semistable family of curves of genus $g \geqq 2$ over a complete non-singular curve Y of genus q, all over an algebraically closed field k of characteristic 0. *Let $F = k(Y)$.*

Inequalities for $(\omega^2_{X/Y})$ or related objects will be called **canonical class inequalities**. Arakelov proved [Ara 71]:

CC 1. *Let g_0 be the dimension of the F/k-*trace *of the Jacobian of a generic fiber X_F. Let s be the number of points of Y where X_F has bad reduction. Then*

$$(\omega^2_{X/Y}) \leqq 6(g - g_0)(2q - 2 + s) - \delta.$$

A variant was obtained by Vojta [Vo 88], who gives

CC 2. $$(\omega^2_{X/Y}) \leqq (2g - 2)(2q - 2 + s),$$

which results from another inequality, arising from the work of van de Ven, Bogomolov, Miyaoka, Yau, Parshin, and Arakelov:

CC 3. $$(\omega^2_{X/Y}) \leqq 3 \sum_y \delta^\#_y + (2g - 2)\max(2q - 2, 0).$$

I find Vojta's proof of **CC 3** in [Vo 88] considerably easier to follow than previous references, e.g. [Ara 71] or [Szp 81]. See also Parshin [Par 89].

On the other hand, Parshin and Arakelov work not only with $(\omega^2_{X/Y})$

but also with the **direct image** $\pi_* \omega_{X/Y}$, which is a vector sheaf of rank g, and with the line sheaf

$$\det \pi_* \omega_{X/Y}.$$

We define the **degree**

$$\deg \pi_* \omega_{X/Y} = \deg \det \pi_* \omega_{X/Y}.$$

We have the relation (a variation of **Noether's formula**)

CC 4. $$(\omega_{X/Y}^2) + \delta = 12 \deg \pi_* \omega_{X/Y}.$$

Furthermore, $(\omega_{X/Y}^2)$ and $\deg \pi_* \omega_{X/Y}$ are of the same order of magnitude, by a result of Xiao Gang [Xi 87], see also Cornalba–Harris [CorH 88], namely:

CC 5. $$\delta \leqq \left(8 + \frac{4}{g}\right) \deg \pi_* \omega_{X/Y}.$$

For $\pi_* \omega_{X/Y}$, Arakelov obtains

CC 6. $$\deg \pi_* \omega_{X/Y} \leqq \tfrac{1}{2}(g - g_0)(2q - 2 + s).$$

Parshin observed that the canonical class inequalities imply **height inequalities**. Several variations of these have been obtained, notably:

H 1. $h_K(P) \leqq 8 \cdot 3^{3g+1}(g-1)^2\left(s + 1 + \dfrac{d(P)}{3^g} + \dfrac{1}{3^{3g}}\right)$ in [Szp 81],

H 2. $h_K(P) \leqq \dfrac{8g - 6}{3} d(P) + O(1)$ in [Vo 88] and [Vo 90b].

On the other hand, from quite another direction, one has:

H 3. $h_K(P) \leqq 2(2g-1)^2(d(P) + s)$ in [EsV 90].

The inequalities **H 1** from Szpiro and **H 3** from Esnault–Viehweg are not only effective but completely explicit. Szpiro's proof is completely algebraic (valid even in characteristic p), whereas Esnault–Viehweg approach the problem from the direction of semipositive sheaves and their proof uses certain aspects of complex analysis for which no immediate substitute over number fields, even conjecturally, is known today. Note that **H 3** improves **H 1** in that at least the factor involving g is quadratic in g, rather than exponential. Vojta's inequality **H 2** has such a factor of $d(P)$ linear in g, but it is not clear from Vojta's proof how effective is the term $O(1)$. Vojta argues by using a refinement of the Parshin construction on the canonical class inequality **CC 2**, whence the problem about $O(1)$. It is already a problem to get an inequality linear in g with explicit $O(1)$.

Of course, Vojta's inequality which we gave in §2, namely

$$h_K(P) \leqq (2 + \varepsilon)\, d(P) + O_\varepsilon(1),$$

is even better in so far as the coefficient $2 + \varepsilon$ of $d(P)$ occurs, independently of the genus. On the other hand, again the term $O_\varepsilon(1)$ is not made explicit, and is even ineffective as things stand today. So the final word on all these inequalities is not yet at hand. From this point of view, the Vojta conjecture does *not* supercede other inequalities involving functions of the genus as coefficients. Indeed, a fundamental problem is to determine simultaneously constants b_1, b_0 so that the inequality

$$h_K(P) \leqq b_1\, d(P) + b_0 \qquad \text{or possibly } b_1\big(d(P) + s\big) + b_0$$

holds for all $P \in X(\mathbf{Q}^{\mathrm{a}})$. As b_1 decreases from a function of the genus to $1 + \varepsilon$, the number b_0 increases, and the problems is to determine the best possible set of pairs (b_1, b_0) in $\mathbf{R}^2$. From the point of view of keeping b_0 an absolute constant, the question then arises whether a linear function of the genus is best possible for b_1.

VI, §4. MANIN'S METHOD WITH CONNECTIONS

The method of this section historically gave the first proof for the Mordell conjecture in the function field case [Man 63]. It is based on horizontal differentiation, and to this day no analogue is known for it in the number field case. But geometrically, it is still giving rise to many investigations. We shall now describe this method, relying on Coleman's account [Col 90]. I am much indebted to Coleman for his useful suggestions.

Let Y be an affine non-singular curve over an algebraically closed field k of characteristic 0. *Let*

$$\pi\colon X \to Y$$

be a proper, smooth scheme over Y. Let $F = k(Y)$. We usually write Ω^1_X and Ω^1_Y instead of $\Omega^1_{X/k}$ and $\Omega^1_{Y/k}$.

De Rham cohomology

We first give the definition for the de Rham cohomology group $H^1_{\mathrm{DR}}(X/Y)$. By a 1-**cocycle** $\gamma = \{(\omega_U, f_{U,V})\}$ we mean a family of elements $\omega_U \in \Omega^1_{X/Y}(U)$

indexed by the open sets of an open covering, so ω_U is a 1-form over U, relative to Y; and a family of functions $f_{U,V} \in \mathcal{O}_X(U \cap V)$, indexed by pairs of open sets in the covering satisfying the following conditions:

(1) $d\omega_U = 0$, so ω_U is closed;
(2) For all pairs U, V we have $\omega_U - \omega_V = df_{U,V}$;
(3) We have $f_{U,V} + f_{V,W} = f_{U,W}$ on $U \cap V \cap W$, and $f_{U,U} = 0$.

By a **coboundary**, we mean a cocycle such that for each U there is a function $f_U \in \mathcal{O}_X(U)$ satisfying

$$\omega_U = df_U \qquad \text{and} \qquad f_{U,V} = f_U - f_V.$$

The factor group of 1-cocycles modulo the subgroup of coboundaries is the **de Rham group** $H^1_{\mathrm{DR}}(X/Y)$. We view $H^1_{\mathrm{DR}}(X/Y)$ as a $k[Y]$-module.

For simplicity, we let $t \in k[Y]$ be a non-constant function, and we assume that

$$H^0\Omega^1_Y = k[Y]\,dt.$$

This can always be achieved after localizing, which won't affect the theorem.

Let Z be a Zariski closed subset of X, finite and smooth over Y under π, such that $\pi(Z) = Y$. We define the **relative cohomology group** $H^1_{\mathrm{DR}}(X/Y, Z)$ in the same way, but with the additional restriction on the functions $f_{U,V}$ and f_U that they vanish on Z.

Connections

Let M be a $k[Y]$-module. By a **connection** on M we mean a k-linear map

$$\nabla\colon M \mapsto \Omega^1_{Y/k}(Y) \otimes M$$

satisfying the Leibniz rule

$$\nabla(u\omega) = du \otimes \omega + u\nabla(\omega)$$

for $u \in k[Y]$ and $\omega \in M$. In the applications, M will consist of modules of differential forms of various sorts. We shall define the **Gauss–Manin** connection

$$\nabla\colon H^1_{\mathrm{DR}}(X/Y) \to \Omega^1_{Y/k}(Y) \otimes H^1_{\mathrm{DR}}(X/Y).$$

We shall give the definition in such a way that it applies later to another situation. Since X is smooth over Y, we have an exact sequence

ES 1. $$0 \to \mathcal{O}_X \otimes \Omega^1_Y \to \Omega^1_X \to \Omega^1_{X/Y} \to 0.$$

Let $\omega \in H^1_{\mathrm{DR}}(X/Y)$. For a sufficiently fine open covering $\{U, V, \ldots\}$ we can find a cocycle $\{(\omega_U, f_{U,V})\}$ representing ω such that by the surjective map in the above exact sequence, we can lift ω_U to a form $\omega_U^X \in \Omega^1_X(U)$. The collection $\{d\omega_U^X\}$ is a family of local sections of the sheaf Ω^2_X, which occurs as the middle term of the exact sequence

ES 2. $$0 \to \Omega^1_{X/Y} \otimes_{\mathcal{O}_Y} \Omega^1_Y \to \Omega^2_X \to \Omega^2_{X/Y} \to 0.$$

Furthermore, $d\omega_U^X$ maps to 0 in the arrow $\Omega^2_X \to \Omega^2_{X/Y}$ and

$$\omega_U^X - \omega_V^X - df_{U,V}$$

maps to 0 in the map $\Omega^1_X \to \Omega^1_{X/Y}$. Hence for each pair U, V there exists $g_{U,V} \in \mathcal{O}_X(U \cap V)$ such that

$$\omega_U^X - \omega_V^X - df_{U,V} = g_{U,V}\, dt,$$

and there exists $\eta_U \in \Omega^1_{X/Y}(U)$ such that

$$d\omega_U^X = \eta_U \otimes dt.$$

Then $\{(\eta_U, g_{U,V})\}$ is a cocycle, which represents $\nabla\omega$ by definition.

Next suppose Z is a Zariski closed subset of X, smooth and finite over Y, so we have the relative group $H^1_{\mathrm{DR}}(X/Y, Z)$ previously defined. We may then define the **relative Gauss–Manin connection**

$$\nabla_Z\colon H^1_{\mathrm{DR}}(X/Y, Z) \to \Omega^1_{Y/X}(Y) \otimes H^1_{\mathrm{DR}}(X/Y, Z).$$

Exactly the same construction using functions which vanish on Z defines $\nabla_Z\omega$ if ω lies in $H^1_{\mathrm{DR}}(X/Y, Z)$. One uses the sheaves:

$$\mathcal{O}_{X,Z} = \text{sheaf of functions vanishing on } Z,$$
$$\Omega^1_{X,Z} = \text{sheaf of differential forms vanishing on } Z,$$

instead of $\mathcal{O}_X$ and Ω^1_X respectively. We also use the exact sequence

ES$_Z$ 1. $$0 \to \mathcal{O}_{X,Z} \otimes \Omega^1_Y \to \Omega^1_{X,Z} \to \Omega^1_{X,Y} \to 0$$

instead of **ES 1**.

We apply these connections to the case when Z is constructed as follows.

Let s_1, s_2 be sections of π and let Z be the Zariski closure of the images of s_1 and s_2. After localizing on Y, we can achieve that the images of s_1 and s_2 do not intersect, and that Z is smooth, so we have the Gauss–Manin connection in this situation. We shall define an injection

$$k[Y] \hookrightarrow H^1_{\mathrm{DR}}(X/Y, Z)$$

making the following diagram exact and commutative.

ES 3.

$$\begin{array}{ccccccccc}
0 & \longrightarrow & k[Y] & \longrightarrow & H^1_{\mathrm{DR}}(X/Y, Z) & \longrightarrow & H^1_{\mathrm{DR}}(X/Y) & \longrightarrow & 0 \\
 & & \Big\downarrow{\scriptstyle d} & & \Big\downarrow{\scriptstyle \nabla_Z} & & \Big\downarrow{\scriptstyle \nabla} & & \\
0 & \longrightarrow & \Omega^1_{Y/X}(Y) & \longrightarrow & \Omega^1_{Y/X}(Y) \otimes H^1_{\mathrm{DR}}(X/Y, Z) & \longrightarrow & \Omega^1_{Y/X}(Y) \otimes H^1_{\mathrm{DR}}(X/Y) & \longrightarrow & 0
\end{array}$$

To define the map of $k[Y]$ into the de Rham group, let $\varphi \in k[Y]$. For each point of X pick an open neighborhood U and function $f_U \in \mathcal{O}_X(U)$ such that

$$f_U \circ s_1 = \varphi \qquad \text{and} \qquad f_U \circ s_2 = 0.$$

Then the families $\{(df_U, f_U - f_V)\}$ define a cocycle, whose class lies in $H^1_{\mathrm{DR}}(X/Y, Z)$, and the association which sends φ to this class is an injective homomorphism

$$0 \to k[Y] \to H^1_{\mathrm{DR}}(X/Y, Z).$$

Taking the vertical maps (connections) into account, we may speak of the exact and commutative diagram **ES 3** as an exact sequence of connections.

Horizontal differentiation

A connection allows us to differentiate horizontally in a manner which we now describe. We fix a derivation $\partial\colon k[Y] \to k[Y]$ such that

$$\mathrm{Der}_{k[Y]/k} = k[Y]\partial.$$

For instance, we could take $\partial = d/dt$. Let $\mathcal{D}$ be the algebra of differential operators which are finite sums

$$\sum f_i \partial^i$$

with $f_i \in k[Y]$. We have two natural injections giving rise to a commutative diagram:

$$\begin{array}{ccc} & & H^1_{\mathrm{DR}}(X/Y, Z) \\ & \nearrow & \downarrow \\ H^0\Omega^1_{X/Y} & & \\ & \searrow & \\ & & H^1_{\mathrm{DR}}(X/Y) \end{array}$$

Namely, a global differential form ω gives rise to a cocycle $\{(\omega, 0)\}$, such that for all open sets U, V we have $\omega_U = \omega$ and $f_{U,V} = 0$. We associate to ω the class of this cocycle in the de Rham group.

There are also homomorphisms (depending on a connection ∇):

$$\nabla(\partial)\colon H^1_{\mathrm{DR}}(X/Y) \to H^1_{\mathrm{DR}}(X/Y),$$
$$\nabla_Z(\partial)\colon H^1_{\mathrm{DR}}(X/Y, Z) \to H^1_{\mathrm{DR}}(X/Y, Z),$$

defined as follows. Let $\alpha \in H^1_{\mathrm{DR}}(X/Y, Z)$, say. They from the duality between derivations and differentials, we get the operators

$$\nabla_Z(\partial)\alpha = \langle \nabla_Z \alpha, \partial \rangle \in H^1_{\mathrm{DR}}(X/Y, Z),$$

and similarly without the Z. We define the **Kodaira–Spencer map**

$$\mathrm{KS}_\partial\colon H^0\Omega^1_{X/Y} \to H^1_{\mathrm{DR}}(X/Y) \bmod H^0\Omega^1_{X/Y}$$

by

$$\omega \mapsto \nabla(\partial)\omega.$$

We shall see later the significance of the Kodaira–Spencer map.

By linearity and by the natural injection of $H^0\Omega^1_{X/Y}$ into the de Rham groups, we obtain homomorphisms

$$\mathscr{D} \otimes H^0\Omega^1_{X/Y} \to H^1_{\mathrm{DR}}(X/Y) \quad \text{and} \quad \mathscr{D} \otimes H^0\Omega^1_{X/Y} \to H^1_{\mathrm{DR}}(X/Y, Z).$$

We define the **Picard–Fuchs** group PF to be the kernel of the first pairing, independent of Z, that is

$$\mathrm{PF} = \ker(\mathscr{D} \otimes H^0\Omega^1_{X/Y} \to H^1_{\mathrm{DR}}(X/Y)).$$

Then PF has a natural image in $H^1_{\mathrm{DR}}(X/Y, Z)$, and it follows from **ES 3** that the image of PF lies in $k[Y]$, so we obtain a homomorphism

$$\mathrm{PF} \to k[Y] \quad \text{which we denote by} \quad D \mapsto f_{D,s}.$$

The image of D in $k[Y]$ depends on the original choice $s = (s_1, s_2)$, so we indicated s in the notation.

For each choice of sections (s_1, s_2) we had to localize on Y to insure the smoothness of Z. Hence we shall view the function $f_{D,s}$ from now on as an element of the function field $k(Y)$, and the association

$$D \mapsto f_{D,s}$$

then applies to all choices of sections (s_1, s_2) as a homomorphism of PF into $k(Y)$.

Abelian varieties

We apply **ES 3** to the case when X is a family of abelian varieties over Y. In that case, we let s_1 be a section and we let s_2 be the zero section, and we use s to denote a section.

Theorem 4.1. *Let A/Y be an abelian scheme. Let (B, τ) be the F/k-trace of the generic fiber A_F. Let s be a section, giving rise to the pair $(s, 0)$. The sequence* **ES 3** *(as a sequence of connections) splits if and only if there exists an integer $m > 0$ such that $ms \in \tau B(k)$.*

We may further describe an imbedding of the group $A(F)/(\tau B(k) + A(F)_{\mathrm{tor}})$ in a finite product

$$F \times F \times \cdots \times F.$$

For this we use the horizontal differentiation and the functions $f_{D,s}$ in the context of an abelian scheme when the sections form a group.

Theorem 4.2 (Theorem of the kernel). *The association*

$$(D, s) \mapsto f_{D,s}$$

is a bilinear map

$$\mathrm{PF} \times A_F(F) \to k(Y)$$

The set of sections s such that $f_{D,s} = 0$ for all $D \in \mathrm{PF}$ is precisely

$$A_F(F)_{\mathrm{tor}} + \tau B(k).$$

Manin claimed to have proved the theorem of the kernel in [Man 63], but a quarter of a century later, Coleman found a gap in Manin's proof [Col 90]. See also Manin's letter in *Izvestia Akad. Nauk*, 1990. For the application to Mordell's conjecture in the function field case, only Theorem 4.1 was needed, and could be proved. On the other hand, Manin's work gave rise to further work by Deligne. Using this work Chai was able to prove the theorem of the kernel completely [Chai 90].

Since by the Lang–Néron theorem, the factor group of $A_F(F)$ by its torsion group and $\tau B(k)$ is finitely generated, we get:

Corollary 4.3. *There exist a finite number of differential operators $D_1, \ldots, D_m$ such that the association*

$$s \mapsto (f_{D_1,s}, \ldots, f_{D_m,s})$$

gives an imbedding of $A_F(F) \bmod A(F)_{\text{tor}} + \tau B(k)$ into $k(Y)^m$.

Curves

We pass on to a smooth family of curves $X \to Y$ of genus $\geqq 2$. We say that the family is **stably split** if there exists a finite morphism $Y' \to Y$ such that the base change $X_{Y'}$ is split, that is there exists a curve X_0 over k such that $X_0 \times Y'$ is birationally equivalent to $X_{Y'}$. Following Coleman, we shall use a differential criterion for the family to split.

We return to the derivation ∂ such that $\text{Der}_{k[Y]/k} = k[Y]\partial$, and to the Kodaira–Spencer map in the present context.

Theorem 4.4. *The following conditions are equivalent:*

(1) *There exists a derivation $\partial^X \in (H^0\Omega^1_{X/k})^\vee$ which lifts ∂ to a derivation in the dual of $H^0\Omega^1_{X/k}$.*
(2) *The Kodaira–Spencer map is 0.*
(3) *X is stably split.*

The next proposition gives a seemingly weaker criterion for the Kodaira–Spencer map to be 0.

Proposition 4.5. *Let $\pi: X \to Y$ be a proper smooth family of curves of genus $\geqq 2$. Suppose that:*

(a) *There exist sections of arbitrarily large height.*
(b) *There exist forms $\omega_1, \omega_2 \in H^0\Omega^1_{X/Y}$ linearly independent over $k[Y]$ such that if ∇ is the Gauss–Manin connection, then*

$$\nabla(\partial)\omega_i \in H^0\Omega^1_{X/Y} \quad \text{for} \quad i = 1, 2.$$

In other words, the Kodaira–Spencer map is 0 on a 2-dimensional space.

Then the Kodaira–Spencer map is 0, and hence ∂ lifts to a derivation in $(H^0\Omega^1_{X/Y})^\vee$, so the family is stably split.

Manin's key lemma used both for the above proposition, and the subsequent step, is used to bound heights, namely:

Lemma 4.6. *Let V be a finite dimensional vector subspace of the function field $k(X)$ over k. Then the set of sections s for which there exists a function $f \in V$ such that $f \circ s = 0$ has bounded height.*

When one cannot apply the situation of Proposition 4.5, one then uses a tower of coverings similar to those used in the theory of integral points which we shall encounter in Chapter IX, Proposition 3.4. Namely given an integer $m \geqq 2$, suppose that there are infinitely many rational points in $X_F(F)$. Then infinitely many points lie in the same coset of $J(F)/mJ(F)$ where J is the Jacobian of X_F over F. Then there is a point $P_0 \in J(F)$ such that all the points P in that coset can be written in the form

$$P = mQ + P_0.$$

We restrict the covering $J \to J$ given by $x \mapsto mx + P_0$ to the curve $X_F = C_0$ to obtain a covering C_1. We then iterate this process to obtain a tower of curves $C_n \to C_{n-1}$.

$$C_n \to C_{n-1} \to \cdots \to C_0 = X_F.$$

By Proposition 4.5, we are reduced to the case when the dimension of the kernel of the Kodaira–Spencer map is $\leqq 1$. Then we can pick the curves C_n so that the kernels of the Kodaira–Spencer maps have the same dimension for all n. Let J_n be the Jacobian of C_n. Then the Kodaira–Spencer maps for C_n and J_n are compatible. Let $B_n = J_n/J$. In the present case, the Kodaira–Spencer maps are injective, and it follows that

$$H^0\Omega^1_{B_n/Y} + \nabla(\partial)H^0\Omega^1_{B_n/Y} = H^1_{\mathrm{DR}}(B_n/Y).$$

Given $\omega \in H^0\Omega^1_{B_n/Y}$ there exist $\omega_1, \omega_2 \in H^0\Omega^1_{B_n/Y}$ such that

$$\nabla(\partial)^2\omega + \nabla(\partial)\omega_1 + \omega_2 = 0.$$

We pull back this relation to the curve C_n. Let η, η_1, η_2 be the pull backs of $\omega, \omega_1, \omega_2$ respectively. They by the definition of PF, which is the kernel of $\mathscr{D} \otimes H^0\Omega^1 \to H^1_{\mathrm{DR}}$, if we put

$$\partial^2 \otimes \eta + \partial \otimes \eta_1 + 1 \otimes \eta_2 = D_\eta$$

then $D_\eta \in \mathrm{PF}$. As we have seen we get an associated function

$$D_\eta \mapsto f_{D_\eta, s}.$$

Manin calculates the function $f_{D_\eta,s}$, and from this calculation deduces that the set of sections have bounded height. One would like to see directly that a bound on the heights of these functions gives a bound for the sections. It would be valuable to have a simplification of this part of Manin's paper, especially an analysis which would give a sharp bound for the height of sections. Furthermore, one might reconsider this part with an eye to seeing if, when passing to coverings Y' of Y, one gets anywhere close to the bound for heights conjectured by Vojta as mentioned in §2.

VI, §5. CHARACTERISTIC *p* AND VOLOCH'S THEOREM

After Manin's proof Samuel gave a proof of the Mordell conjecture in characteristic p when the curve is defined over a function field and when it cannot be defined over the constant field [Sam 66]. Following Raynaud's work dealing with my conjecture on the intersection of a curve with the division group of a finitely generated group in the Jacobian, Voloch [Vol 90] gave a 2-page proof of this more general property, under certain circumstances which we shall now make precise.

Let F be a function field over a constant field k, which we assume algebraically closed for simplicity. Let C be a projective non-singular curve defined over F. Recall that we say that C is **stably split** if there exists a curve C_0 defined over k such that for some finite extension E of F we have $C_E \approx C_{0,E}$; in other words C becomes isomorphic to a constant curve over E. We suppose that k has characteristic $p > 0$.

Suppose C is imbedded in its Jacobian J over F. Let Γ be a finitely generated subgroup of $J(F^{\mathrm{a}})$ and let Γ' be its **prime-to-*p* division group**, that is

> Γ' = subgroup of $J(F^{\mathrm{a}})$ consisting of all points x such that $nx \in \Gamma$ for some positive integer n prime to the characteristic p.

Theorem 5.1 ([Vol 90]). *Let C be a projective non-singular curve of genus $g \geqq 2$ defined over the function field F of characteristic $p > 0$. Assume that C is not stably split. Let Γ be a finitely generated subgroup of $J(F^{\mathrm{a}})$ and let Γ' be its prime-to-p division group. Assume that J is* **ordinary** *(that is $J(F^{\mathrm{a}})$ has p^g points of order p). Then $C \cap \Gamma'$ is finite.*

The proof is of independent interest, and involves the following lemma.

Lemma 5.2. *Let $L = F^{\mathrm{s}}$ be the separable closure of F. Let D be a non-trivial derivation of L. Let X be a projective non-singular curve over*

L. Suppose D lifts to X (meaning that D lifts to a derivation of the function field $L(X)$ mapping every local ring of a closed point into itself). Then X may be defined over L^p (i.e. X is isomorphic to a curve lifted to L from L^p).

This lemma is the analogue of Theorem 4.4 used in Manin's proof. If C is as in Theorem 5.1, then first without loss of generality one can replace C by a curve which cannot be defined over L^p. Then Voloch proves a first-order analogue of Raynaud's results in number fields, namely:

Lemma 5.3. *If C cannot be defined over L^p then $C \cap pJ(L)$ is finite.*

Theorem 5.1 then follows rapidly.

Voloch's proof suggests a possibility for a new proof of Manin's theorem in characteristic 0. Indeed, taking a non-stably split curve defined over a function field in characteristic 0, one may reduce the curve mod p after taking a suitable model. By a suitable mixed characteristic version of Hilbert irreducibility ([La 83a], Chapter 9, Corollary 6.3) one obtains a reduction to Voloch's theorem, provided one could prove that there is some prime p (presumably infinitely many) such that the Jacobian of the reduced curve is ordinary. This is true for elliptic curves, since the j-invariant is transcendental over $\mathbf{Q}$, and the curve is always a "Tate curve", cf. [La 72]. See also my *Elliptic Functions*, Chapter 15, §2.

CHAPTER VII

Arakelov Theory

In 1974 Arakelov indicated a way to complete a family of curves over the ring of integers of a number field by including the fibers at infinity. This amounted to the corresponding Riemann surfaces and their differential geometric properties once the number field gets imbedded into the complex numbers. Arakelov showed how one could define a global intersection number for two arithmetic curves on an arithmetic surface, and that this intersection number was actually defined on the rational equivalence classes, thus providing the beginning for the ultimate transposition of all algebraic geometry to this case. This is a huge program, which combines the algebraic side of algebraic geometry, the complex analytic side, complex differential geometry, partial differential equations and Laplace operators with needed estimates on the eigenvalues, in a completely open ended unification of mathematics as far as one can see.

In this chapter, the point is not to give a general summary of Arakelov theory, but to extract some basic definitions to show how that theory is relevant to diophantine applications today. I do cover the two main existing possibilities: the first, conjectural, is Parshin's proposed bound for $(\omega_{X/Y}^2)$, analogous to the one of Chapter VI, §3; the second is Vojta's use of the higher dimensional theory to prove the existence of sections for certain sheaves on the product of a curve with itself, to carry out his vast extension of the older methods of diophantine approximation on curves of higher genus.

We shall be concerned with metrized line sheaves and vector sheaves, so I review some analytic terminology in §1, before globalizing over number fields. Here I shall assume some elementary definitions (of d, d^c, the Chern form), which will however be given later in the more differential geometric context of Chapter VIII, §2.

VII, §1. ADMISSIBLE METRICS OVER C

Since metrics are given at infinity, we here consider a complete non-singular curve X over the complex numbers $\mathbf{C}$, and for this section we identify X with the Riemann surface $X(\mathbf{C})$. We suppose that the genus of X is $g \geqq 1$.

Let φ be a real (1, 1)-form on X *normalized such that*

$$\int_X \varphi = 1.$$

Let D be a divisor on X. By a **Green's function** for D with respect to φ we mean a function

$$\mathbf{g}_D\colon X - \operatorname{supp}(D) \to \mathbf{R}$$

satisfying the following conditions:

GR 1. *If D is represented by a rational function f on an open set U, then there exists a C^∞ function α on U such that for all $P \notin \operatorname{supp}(D)$ we have*

$$\mathbf{g}_D(P) = -\log|f(P)|^2 + \alpha(P).$$

GR 2. $dd^c\mathbf{g}_D = (\deg D)\varphi$ *outside the support of D.*

A function satisfying these two conditions is uniquely determined up to an additive constant, because the difference of two such functions is harmonic on the complement of D, and being sufficiently smooth on D by the continuity of the partial derivatives, it is also harmonic on D, so constant.

Finally we require:

GR 3. $$\int_X \mathbf{g}_D\varphi = 0.$$

A Green's function always exists.

Condition **GR 1** independently defines a quite general notion. On any variety V over $\mathbf{C}$ with a Cartier divisor D one defines a **Weil function** associated with D to be a function

$$\lambda_D\colon V - \operatorname{supp}(D) \to \mathbf{R}$$

which is continuous, and such that if D is represented by a rational function f on an open set U, then there exists a continuous function α

on U such that for all $P \notin \operatorname{supp}(D)$ we have

$$\lambda_D(P) = -\log|f(P)| + \alpha(P).$$

Thus a Green's function is 2 times a Weil function.

We shall also write

$$\mathbf{g}_D(Q) = \mathbf{g}(D, Q) \quad \text{or} \quad \mathbf{g}_P(Q) = \mathbf{g}(P, Q).$$

It is a fact that $\mathbf{g}$ is symmetric, in the sense that

$$\mathbf{g}(P, Q) = \mathbf{g}(Q, P) \quad \text{for} \quad P \neq Q,$$

and can be viewed as a function on $X \times X$ minus the diagonal. As such, the Green's function is a Weil function on $X \times X$ with respect to the diagonal, and is C^∞ outside the diagonal.

Let $\mathscr{L}$ be a line sheaf on X with a metric ρ. We say that this metric is **φ-admissible** if $c_1(\rho)$ equals φ. Such a metric exists, and the quotient of two such metrics is constant.

Let D be a divisor on X. We denote by $[D]$ the line sheaf $\mathcal{O}_X(D)$ with the unique admissible metric ρ such that if 1_D is the section defined by 1 in the function field, then

$$\int_X (-\log|1_D|_\rho)\varphi = 0.$$

Then

$$\mathbf{g}(P, Q) = -\log|1_P(Q)|_\rho^2.$$

Let Δ be the diagonal of $X \times X$. Then Δ is a divisor on $X \times X$, with the line sheaf $\mathcal{O}_{X \times X}(\Delta)$. There is a unique metric ρ on $\mathcal{O}(\Delta)$ such that

$$\mathbf{g}(P, Q) = -\log|1_\Delta(P, Q)|_\rho^2.$$

We call this the **canonical metric** (with respect to φ).

Among all forms φ, there is one which gives rise to special structures about which more explicit theorems have been proved. Although we shall not go specifically into these applications, we define that form because of its importance.

Let $\{\varphi_1, \ldots, \varphi_g\}$ be an orthonormal basis of the space of differentials of first kind with respect to the hermitian product

$$\langle \varphi, \eta \rangle = \frac{\sqrt{-1}}{2} \int_X \varphi \wedge \bar{\eta}.$$

We define the **Arakelov volume form**, or **canonical form**, to be

$$\mu = \frac{\sqrt{-1}}{2g} \sum_{i=1}^{g} \varphi_i \wedge \bar{\varphi}_i .$$

Let $\Omega^1_{X/\mathbf{C}}$ be the sheaf of differential forms on X. Then there is a natural isomorphism

$$\Omega^1_{X/\mathbf{C}} \approx \mathcal{O}(\Delta)^\vee | \Delta,$$

where Δ is the diagonal, whence we get a metric on $\Omega^1_{X/\mathbf{C}}$ via this isomorphism. This metric is admissible (with respect to the canonical Arakelov form) and will be called the **canonical metric**. Furthermore, let $P \in X$. Then the residue gives an isomorphism

$$\Omega^1_{X/\mathbf{C}} \otimes \mathcal{O}_X(P) = \Omega^1_{X/\mathbf{C}}(P)|P \to \mathbf{C},$$

and this isomorphism is an isometry if we give $\mathbf{C}$ the metric of the ordinary absolute value, and if we give $\Omega^1_{X/\mathbf{C}}(P)$ the tensor product metric of the canonical metrics as described above.

VII, §2. ARAKELOV INTERSECTIONS

Let F be a number field with ring of integers $\mathfrak{o}_F$, and let $Y = \operatorname{spec}(\mathfrak{o}_F)$ as before. Let S_∞ be the set of archimedean absolute values on F, and let Σ be the set of imbeddings $\sigma: F \to \mathbf{C}$.

By an **arithmetic surface**, we mean an integral scheme of dimension 2 together with a projective flat morphism

$$\pi: X \to Y.$$

whose generic fiber X_F is geometrically irreducible, so the generic fiber is a projective curve over the field F. We let

$$X_\infty = \coprod_{\sigma \in \Sigma} X_\sigma$$

be the disjoint union of the curves over $\mathbf{C}$ defined by all imbeddings of F into $\mathbf{C}$. We often identify $X_\infty(\mathbf{C})$ with X_∞.

By a **hermitian vector sheaf** on X we mean a vector sheaf $\mathcal{E}$ such that, if $\mathcal{E}_\infty$ denotes the restriction of $\mathcal{E}$ to X_∞, then $\mathcal{E}_\infty$ has been equipped with a hermitian metric h invariant under complex conjugation. The norm on $\mathcal{E}$ coming from the metric will be denoted by

$$|s|_{\sigma,h} = h_\sigma(s).$$

We let

$$h_\infty(s) = \prod_\sigma h_\sigma(s).$$

The collection of norms $\{h_\sigma\}$ on each component $\mathscr{E}$ will be denoted by $h^{\mathscr{E}}$ if we need to refer to $\mathscr{E}$ in the notation. Instead of hermitian vector sheaf, we also write of a **metrized vector sheaf**.

Let $(\mathscr{L}, \rho^{\mathscr{L}})$ and $(\mathscr{M}, \rho^{\mathscr{M}})$ be two metrized line sheaves on an arithmetic surface X. Suppose that $\mathscr{L}$, $\mathscr{M}$ have non-zero sections s, t respectively, and that the divisor (s) of s has no common component with the divisor (t). Let x be a closed point of X, and let $(s, t)_x$ be the ideal of $\mathscr{O}_x$ $(= \mathscr{O}_{x,X})$ generated by s and t after a choice of local trivialization of $\mathscr{L}$, $\mathscr{M}$ at the point x. Then $\mathscr{O}_x/(s, t)_x$ is a finite abelian group, and we define the **finite part** of the **intersection number** by

$$(s, t)_{\text{fin}} = \sum_{x \in X} \log(\mathscr{O}_x : (s, t)_x).$$

The sum on the right-hand side is finite, and only points in the intersection of (s) and (t) make a non-zero contribution to this sum.

On the other hand, suppose that the divisor (s) on $X_\infty(\mathbf{C})$ splits into a finite number of points

$$(s)_\infty = \sum n_j(P_j) \qquad \text{with} \quad P_j \in X_\infty(\mathbf{C}).$$

Let $c_1(\rho^{\mathscr{M}})$ be the first Chern form of the metric $\rho^{\mathscr{M}}$ on $\mathscr{M}$. We define the **intersection number at infinity** to be

$$(s, t)_\infty = \sum - n_j \log|t(P_j)|_\infty + \int_{X_\infty(\mathbf{C})} - (\log|s|_\infty) c_1(\rho^{\mathscr{M}}).$$

We define the **intersection number** to be

$$(\mathscr{L} . \mathscr{M}) = (s, t)_{\text{fin}} + (s, t)_\infty,$$

which is independent of the choice of s and t. As we have done above, we often omit the specific mention of the norms $\rho^{\mathscr{L}}$, $\rho^{\mathscr{M}}$ from the notation.

We define the **Arakelov–Picard group** (or **arithmetic Picard group**) $\text{Pic}_{\text{Ar}}(X)$ to be the group of metrized line sheaves, with admissible metrics, up to metric isomorphisms, with the group operation given by the tensor product. Then a basic theorem is that:

The intersection number $(\mathscr{L} . \mathscr{M})$ extends uniquely to a symmetric bilinear form

$$\text{Pic}_{\text{Ar}}(X) \times \text{Pic}_{\text{Ar}}(X) \to \mathbf{R}.$$

We can also define the notion of a **hermitian vector sheaf** on $Y = \operatorname{spec}(\mathfrak{o}_F)$. Without the hermitian structure, such a sheaf corresponds to a finite module over $\mathfrak{o}_F$ without torsion, i.e. a projective module over $\mathfrak{o}_F$. The hermitian structure simply gives the module a hermitian positive definite scalar product at each imbedding σ of F into $\mathbf{C}$, invariant under complex conjugation. If $(\mathscr{L}, \rho)$ is a hermitian line sheaf on Y then as in Chapter IV, §5 we define its **(Arakelov) degree**

$$\deg(\mathscr{L}, \rho) = \log(\mathscr{L} : s\mathfrak{o}_F) - \sum_{\sigma} \log|s|_{\sigma}$$

for any non-zero section s of $\mathscr{L}$. The expression on the right is independent of the choice of section.

Having the intersection number, one can then try to translate intersection theory on surfaces into the present context. This was done for the adjunction formula, Riemann–Roch (Faltings), and the Hodge index theorem (Faltings–Hriljac). Proofs will be found in [La 88]. A summary of the results, going beyond and including Noether's formula (also due to Faltings) will be found in Soulé [So 89].

To any point $x \in \mathbf{P}^n(F)$ there corresponds a morphism

$$s_x : \operatorname{spec}(\mathfrak{o}_F) \to \mathbf{P}^n_{\mathbf{Z}} = \mathbf{P}.$$

On $\mathbf{P}^n_{\mathbf{Z}}$ we have the line sheaf $\mathscr{O}_{\mathbf{P}}(1)$, and the universal quotient

$$\mathscr{O}_{\mathbf{P}}^{n+1} \to \mathscr{O}_{\mathbf{P}}(1) \to 0.$$

Cf. [Ha 77], Chapter II, Theorem 8.13. On the constant vector sheaf $\mathscr{O}_{\mathbf{P}}^{n+1}$ we have the standard product metric arising from the ordinary absolute value on $\mathbf{C}$, and thus we get a hermitian metric on the line sheaf

$$\mathscr{O}_{\mathbf{P}}(1) \otimes_{\mathbf{Z}} \mathbf{C} \quad \text{on } \mathbf{P}^n_{\mathbf{C}}$$

by taking the quotient of the standard metric.

By pull back, $s_x^* \mathscr{O}_{\mathbf{P}}(1)$ is a metrized line sheaf on $\operatorname{spec}(\mathfrak{o}_F)$, and one can then show that if $h_{\mathbf{P}}(x)$ denotes the height of the point x in projective space defined in Chapter II, §1, then

$$h_{\mathbf{P}}(x) = \frac{1}{[F:\mathbf{Q}]} \deg\big(s_x^* \mathscr{O}_{\mathbf{P}}(1)\big).$$

More generally, let us consider a divisor D on X. Then we have the line sheaf $[D]$, which is just $\mathscr{O}_X(D)$ with the metrics at infinity defined in §1. If P is an algebraic point on the generic fiber, so $P \in X_F(\mathbf{Q}^{\mathrm{a}})$, then we let E_P be the Zariski closure of P in X, and we let $[E_P]$ be the corresponding metrized line sheaf. The height can then be described in Arakelov intersections as in the next proposition.

Proposition 2.1. *Fix a divisor D on X, and let D_F be its restriction to the generic fiber. Then the association*

$$P \mapsto \frac{1}{[F(P):\mathbf{Q}]}([D].[E_P])$$

is a height function in the class of heights mod $O(1)$ *on $X(\mathbf{Q}^a)$ associated with D_F.*

One can define the **canonical sheaf** $\omega_{X/Y}$ just as we did in the function field case of Chapter VI, §1. Indeed, if $j\colon X \to S$ is an imbedding of X into a smooth scheme S over Y, then we have the **conormal sheaf**

$$\mathscr{C} = \mathscr{J}/\mathscr{J}^2$$

as before, where $\mathscr{J}$ is the sheaf of ideals of $\mathscr{O}_S$ defining X, and also as before

$$\omega_{X/Y} = \det j^*\Omega^1_{X/S} \otimes \det \mathscr{C}^\vee_{X/S}.$$

The discussion in Chapter VI, §1 is valid in the present context, no constant field played a role in this discussion.

Note that the restriction $\omega_{X/Y}$ to a complex fiber X_σ is simply

$$\omega_{X/Y}|X_\sigma = \Omega^1_{X_\sigma/\mathbf{C}},$$

and we shall assume throughout that $\omega_{X/Y}$ has the admissible metric which we defined in §2, via the scalar product of forms on a Riemann surface. Thus $\omega_{X/Y}$ is a metrized line sheaf, with an admissible metric.

Furthermore, the restriction of $\omega_{X/Y}$ to the generic fiber X_F, as a line sheaf without metrics, is just

$$\omega_{X/Y}|X_F = \Omega^1_{X/S} \approx \mathscr{O}_{X_F}(K)$$

where K is a canonical divisor on X_F. By Proposition 2.1, we have

$$h_K(P) = \frac{1}{[F(P):\mathbf{Q}]}([K].[E_P]) + O(1) \qquad \text{for} \quad P \in X(\mathbf{Q}^a).$$

One great advantage of the method of Chapter VI, §2 via canonical class inequalities and Parshin's construction to obtain height inequalities, is the possibility they give of being translated to the number field case via Arakelov theory. A theorem of Faltings asserts that in the semistable case, we always have $(\omega_{X/Y}^2) \geqq 0$. Parshin has raised the following question.

Do there exist positive numbers a_0, a_1, a_2, effectively computable, with a_1, a_2 absolute constants, and a_0 depending on g, such that for all number fields F, and all semistable families X/Y with $Y = \mathrm{spec}(\mathfrak{o}_F)$, and generic fiber of genus $g \geqq 2$ the following inequality holds:

$$(\omega_{X/Y}^2) \leqq a_2 \sum_y \delta_y^{\#} \log \#k(y) + a_1(2g-2)[F:\mathbf{Q}]\, d(F) + a_0[F:\mathbf{Q}].$$

(The sum is taken over all closed points of Y.)

Parshin showed how such an inequality would imply bounds for the height of rational points, that is, could be used to prove the Mordell conjecture–Faltings theorem.

Vojta shows in [Vo 88] how the proposed Parshin inequality, and even a weaker form of it, implies **height inequalities** of the same type as **H2** of Chapter VI, §2, by refining the Parshin construction. Vojta's arguments apply both to the function field case and to the number field case. For Parshin's own discussion of these questions see [Par 89].

One may also view the constants a_2, a_1, a_0 as variable. For instance, Vojta conjectures that a_1 can be taken as $1 + \varepsilon$, for every $\varepsilon > 0$, in which case a_0 would depend on ε. The problem about the set of (a_0, a_1, a_2) for which the inequality is true gets raised here in even stronger form than in the function field case. What is the geometric shape of this set in $\mathbf{R}^3$?

Szpiro has also raised the question when is $(\omega_{X/Y}^2) = 0$. In the function field case, this happens only if the family is **stably split**, i.e. the family becomes birationally equivalent to a product over a finite extension of the base Y. In the number field case, if $(\omega_{X/Y}^2) = 0$, this is presumably a rare occurrence. I suggested that this may happen only if there is complex multiplication. Bost, Mestre and Moret-Bailly [BMM] have done some computations in the case of one of the irreducible factors of the Fermat curve, and found a non-zero numerical value for $(\omega_{X/Y}^2)$ in this case.

In addition, if $(\omega_{X/Y}^2) > 0$ (strict positivity!), and under some additional hypothesis (for instance if X_F has good reduction everywhere), Szpiro gave another proof for Raynaud's theorem that the intersection of the curve with the group of torsion points in the Jacobian is finite [Szp 84]. As usual, a proof of such a diophantine result using Arakelov theory techniques may ultimately lead to effective upper bounds for the heights of solutions, going beyond the original finiteness proof.

In the present context of Arakelov theory, we also mention the extent to which Vojta has gone toward proving his conjecture bounding the height of algebraic points. Vojta defines the **arithmetic discriminant** of an algebraic point P to be

$$d_a(P) = \frac{1}{[F(P):\mathbf{Q}]}([E_P].[E_P] + \omega_{X/Y}).$$

This arithmetic discriminant bears to the (logarithmic) discriminant $d(P)$ the same relation as the arithmetic genus of the singular curve in algebraic geometry bears to the genus of a desingularized curve. In particular, it is usually large compared to $d(P)$. But using it, Vojta proves [Vo 90d]:

Theorem 2.2. *Let $\pi: X \to Y$ be an arithmetic surface as above. Fix an integer n and $\varepsilon > 0$. Then for all points $P \in X(\mathbf{Q}^a)$ of degree $\leqq n$ we have*

$$h_K(P) \leqq (1 + \varepsilon)\, d_a(P) + O(1).$$

Although nothing like such a bound was known up to now on curves, it is still too weak to prove even a weak form of the *abc* conjecture, for instance, because the arithmetic discriminant is too large compared to the discriminant.

VII, §3. HIGHER DIMENSIONAL ARAKELOV THEORY

This higher dimensional theory was developed by Gillet–Soulé [GiS 88], who define the arithmetic analogue of intersections for cycles of all dimensions, and also the usual objects entering into the Hirzebruch–Grothendieck Riemann–Roch formula. It would take too long here to give all the definitions, but I want to deal with one aspect of the theory having to do with the existence of sections for a vector sheaf or a line sheaf. The existence of such sections was used in a spectacular way by Vojta in his new proof of Faltings' theorem, and it is worth while to see more precisely how the higher dimension enters into that picture.

We let again $Y = \operatorname{spec}(\mathfrak{o}_F)$ where F is a number field. By an **arithmetic variety**

$$\pi: X \to Y$$

we mean a regular integral scheme X, projective and flat over Y, and such that the generic fiber X_F is a variety. We often view X over $\operatorname{spec}(\mathbf{Z})$, since Y itself is over $\operatorname{spec}(\mathbf{Z})$. For each imbedding $\sigma: F \to \mathbf{C}$ we get a variety

$$X_\sigma = X \times_\sigma \mathbf{C}.$$

We let X be of relative dimension d over Y, so of absolute dimension $d+1$.

Let $\mathscr{E}$ be a vector sheaf on X. For each $\sigma: F \to \mathbf{C}$ we get the vector sheaf $\mathscr{E}_\sigma$ on X_σ. For each σ we suppose given:

a hermitian metric h_σ on $\mathscr{E}_\sigma$;
a Kähler form ω_σ on X_σ, with corresponding metric g_σ;
we let $\mu_\sigma = \omega_\sigma^d/d!$.

Let $\bigwedge^{0,q}(\mathscr{E}_\sigma)$ be the sheaf of differential forms of type $(0, q)$ with coefficients in $\mathscr{E}_\sigma$, and let $A^{0,q}(\mathscr{E}_\sigma)$ be the vector space of C^∞ sections. Via g_σ and h_σ there is a hermitian scalar product on $\bigwedge^{0,q}(\mathscr{E}_\sigma)$. See [GrH 78], Chapter 0, §6. Therefore we get a scalar product on $A^{0,q}(\mathscr{E}_\sigma)$ by the formula

$$(*) \qquad \langle \eta, \eta' \rangle = \int_{X_\sigma(\mathbf{C})} \langle \eta(x), \eta'(x) \rangle \mu_\sigma(x).$$

The Dolbeault operator

$$\bar{\partial}: A^{0,q}(\mathscr{E}_\sigma) \to A^{0,q+1}(\mathscr{E}_\sigma)$$

has an adjoint $\bar{\partial}^*$ relative to this product, and as shown in [GrH 78] p. 84, there is an orthogonal direct sum decomposition

$$A^{0,q}(\mathscr{E}_\sigma) = \operatorname{Im} \bar{\partial} \oplus \operatorname{Im} \bar{\partial}^* \oplus H^{0,q}(X_\sigma, \mathscr{E}_\sigma).$$

We can form the derived functor (cohomology functor) $\mathscr{R}^q\pi_*\mathscr{E}$, which is a coherent sheaf on Y, and we denote its module of global sections by $R^q\pi_*\mathscr{E}$. This is a finitely generated module over $\mathfrak{o}_F$. We also have

$$R^q\pi_*\mathscr{E} \otimes_\sigma \mathbf{C} \approx H^{0,q}(X_\sigma, \mathscr{E}_\sigma).$$

Recall that if M is a finitely generated module over $\mathbf{Z}$, given a volume on $M_\mathbf{R} = \mathbf{R} \otimes M$, we define the **Euler characteristic**

$$\chi(M) = -\log \operatorname{Vol}(M_\mathbf{R}/M) + \log \#(M_{\text{tor}}).$$

Note the role of torsion in this definition. This definition can be applied to the cohomology groups over $\mathbf{Z}$ of some sheaf, and then one defines the Euler characteristic of the sheaf as the alternating sum of the Euler characteristics of its cohomology.

Similarly, choose elements $s_1, \ldots, s_m \in R^q\pi_*\mathscr{E}$ which are linearly independent over $\mathfrak{o}_F$ and are maximal such. Then $R^q\pi_*\mathscr{E}/(\Sigma\mathfrak{o}_F s_j)$ is finite. Furthermore, for each σ, the elements $s_1, \ldots, s_m$ under the imbedding σ

form a basis of $H^{0,q}(X_\sigma, \mathscr{E}_\sigma)$. Let H_σ be matrix representing the scalar product (*) with respect to this basis. Then det H_σ is a positive real number. We define the **L^2-degree** of $R^q\pi_*\mathscr{E}$ by the formula

$$\deg_{L^2} R^q\pi_*\mathscr{E} = -\frac{1}{2}\sum_\sigma \log \det H_\sigma + \log(R^q\pi_*\mathscr{E} : \Sigma\mathfrak{o}_F s_j)$$

The right-hand side is independent of the choice of basis.

In a purely geometric context, we would define the Euler characteristic to be

$$\sum_{q=0}^{d} (-1)^q \deg_{L^2} R^q\pi_*\mathscr{E},$$

but then we would be missing the "torsion" at infinity, which we must therefore introduce.

Let

$$\Delta_{\sigma,q} = \bar{\partial}\bar{\partial}^* + \bar{\partial}^*\bar{\partial}$$

viewed as an operator on Im $\bar{\partial} \oplus$ Im $\bar{\partial}^*$ in the orthogonal decomposition of $A^{0,q}(\mathscr{E}_\sigma)$. Then $\Delta_{\sigma,q}$ has eigenvalues $0 < \lambda_1 \leqq \lambda_2 \ldots$ and the associated **zeta function**

$$\zeta_{\sigma,q}(s) = \sum \lambda_i^{-s},$$

which converges absolutely for $\mathrm{Re}(s) > d$. The operator, and hence the zeta function, depend on $\mathscr{E}$ and the metrics h, g. By a basic theorem of Seeley [See 67], the zeta function has an analytic continuation as a meromorphic function on **C**, holomorphic at $s = 0$. We define the **analytic torsion**

$$\tau(\mathscr{E}_\sigma) = \sum_{q \geqq 0} (-1)^q q \zeta'_{\sigma,q}(0).$$

Then we define the **arithmetic Euler characteristic** of Gillet–Soulé to be

$$\chi_{\mathrm{Ar}}(\mathscr{E}) = \sum_{q=0}^{d} (-1)^q \deg_{L^2} R^q\pi_*\mathscr{E} + \sum_\sigma \tfrac{1}{2}\tau(\mathscr{E}_\sigma).$$

We have the relative **tangent sheaf** $\mathscr{T}_{X/Y}$, with a hermitian metric $h_{X/Y}$ corresponding to the chosen Kähler form ω (at each imbedding σ).

Gillet–Soulé define an **arithmetic Todd class** $\mathrm{Td}_{\mathrm{Ar}}(\mathscr{T}_{X/Y}, h_{X/Y})$, but we do not reproduce this definition (cf. the critical comments in [Weng 91]). They also define an **arithmetic Chern character**

$$\mathrm{ch}_{\mathrm{Ar}}\colon \text{metrized vector sheaves} \to \mathbf{Q} \otimes \mathrm{CH}_{\mathrm{Ar}}(X),$$

where

$$\mathbf{Q} \otimes \mathrm{CH}_{\mathrm{Ar}}(X) = \bigoplus \mathbf{Q} \otimes \mathrm{CH}^p_{\mathrm{Ar}}(X)$$

is the direct sum of the **arithmetic Chow groups**, tensored with **Q**. Then Gillet–Soulé have announced that the usual Hirzebruch formula is valid, namely:

Statement 3.1 (Gillet–Soulé [GiS 89]). *Let $\pi: X \to Y$ be an arithmetic variety of relative dimension d. Let $(\mathscr{E}, h)$ be a hermitian vector sheaf on X. Then*

$$\chi_{\mathrm{Ar}}(\mathscr{E}_h) = \deg \mathrm{ch}_{\mathrm{Ar}}(\mathscr{E}_h) \cdot \mathrm{Td}_{\mathrm{Ar}}(\mathscr{T}_{X/Y}, h_{X/Y}).$$

However, Weng [Weng 91] has pointed out that the formula is not correct, and he has given a modification of the definition of the arithmetic Todd genus for which the formula is conjecturally valid. Let $(\mathscr{L}, \rho)$ be a metrized line sheaf. In the applications described below, one needs the formula only for $\mathscr{E} \otimes \mathscr{L}^{\otimes n}$, and only asymptotically with an error term $O(n^d \log n)$ for $n \to \infty$. In this case this formula is proved. One of the main applications of such an expression is to provide sections for, say, powers of a line sheaf. The simplest context is that of a line sheaf $\mathscr{L}$ with hermitian metric ρ_σ for each σ. Let us define

$$h^0(X, \mathscr{E}_h) = \log \#\{\text{sections } s \in H^0(X, \mathscr{E}) \text{ such that } |s|_{L^2} \leqq 1\}.$$

Then in particular, we also have $h^0(X, \mathscr{E}_h \otimes \mathscr{L}_\rho^{\otimes n})$ for all positive integers n. By Gillet–Soulé's arithmetic intersection theory, the maximal power $(\mathscr{L}_\rho^{d+1})$ is defined as a real number. We let r be the rank of $\mathscr{E}$.

Theorem 3.2 (Gillet–Soulé [GiS 88d]). *If $(\mathscr{L}_\rho^{d+1}) > 0$ and $c_1(\rho_\sigma) > 0$ for all σ, then*

$$h^0(X, \mathscr{E}_h \otimes \mathscr{L}_\rho^{\otimes n}) \geqq n^{d+1} \frac{r}{(d+1)!} (\mathscr{L}_\rho^{d+1}) - O(n^d \log n)$$

for $n \to \infty$.

In particular, the first term on the right-hand side dominates, and gives the existence of sections for n large. However, Gillet–Soulé's theorem is proved under the assumption that the metric on $\mathscr{L}$ is positive, and Vojta used metrics which are not necessarily positive. Consequently, Theorem 3.2 as stated was of no use, and Vojta had to go back to the ingredients which went into its proof, notably the analysis of Bismut–Vasserot [BiV 88a, b], some of which applied to non-positive metrics.

Specifically, Vojta works in the case of an arithmetic variety $\pi: X \to Y$ such that X has a birational morphism $X \to X_1 \times_Y X_2$ on the product of two arithmetic surfaces X_1 and X_2 over Y. We take the trivial vector sheaf $\mathscr{E}$ and some line sheaf $\mathscr{L}$ with metrics. We go back to Statement

3.1, which is proved asymptotically to get

$$\sum_{i=0}^{2} (-1)^i \deg_{L^2} R^i \pi_* \mathscr{L}^{\otimes n} + \sum_{\sigma} \tau_\sigma(\mathscr{L}_\sigma^{\otimes n}) = \frac{n^3}{6}(\mathscr{L}^3) + O(n^2 \log n)$$

and we want to use this formula to show the existence of sections for large n. The arguments of ordinary algebraic geometry show that the terms with R^i for $i = 1, 2$ are $O(n^2 \log n)$, and can thus be absorbed into the error term. The problem is then with the analytic torsion, and one has to show that these terms also can get absorbed into the error term. This requires certain upper and lower bounds on $\zeta'_{q,n}(0)$, which we state in the special case used in [Vo 90a], see Chapter IX, §6.

Lemma 3.3 *Let M be a compact Kähler manifold of dimension d. Let $\mathscr{E}_h$ be a metrized vector sheaf on M, let $\mathscr{L}_\rho$ be a metrized line sheaf, and let $\zeta_{q,n}$ be the zeta function of the Laplacian on $A^{0,q}(\mathscr{E} \otimes \mathscr{L}^{\otimes n})$. Then*

$$\zeta'_{q,n}(0) \geqq -O(n^d \log n) \qquad \textit{for} \quad n \to \infty.$$

Lemma 3.4. *Let $M = X_1 \times X_2$ be a product of two compact connected Riemann surfaces. For $i = 1, 2$ let μ_i be a volume form on X_i, normalized to have volume 1. Let $p_i\colon M \to X_i$ be the projection. Let the Kähler form on M be $p_1^*\mu_1 + p_2^*\mu_2$. Let $\mathscr{L}$ be a metrized line sheaf on M. Assume that for each fiber $X_1 \times \{P_2\}$ the restriction of $\mathscr{L}$ to this fiber is a positive metrized line sheaf whose metric is admissible with respect to μ_1. Then*

$$\zeta'_{2,n}(0) = O(n^2 \log n) \qquad \textit{for} \quad n \to \infty.$$

The bound in Lemma 3.4 amounts to finding a lower bound for the first eigenvalue of the Laplacian. The two lemmas are purely analytic, and I wanted to show one example of how estimates at infinity are needed in the Arakelov geometry to obtain number theoretic results. The estimates showing how the analytic torsion can be absorbed into the error term are particularly striking, and one then sees how

$$\deg_{L^2} \pi_* \mathscr{L}^{\otimes n} = \frac{n^3}{6}(\mathscr{L}^3) + O(n^2 \log n),$$

whence the first term on the right dominates if $(\mathscr{L}^3) > 0$, but without the assumption that the metric is positive.

CHAPTER VIII

Diophantine Problems and Complex Geometry

Complex differential geometry intervenes in diophantine problems through several factors. First, if one considers holomorphic families of varieties, the problem of determining whether there exist only finitely many sections can be studied from a complex geometric point of view. But it also turns out (conjecturally at the moment) that the property of being Mordellic for a projective variety can be characterized in terms of purely complex differential geometric invariants, or complex analytic invariants. For instance, I conjectured that a projective variety X defined over a subfield of $\mathbf{C}$ finitely generated over the rationals is Mordellic if and only if every holomorphic map of $\mathbf{C}$ into $X(\mathbf{C})$ is constant. It is known in many cases that certain projective varieties have this holomorphic property, but except for curves of genus $\geqq 2$ or subvarieties of abelian varieties which do not contain translations of abelian subvarieties of dimension > 0 (Faltings' theorems) or varieties derived from those by products or unramified coverings or quotients, it is not known that they are Mordellic. Thus one obtains complex analytic criteria for a variety to be Mordellic. Similarly, in Chapter IX, we shall get quantitative diophantine criteria by inequalities at one absolute value.

By now, we see a pattern emerging, that *certain global diophantine properties* of a variety are controlled conjecturally, in first approximation and qualitatively, by *geometric conditions* (the Mordellicity of the complement of the special set), and also by conditions at *one* archimedean absolute value, which is a *local condition.*

In addition, to deal with quantitative estimates, Vojta has taught us that not only was there a classical analogy between algebraic numbers and algebraic functions, but also there is an equally deep seated analogy between algebraic numbers and holomorphic functions, via Nevanlinna theory. In this chapter, we touch on all these themes.

A fairly complete exposition with proofs is given in [La 87]. Hence the account in this chapter will be rapid, and is intended only as a brief guide.

VIII, §1. DEFINITIONS OF HYPERBOLICITY

We shall work not only with complex manifolds, but with complex spaces. Just as an algebraic space is defined locally by a finite number of polynomial equations in affine space, a complex space is defined locally by a finite number of holomorphic equations.

We let $\mathbf{D}$ be the unit disc in $\mathbf{C}$, centered at the origin. If $z \in \mathbf{D}$, the tangent plane $T_z\mathbf{D}$ can be identified with $\mathbf{C}$ itself, and a tangent vector $v \in T_z\mathbf{D}$ can be identified with a complex number. We have the **hyperbolic metric** on $T_z\mathbf{D}$, defined on a tangent vector v by the formula

$$|v|_{z,\mathrm{hyp}} = \frac{|v|_{\mathrm{euc}}}{1 - |z|^2},$$

where $|v|_{\mathrm{euc}}$ is the euclidean norm on $\mathbf{C}$. Note that for $z = 0$, the hyperbolic metric is the same as the euclidean metric.

For any positive number r we let $\mathbf{D}(r)$ be the open disc of radius r. The **hyperbolic metric** on $\mathbf{D}(r)$ is defined by

$$|v|_{\mathrm{hyp},r,z} = \frac{|v/r|_{\mathrm{euc}}}{1 - |z/r|^2}.$$

Thus multiplication by r

$$\mathbf{m}_r \colon \mathbf{D} \to \mathbf{D}(r)$$

gives an metric holomorphic isomorphism between $\mathbf{D}$ and $\mathbf{D}(r)$.

Let X be a complex manifold with a hermitian metric, or as we shall also say, a complex **hermitian manifold**. We can define the **distance** between two points x, y with respect to this metric by

$$d(x, y) = \inf_{\gamma} \int_a^b |d\gamma(t)|\, dt$$

where the inf is taken over all C^1 curves $\gamma\colon [a, b] \to X$ such that $\gamma(a) = x$ and $\gamma(b) = y$. The inf could also be taken over piecewise C^1 curves. In particular, we let d_{hyp} denote the hyperbolic distance on the disc $\mathbf{D}$ or $\mathbf{D}_r$, to distinguish it from the euclidean distance d_{euc}.

Next, let X be a connected complex space. Let $x, y \in X$. We consider sequences of holomorphic maps

$$f_i: \mathbf{D} \to X, \qquad i = 1, \ldots, m,$$

and points $p_i, q_i \in \mathbf{D}$ such that $f_1(p_1) = x$, $f_m(q_m) = y$, and

$$f_i(q_i) = f_{i+1}(p_{i+1}).$$

In other words, we join x to y by what we call a **Kobayashi chain** of discs. We add the hyperbolic distances between p_i and q_i, and take the inf over all such choices of f_i, p_i, q_i to define the **Kobayashi semidistance**

$$d_X(x, y) = \inf \sum_{i=1}^{m} d_{\text{hyp}}(p_i, q_i).$$

Then d_X satisfies the properties of a distance, except that $d_X(x, y)$ may be 0 if $x \neq y$, so we call d_X a **semidistance.**

If $X = \mathbf{D}$ then $d_{\text{hyp}} = d_{\mathbf{D}}$, in other words, the Kobayashi semidistance is the hyperbolic distance.

If $X = \mathbf{C}$ with the euclidean metric, then $d_X(x, y) = 0$ for all $x, y \in \mathbf{C}$.

Let $f: X \to Y$ be a holomorphic map of complex spaces. Then f is distance decreasing for the Kobayashi semidistance, that is

$$d_Y(f(x), f(x')) \leqq d_X(x, x') \qquad \text{for} \quad x, x' \in X.$$

The Kobayashi semidistance is continuous for the topology of X.

We define X to be **Kobayashi hyperbolic** if the semidistance d_X is a distance, that is, $x \neq y$ in X implies $d_X(x, y) > 0$. **Hyperbolic** will always mean Kobayashi hyperbolic. All other types of hyperbolic properties which we encounter will be subjected to a prefix to distinguish them.

Directly from the definition, we note that to be hyperbolic is a biholomorphic invariant. Furthermore, if X, Y are hyperbolic, so is $X \times Y$. A complex subspace of a hyperbolic space is hyperbolic. Discs and polydiscs in $\mathbf{C}^n$ are hyperbolic. A bounded domain in $\mathbf{C}^n$ is hyperbolic, since it is an open subset of a product of polydiscs. A quotient of a bounded domain by a discrete group of automorphisms without fixed points is also hyperbolic. More generally, let $X' \to X$ be an unramified covering. Then X is hyperbolic if and only if X' is hyperbolic.

There is another notion of hyperbolicity which will be relevant. We define a complex space X to be **Brody hyperbolic** if every holomorphic map $f: \mathbf{C} \to X$ is constant. It is trivial that:

Kobayashi hyperbolic implies Brody hyperbolic.

The converse holds under various compactness conditions, for instance:

Theorem 1.1 (Brody). *Let X be a compact complex space. Then X is Kobayashi hyperbolic if and only if X is Brody hyperbolic.*

For proofs of this and related properties, see [La 87].

We now return to the considerations of Chapter I, §3 in light of hyperbolicity. I conjectured [La 74] and [La 86]:

Conjecture 1.2. *The following conditions are equivalent for a projective variety X, defined over a subfield of the complex numbers finitely generated over the rationals.*

$X(\mathbf{C})$ is hyperbolic;
X is Mordellic;
Every subvariety of X is pseudo canonical.

Either one of the second or third condition would show that the property of being hyperbolic is algebraic. See also Chapter I, §3, Conjecture 3.6. In particular, we have the subsidiary conjecture:

Let X be a projective variety defined over a field F finitely generated over $\mathbf{Q}$. If $\sigma: F \to \mathbf{C}$ is one imbedding of F into the complex numbers, if X_σ denotes the resulting complex variety, and $X_\sigma(\mathbf{C})$ is hyperbolic, then for every imbedding $\sigma: F \to \mathbf{C}$ the complex space $X_\sigma(\mathbf{C})$ is hyperbolic.

Just by itself, this constitutes an unsolved problem today, independently of any connections with diophantine properties, or the algebraic geometric condition of being pseudo canonical.

In addition, we recall the algebraic special set defined in Chapter I, §3, which we now write as $\mathrm{Sp}_{\mathrm{alg}}(X)$ because we introduce the **holomorphic special set** $\mathrm{Sp}_{\mathrm{hol}}(X)$ to be the Zariski closure of the union of all images of non-constant holomorphic maps $f: \mathbf{C} \to X$. But I conjectured [La 86]:

Conjecture 1.3. *The algebraic and holomorphic special sets are equal.*

Thus the conjecture that X is pseudo canonical if and only if the special set is a proper subset now applies also to the holomorphic special set. A question also arises as to the extent it is necessary to take the Zariski closure in the above definition. The answer is known for abelian varieties, see Theorem 1.10.

Previously, there was a weaker **conjecture of Green–Griffiths**, implicit in [GrG 80].

Let X be a pseudo-canonical projective variety over **C**. *Let $f: \mathbf{C} \to X$ be holomorphic. Then the image of f is contained in a proper Zariski closed subset.*

For a result in this direction, see Lu–Yau [LY 90]. Note that the property of a variety expressed in the Green–Griffiths conjecture, i.e. that every holomorphic map of **C** into the variety is not Zariski dense, is *not equivalent* to X being pseudo canonical. The example $X = C \times \mathbf{P}^1$ where C is a curve of genus $\geqq 2$ shows that there may be a surface covered by holomorphic images of **C**, without the surface being pseudo canonical. The special set in this case is the whole surface. Nevertheless, the image of every non-constant holomorphic map $\mathbf{C} \to X$ is contained in one of the fibers of the projection $C \times \mathbf{P}^1 \to C$.

One can also **pseudofy** the notion of hyperbolicity. We say that X is **pseudo-Kobayashi hyperbolic** if there exists a proper algebraic subset Y such that if $x, x' \in X$ and $d_X(x, x') = 0$ then $x = x'$ or $x, x' \in Y$. We say that X is **pseudo-Brody hyperbolic** if the holomorphic special set is a proper subset. I conjectured:

Conjecture 1.4. *The following conditions are equivalent for a projective variety X defined over a subfield of the complex numbers finitely generated over* **Q**:

X is pseudo-Kobayashi hyperbolic;
X is pseudo-Brody hyperbolic;
X is pseudo canonical;
X is pseudo Mordellic.

Furthermore the set Y mentioned above can be taken to be the special set.

Even the equivalence of the first two conditions is not known today.

In parallel with the conjecture that the complement of the special set is Mordellic, I also conjecture that the complement of the special set is hyperbolic.

Example 1.5 (Hyperbolic hypersurfaces and complete intersections). Brody proved a conjecture of Kobayashi that the property of being hyperbolic is open, say in the following sense. Let

$$f: X \to Y$$

be a proper holomorphic map of complex spaces. If $f^{-1}(y_0)$ is hyperbolic for some point $y_0 \in Y$, then $f^{-1}(y)$ is hyperbolic for all y in some open neighborhood of y_0. However, the property is not closed. An example was first given by **Brody–Green**, namely the family of hypersurfaces

$$x_0^d + \cdots + x_3^d + (tx_0x_1)^{d/2} + (tx_0x_2)^{d/2} = 0.$$

They proved that these varieties are hyperbolic for d even $\geqq 50$, and all but a finite number of $t \neq 0$. But for $t = 0$ the variety is a Fermat hypersurface which contains lines, and so is not hyperbolic.

Kobayashi has raised the question whether the **generic hypersurface** of degree d in $\mathbf{P}^n$, with $d \geqq n + 2$ is hyperbolic [Kob 70], and similarly for the **generic complete intersection** of hypersurfaces of degrees $d_1, \ldots, d_r$ if

$$d_1 + \cdots + d_r \geqq n + 2.$$

Since a non-singular hypersurface is simply connected for $n \geqq 3$, one sees that hyperbolic spaces include a lot more than those which have bounded domains as universal covering spaces.

Suppose that X is a projective non-singular variety over $\mathbf{C}$. Then we have the canonical class K_X and also the cotangent bundle $T_X^\vee$. The class K_X is the divisor class associated with the maximal exterior power $\bigwedge^{\max} T_X^\vee$. Kobayashi proved [Kob 75]:

Theorem 1.6. *If $T_X^\vee$ is ample, then X is hyperbolic.*

Kobayashi–Ochiai [KoO 75] conjectured that if X is hyperbolic then the canonical class K_X is pseudo-ample, but I would make the stronger conjecture:

Conjecture 1.7. *If X is hyperbolic then K_X is ample.*

The converse of this last statement is not always true. A Fermat hypersurface of high degree has ample canonical class, but contains complex lines, so is not hyperbolic. In any case we have (with a conjecture in the middle):

$$T_X^\vee \text{ ample} \quad\Rightarrow\quad X \text{ hyperbolic} \quad\Rightarrow\quad K_X \text{ ample} \quad\Rightarrow\quad K_X \text{ pseudo ample.}$$

Subvarieties of abelian varieties

My conjecture that a subvariety of an abelian variety is Mordellic unless it contains the translation of an abelian subvariety of dimension > 0 led me to conjecture its hyperbolic analogue, which was proved by Mark Green [$\mathrm{Gr_M}$ 78], namely:

Theorem 1.8. *Let X be a closed complex subspace of a complex torus. Then X is hyperbolic if and only if X does not contain a translated complex subtorus $\neq 0$.*

In addition, the study of complex lines in an abelian variety from the point of view of transcendental numbers led me to conjecture the following statement, proved in [Ax 72]:

Theorem 1.9. *Let A be an abelian variety imbedded in projective space over $\mathbf{C}$, and let X be a hyperplane section. Let $g: \mathbf{C} \to A$ be a one parameter subgroup, i.e. a holomorphic homomorphism. Then X contains a translation of $g(\mathbf{C})$ or the intersection of X and $g(\mathbf{C})$ is not empty.*

For subvarieties of abelian varieties over the complex numbers, the Ueno–Kawamata fibrations of Chapter I, §6 have the stronger property to take into account holomorphic maps of $\mathbf{C}$ into X.

Theorem 1.10. *Let X be a subvariety of an abelian variety over $\mathbf{C}$. Let $f: \mathbf{C} \to X$ be a non-constant holomorphic map. Then the image of f is contained in the translate of an abelian subvariety, contained in X.*

Several people contributed to this theorem. Bloch in 1926 was the first to make the conjecture that if X is not the translation of an Abelian subvariety then a holomorphic map of $\mathbf{C}$ into X is **degenerate**, in the sense of being contained in a proper algebraic subset. Ochiai [Och 77] made major progress toward this conjecture. Then simultaneously Green–Griffiths and Kawamata proved Bloch's conjecture. Specifically, Green–Griffiths proved [GrG 80], §3, Theorem I′:

Let X be a closed complex subspace of a complex torus A. If X is not the translate of a subtorus of A, then the image of a non-constant holomorphic map $f: \mathbf{C} \to X$ lies in the translate of a proper complex subtorus in X.

On the other hand, Kawamata [Ka 80] proved not only Theorem 1.10, but also further results which combined with Ochiai's criterion yielded the full fibration theorem recalled in Chapter I, §6. In particular:

Theorem 1.11. *For a subvariety of an abelian variety, the algebraic and holomorphic special sets are equal, and one does not have to take the Zariski closure to define them.*

However, note that there may exist countably many translates of abelian subvarieties which are not contained in the fibers of the Ueno–Kamawata fibration. They may occur as sections, but their images are contained in the union of those fibers.

For the analogous structure theorem concerning semiabelian varieties, see Noguchi [No 81a].

Non-compact spaces

The above examples concern compact complex manifolds. There are also results concerning non-compact manifolds or spaces, but some new subtleties arise. Green has given an example of a Zariski open subset of a projective variety which is Brody hyperbolic but not Kobayashi hyperbolic. However, the possibility remains open that for an affine complex variety, the two are equivalent. For a discussion of this and connections with the diophantine properties of integral points, see [La 87].

The most natural way of obtaining non-compact spaces is to take away some proper algebraic subset in a projective variety over $\mathbf{C}$. In that line we have Borel's theorem. Recall that hyperplanes in $\mathbf{P}^n$ are said to be in general position if any $n + 1$ of them or fewer are linearly independent. Also if we pick $n + 2$ hyperplanes such that they are given by the equations

$$x_j = 0 \text{ for } j = 0, \ldots, n, \qquad \text{and} \qquad x_0 + \cdots + x_n = 0,$$

where x_j is the j-th projective coordinate, then for each subset I of $\{0, \ldots, n\}$ which consists of at least two elements and not more than n elements, we let the corresponding **diagonal hyperplane** be

$$D_I = \text{solutions of the equation } \sum_{i \in I} x_i = 0.$$

Then we have one form of

Theorem 1.12 (Borel's theorem). *Let Y be the complement of $n + 2$ hyperplanes of $\mathbf{P}^n(\mathbf{C})$ in general position. Then every holomorphic map $\mathbf{C} \to Y$ is either constant, or its image is contained in the diagonals.*

Borel's theorem was complemented by Green and Fujimoto as follows:

Theorem 1.13. *Let Y be the complement of q hyperplanes of $\mathbf{P}^n(\mathbf{C})$ in general position. Assume $q \geqq 2n + 1$. Then Y is Brody hyperbolic, that is every holomorphic map $\mathbf{C} \to Y$ is constant.*

Not much is known about the complement of hypersurfaces rather than the complement of hyperplanes. For a fuller discussion, examples, and bibliography of results of Fujimoto, Green, Noguchi, cf. [La 87] in addition to [Kob 70], especially IX, §3. We shall return to these questions from the affine diophantine point of view of integral points in Chapter IX.

VIII, §2. CHERN FORM AND CURVATURE

Let X be a complex manifold, and let L be a holomorphic line bundle over X. A hermitian metric is given by a positive definite hermitian product on each fiber, varying C^∞ over x. If U is an (ordinary) open set over which L admits a trivialization, and s is a section of L over U, represented by a function $s_U: U \to \mathbf{C}$ in the trivialization, then the metric ρ is given by

$$|s(P)|_\rho^2 = \frac{|s_U(P)|^2}{h_U(P)}$$

where $h_U: U \to \mathbf{R}_{>0}$ is a C^∞ function.

We let d be the usual exterior derivative, $d = \partial + \bar{\partial}$. We let

$$d^c = \frac{1}{2\pi}\left(\frac{\partial - \bar{\partial}}{2\sqrt{-1}}\right).$$

Then d^c is a real operator, i.e. maps real functions to real functions. We have

$$dd^c = \frac{\sqrt{-1}}{2\pi}\partial\bar{\partial}.$$

The **Chern form** of a metric ρ is the unique form, denoted by $c_1(\rho)$, such that on an open set U as above, we have

$$c_1(\rho)|U = -dd^c \log|s|_\rho^2 = dd^c \log h_U.$$

If a form is expressed in terms of complex coordinates $z_1, \ldots, z_n$ as

$$\omega = \sum_{\substack{|I|=p \\ |J|=q}} f_{IJ}(z, \bar{z})\, dz_I \wedge d\bar{z}_J,$$

where $dz_I = dz_{i_1} \wedge \cdots \wedge dz_{i_p}$ and similarly for $d\bar{z}_J$, then we say that the form is of **type** (p, q). The integers p, q are independent of the choice of holomorphic coordinates. The Chern form is of type (1, 1). We say that a (1, 1)-form

$$\omega = \sum h_{ij}(z)\frac{\sqrt{-1}}{2\pi}dz_i \wedge d\bar{z}_j$$

is **positive** and we write $\omega > 0$ if the matrix $h = (h_{ij})$ is hermitian positive definite for all values of z. A metric ρ is called **positive** if $c_1(\rho)$ is positive. Kodaira's imbedding theorem states that a compact complex manifold admits a projective imbedding if and only if it has a holomorphic line bundle with a positive metric.

We call

$$\Phi(z) = \bigwedge_{i=1}^{n} \frac{\sqrt{-1}}{2\pi} dz_i \wedge d\bar{z}_i$$

the **euclidean form**. Let Ψ be a form of type (n, n), where $n = \dim X$. We can write Ψ locally in terms of complex coordinates

$$\Psi(z) = h(z)\Phi(z)$$

with a C^∞ function h. If $h > 0$ everywhere then we say that Ψ is a **volume form**. A volume form Ψ as above determines its **Ricci form**

$$\mathrm{Ric}(\Psi) = dd^c \log h(z) \quad \text{in terms of the coordinates } z.$$

Let $K_X = \bigwedge^{\max} T_X$ be the **canonical bundle**. A volume form Ψ as above determines a metric κ on K_X (via the local functions h), and by definition,

$$c_1(\kappa) = \mathrm{Ric}(\Psi).$$

A 2-form commutes with all forms. By the n-th power $\mathrm{Ric}(\Psi)^n$ we mean the exterior n-th power. Since Ψ is a volume form, there is a unique function G on X such that

$$\frac{1}{n!} \mathrm{Ric}(\Psi)^n = G\Psi.$$

We call G the **Griffiths function**. In dimension 1, G is minus the **Gauss curvature**, by definition. We write $G = G(\Psi)$ to denote the dependence on Ψ. A $(1, 1)$ form ω will be called **strongly hyperbolic** if it is positive and if there exists a constant $B > 0$ such that, for all holomorphic imbeddings $f: \mathbf{D} \to X$ the Griffiths function of $f^*\omega$ is $\geqq B$, that is

$$\mathrm{Ric}(f^*\omega) = Gf^*\omega \qquad \text{with} \quad G \geqq B.$$

It is easy to show (by what is called the Ahlfors–Schwarz lemma) that if a strongly hyperbolic form exists, then X is hyperbolic. The converse is a major question of Kobayashi [Ko 70]:

Let X be a compact projective complex variety. If X is hyperbolic, does there exist a strongly hyperbolic $(1, 1)$*-form on X?*

Given the problematic status of this question, when I conjectured the equivalence of hyperbolicity and Mordellicity, I was careful not to take as definition of hyperbolicity this property involving (1, 1)-forms. However, weaker objects than (1, 1)-forms may exist as substitutes, in the direction of jet metrics as in Green–Griffiths [GG 80], and length functions which generalize the notion of hermitian metric. It is not known if the Brody–Green hypersurfaces have a hyperbolic (1, 1)-form on them.

The existence of a hyperbolic (1, 1)-form, however, gives an easy way of estimating from below the Kobayashi distance between points as follows. If

$$\omega = \sum h_{ij} \frac{\sqrt{-1}}{2\pi} dz_i \wedge d\bar{z}_j,$$

then the matrix (h_{ij}) defines a hermitian metric. We call the pair (X, ω) a **hermitian manifold**, meaning the manifold endowed with this metric. We let d_ω be the hermitian distance obtained from this metric. For example on the unit disc $\mathbf{D}$ we have the hyperbolic form

$$\omega_{\mathbf{D}} = \frac{2}{(1-|z|^2)^2} \frac{\sqrt{-1}}{2\pi} dz \wedge d\bar{z}.$$

The factor of 2 in the numerator is placed there so that $\mathrm{Ric}(\omega_{\mathbf{D}}) = \omega_{\mathbf{D}}$, so $G(\omega_{\mathbf{D}}) = 1$ (negative curvature -1).

Proposition 2.1. *Let (X, ω) be a hermitian manifold. Assume that there exists a constant $B > 0$ such that for every complex submanifold Y (not necessarily closed) of dimension* 1 *we have*

$$G(\omega | Y) \geqq B.$$

Then X is hyperbolic. Furthermore

$$Bf^*\omega \leqq \omega_{\mathbf{D}} \qquad \text{and} \qquad \sqrt{B}\, d_\omega \leqq d_X.$$

The first inequality is the infinitesimal version of the second. Thus we see that the existence of the hyperbolic (1, 1)-form gives a measure of hyperbolicity for the Kobayashi distance.

The above properties are the ones which are most important for us. To keep as sharp a focus here as possible, we have omitted other properties, which will be found in [Kob 70], [La 87] and the survey article [La 86]. It is also possible to define analogous objects for (n, n)-forms, and to define a weaker notion than hyperbolicity, namely **measure hyperbolicity**, using *equidimensional* holomorphic mappings

$$f\colon \mathbf{C}^n \to X \text{ (of dimension } n) \qquad \text{or} \qquad f\colon \mathbf{D}^n \to X.$$

This notion is less important for us than hyperbolicity, and we refer to the expositions in [Kob 70] and [La 87] for a more complete treatment.

VIII, §3. PARSHIN'S HYPERBOLIC METHOD

In [Pa 86], Parshin gave another proof for the function field case of Mordell's conjecture, actually in a slightly weaker form. We need a definition. Let Y be a complete non-singular curve over $\mathbf{C}$, let $F = \mathbf{C}(Y)$, and let

$$f: X \to Y$$

be a projective morphism from a non-singular surface X to Y, such that the generic fiber X_F is a non-singular curve of genus $\geqq 2$. We say that the family X over Y is **stably non-split** if for every finite extension of the base $Y' \to Y$, the curve $X_{F'}$ obtained by base change from F to the function field $F' = \mathbf{C}(Y')$ cannot be defined over $\mathbf{C}$. The weaker form of the theorem then reads:

Theorem 3.1. *Suppose $f: X \to Y$ as above is stably non-split. Then $X_F(F)$ is finite.*

Parshin's arguments use a mixture of hyperbolicity and topology as follows. Let

S = finite set of points $y \in Y(\mathbf{C})$ such that X_F has bad reduction at y;

U = union of a finite number of discs centered at the points of S in some chart;

$Y_0 = Y - U$ and $y_0 \in Y_0$;

$X_0 = X - f^{-1}(U)$ and $x_0 \in f^{-1}(y_0)$;

Z = fiber $f^{-1}(y_0)$.

The points y_0 and x_0 as above are fixed for the remainder of the discussion. From the smooth fibration

$$f: X_0 \to Y_0$$

we obtain an exact sequence of fundamental groups

$$(*) \qquad 1 \longrightarrow \pi_1(Z, x_0) \longrightarrow \pi_1(X_0, x_0) \underset{\overset{\alpha_s}{\dashleftarrow}}{\longrightarrow} \pi_1(Y_0, y_0) \longrightarrow 1.$$

The goal is to prove that there is only a finite number of sections of f. If a section s goes through x_0, then s defines a splitting α_s of the sequence $(*)$. If not, let x_1 be the point of intersection of $s(Y_0)$ and Z, and connect x_1 with x_0 by some path γ on Z. Then s defines a splitting of a sequence like $(*)$, but with new base points x_1 and $y_1 = f(x_1)$. Then the path γ induces an isomorphism of the new exact sequence with the previous one, and therefore s induces a conjugacy class of splittings $[\alpha_s]$. The finiteness of rational points follows from two statements:

Proposition 3.2. *There is only a finite number of sections s giving rise to the same class $[\alpha_s]$.*

Proposition 3.3. *The set of conjugacy classes of splittings of $(*)$ coming from sections of f is finite.*

The proof of the first proposition consists of general considerations of intersection theory and homology. We give the basic ideas. We consider the exact and commutative diagram:

$$\begin{array}{ccccc} H_2(f^{-1}(U)) & \longrightarrow & H_2(X) & \longrightarrow & H_2(X, f^{-1}(U)) \\ \uparrow & & \uparrow & & \\ \cup & & \cup & & \\ \mathrm{NS}_S(X) & \longrightarrow & \mathrm{NS}(X) & & \end{array}$$

where $\mathrm{NS}_S(X)$ is the subgroup of the Néron–Severi group generated by the classes of components of the bad fibers over points of S. If C is a curve in X, we let $[C]$ denote its image in $\mathrm{NS}(X)$, whence in the homology groups according to the arrows in the above diagram. We write $[s]$ for $[s(Y)]$.

Then in the first place, the conjugacy class $[\alpha_s]$ defines the image of $[s]$ in $H_2(X, f^{-1}(U))$. Furthermore, by intersection theory, if s, s' are sections such that

$$[s] \equiv [s'] \bmod \mathrm{NS}_S(X),$$

then $[s] = [s']$ except possibly for a finite number of sections. Finally, the image of $[s]$ in $H_2(X)$ determines the section s up to a finite number of possibilities because a theorem of Arakelov implies that $([s]^2) < 0$ under the hypothesis that X is stably non-split. Hence $[s]$ does not lie in an algebraic family of dimension > 0, and Proposition 3.2 follows.

Hyperbolicity comes in the proof of Proposition 3.3. In the first place, by general criteria of hyperbolicity, the space X_0 is hyperbolic, essentially because Y_0 is hyperbolic and the fibers are hyperbolic, and there is enough relative compactness involved in the definition of Y_0, X_0 by taking out U and $f^{-1}(U)$. By the distance decreasing property of holomorphic maps, one also sees that for any section s, $s(Y_0)$ is **totally geodesic**, meaning that for any points $x, x' \in s(Y_0)$ we have

$$d_{s(Y_0)}(x, x') = d_{X_0}(x, x').$$

We then choose loops $\gamma_1, \ldots, \gamma_n$ representing generators of $\pi_1(Y_0, y_0)$. Then representatives for a class α_s corresponding to a section s are given in the form

$$\gamma\tilde{\gamma}_j\gamma^{-1}, \qquad j = 1, \ldots, n,$$

where we write $\tilde{\gamma}_j$ to denote the loop γ_j considered on $s(Y_0)$, and γ is a path in the fiber connecting x_0 with some point $x_1 \in s(Y_0) \cap f^{-1}(y_0)$. Since $s(Y_0)$ is totally geodesic, it follows that the lengths of the loops $\tilde{\gamma}_j$ are bounded in the Kobayashi metric of X_0, and the same is true for the set of loops $\gamma\tilde{\gamma}_j\gamma^{-1}$ by the compactness of the fiber. Also by a compactness argument, the set of elements of $\pi_1(X_0, x_0)$ represented by loops of bounded length is finite, whence Proposition 3.3 follows.

By similar arguments, mixing the properties of the fundamental group and hyperbolicity, Parshin gave another proof for Raynaud's theorem concerning subvarieties of abelian varieties in the function field case, which we mentioned as Theorem 6.7 of Chapter I. Also by similar arguments, Parshin proved the function field case of my conjecture concerning integral points on affine subsets of abelian varieties under the restriction that the hyperplane at infinity does not contain the translation of an abelian subvariety of dimension $\geqq 1$. We shall mention this again in the context of integral points in Chapter IX.

VIII, §4. HYPERBOLIC IMBEDDINGS AND NOGUCHI'S THEOREMS

Consider again as in §3 our standard situation of a proper morphism

$$f: X \to Y$$

where Y is a complete non-singular curve of genus $\geqq 2$ over $\mathbf{C}$, and X is a non-singular surface, so we get a fibering. Let $y_0 \in Y$ be a point where X_F has good reduction, $F = \mathbf{C}(Y)$, and let U be an open neighborhood of y_0 which is a disc in a chart. Then U is hyperbolic, and if we take U small enough, then X_F has good reduction for all $y \in U$, so $f^{-1}(U)$ is hyperbolic. Let $s: Y \to X$ be a section. Then the restriction s_U of s to U is a holomorphic map

$$s_U: U \to f^{-1}(U),$$

which is Kobayashi-distance decreasing. If we let Y_0 be the open set of points where X_F has good reduction, then

$$s\colon Y_0 \to f^{-1}(Y_0) = X_0$$

is also distance decreasing, and the set of sections restricted to Y_0 is equicontinuous.

If $Y = Y_0$ then the set of sections is compact by Ascoli's theorem. Since the degrees of sections in some projective imbedding depends continuously on the sections, it follows that these degrees are bounded, whence the heights of sections are bounded, and we have another proof of the Mordell conjecture in the function field case. Since in fact there usually are points of Y above which the fiber is degenerate, we must look more closely at that possibility, and how the Kobayashi distance degenerates in the neighborhood of such a fiber.

So let X be a complex manifold (for simplicity) and let X_0 be a relatively compact open subset (for the ordinary topology). Following Kobayashi, we shall say that X_0 is **hyperbolically imbedded** in X if there exists a positive (1, 1) form ω on X (or equivalently a hermitian metric) and a constant $C > 0$ such that

$$d_{X_0} \geqq C d_\omega.$$

There are other definitions (Cf. [La 87]) which are equivalent to this one, but we have picked the most convenient one. In particular, if X_0 is hyperbolically imbedded in X, then X_0 is hyperbolic, both Brody and Kobayashi.

Apply the above notion to our standard situation of a fibering $f\colon X \to Y$, in the neighborhood of a point $y_0 \in Y$ where X_F has bad reduction. To make the argument using the compactness of the set of sections go through, Noguchi proved the following result [No 85].

Theorem 4.1. *Given a complete non-singular curve Y over* **C** *and a complete non-singular curve X_F over the function field $F = \mathbf{C}(Y)$, of genus $\geqq 2$, there exists a proper morphism*

$$f\colon X \to Y$$

from a non-singular surface X onto Y whose generic fiber is X_F, and such that if Y_0 is the subset of Y over which f is smooth, then $f^{-1}(Y_0)$ is hyperbolically imbedded in X. For such a model X the set of sections of f is compact, that is, every sequence of sections has a subsequence which converges uniformly on compact subsets of Y.

Thus the condition of being hyperbolically imbedded insures that in the neighborhood of a bad point of Y, the set of sections is still locally compact, and the same argument as before works to get the conclusion that the projective degrees (or heights) of sections are bounded.

Noguchi also has a higher dimensional version, assuming the hyperbolic imbedding of the open set where f is smooth in its compactification [No 85], [No 87], but Noguchi has shown in general that there does not exist a good compactification in the higher dimensional case, namely he has shown that the following statement is not true in general:

Let Y be a complete non-singular curve over $\mathbf{C}$, let $F = \mathbf{C}(Y)$, and let X_F be a non-singular projective variety, such that over some point y_0 where X_F has good reduction, the fiber X_{y_0} is hyperbolic; or alternatively, assume X_F algebraically hyperbolic. Then there is a projective morphism

$$f\colon X \to Y$$

from some non-singular variety X onto Y, such that the generic fiber is X_F, and such that, if Y_0 is the open subset of Y over which f is smooth, then $f^{-1}(Y_0)$ is hyperbolically imbedded in X.

Once one has such a family $f\colon X \to Y$, then the compactness of the space of sections follows. We describe the situation somewhat more generally. Let M be a complex manifold and D an effective divisor on M. We say that D has **normal crossings** if in the neighborhood of each point there exist complex coordinates $z_1, \ldots, z_m$ such that in that neighborhood, there exists a positive integer r with $1 \leqq r \leqq m$ such that the divisor is defined by the equation

$$z_1 \ldots z_r = 0.$$

Theorem 4.2. *Let:*

$X_0 \subset X$ be a relatively compact, hyperbolically imbedded complex subspace;
M be a complex manifold of dimension m;
D a divisor with normal crossings on M.

Let $f_n\colon M - D \to X_0$ be a sequence of holomorphic maps, which converge uniformly on compact subsets of $M - D$ to a holomorphic map

$$f\colon M - D \to X_0.$$

Then there exist holomorphic extensions $\bar{f}_n$ and $\bar{f}$ from M into X, and the sequence of extensions $\bar{f}_n$ converges uniformly to $\bar{f}$ on every compact subset of M itself.

The existence of the extensions $\bar{f}_n$ and $\bar{f}$ is due to Kwack–Kobayashi–Kiernan, and the fact that the sequence of extensions $\bar{f}_n$ converges uniformly to $\bar{f}$ is due to Noguchi [No 87]. For proofs, Cf. also [La 87], Chapter II, Theorems 5.2 and 5.4. Also compare the preceding and following theorem with the Kobayashi–Ochiai Theorem 3.7 of Chapter I.

From Noguchi [No 81b] we have the following higher dimensional version of Theorem 3.1.

Theorem 4.3. *Let F be a function field over an algebraically closed field k of characteristic 0. Let X be a projective non-singular variety defined over F. Assume that the cotangent bundle of X is ample. Then:*

(1) *the set $X(F)$ is not Zariski dense; or*
(2) *there is a variety X_0 over k which is isomorphic to X over F, and $X_0(F) - X_0(k)$ is finite.*

Note that the condition for $T^\vee(X)$ to be ample is the condition under which Kobayashi proved that X_σ is hyperbolic for any imbedding σ of F into $\mathbf{C}$ [Kob 75].

VIII, §5. NEVANLINNA THEORY

This theory gives a quantitative version for the qualitative property concerning the existence or non-existence of a non-constant holomorphic map of $\mathbf{C}$ into a complex non-singular variety.

Let X be a projective variety over $\mathbf{C}$. Let D be a Cartier divisor on X. By a **Weil function** for D we mean a function

$$\lambda_D: X - \operatorname{supp}(D) \to \mathbf{R}$$

which is continuous, and is such that if D is represented by φ on a Zariski open set U, then there exists a continuous function $\alpha: U \to \mathbf{R}$ such that for all $P \notin \operatorname{supp}(D)$ we have

$$\lambda_D(P) = -\log|\varphi(P)| + \alpha(P).$$

The difference of two Weil functions is the restriction to $X - \operatorname{supp}(D)$ of a continuous function on X, and so is bounded. Thus two Weil functions differ by $O(1)$.

If L is a line bundle which has a meromorphic section s such that the divisor of s is $(s) = D$, and ρ is a metric on L, then we can take

$$\lambda_D(P) = -\log|s(P)|_\rho.$$

Let

$$f: \mathbf{C} \to X$$

be a holomorphic map. *Suppose $f(\mathbf{C})$ is not contained in D.* This is equivalent to the fact that f meets D discretely, i.e. in any disc $\mathbf{D}(r)$ there are only a finite number of points $a \in \mathbf{D}(r)$ such that $f(a) \in D$. Given a Weil function λ_D, we define the **proximity function**

$$m_{f,D}(r) = m_f(r, D) = \int_0^{2\pi} \lambda_D\big(f(re^{i\theta})\big)\frac{d\theta}{2\pi}.$$

For a real number $\alpha > 0$ we let as usual $\log^+(\alpha) = \max(0, \log \alpha)$. Similarly, let

$$\lambda_D^+ = \max(0, \lambda_D).$$

If D is effective, we could also use λ_D^+ instead of λ_D in the definition of $m_{f,D}$.

The association $D \mapsto \lambda_D \bmod O(1)$ is a homomorphism, and hence so is the association

$$D \mapsto m_{f,D} \bmod O(1).$$

Given $a \in \mathbf{C}$, let D be represented by the pair (U, φ) on an open set U containing $f(a)$. We define

$$v_f(a, D) = \operatorname{ord}_a(\varphi \circ f) \qquad (\geqq 0 \text{ if } D \text{ is effective}),$$

$$N_{f,D}(r) = N_f(r, D) = \sum_{\substack{a \in \mathbf{D}(r) \\ a \neq 0}} v_f(a, D) \log\left|\frac{r}{a}\right| + v_f(0, D) \log r.$$

We call $N_{f,D}$ the **counting function**, which counts how many times f hits D in the disc of radius r. We define the **height** $T_{f,D} \bmod O(1)$ to be

$$T_{f,D} = m_{f,D} + N_{f,D}.$$

Following Vojta, note that the definition of the height is entirely analogous to the definition in the algebraic case that we met in Chapter II. Indeed, for each θ, r we have something like an absolute value defined on the field of meromorphic functions on $\mathbf{D}(r)$ by

$$\|g\|_{\theta,r} = |g(re^{i\theta})|,$$

and for $a \in \mathbf{D}(r)$ we have the absolute value defined by

$$-\log\|g\|_{a,r} = v_{a,r}(g) = (\operatorname{ord}_a g) \log\left|\frac{r}{a}\right| \qquad \text{for} \quad a \neq 0.$$

As Vojta observed, the Jensen formula from elementary complex analysis is the analogue of the product formula (written additively). Of course, here, the places corresponding to θ vary continuously, so instead of a sum we have to use an integral.

Suppose $X = \mathbf{P}^n(\mathbf{C})$ is projective space, and D is a hyperplane. Let

$$f: \mathbf{C} \to \mathbf{P}^n$$

be a holomorphic map. We can always represent f by entire functions

$$f = (f_0, \ldots, f_n)$$

without common zero, by using the Weierstrass factorization theorem. Then it is easy to see that the height is given by the **Cartan–Nevanlinna** expression

$$T_{f,D}(r) = \int_0^{2\pi} \log \max_i \|f_i\|_{\theta,r} \frac{d\theta}{2\pi} + O(1).$$

Thus the height is entirely similar to the height defined previously for algebraic points in $\mathbf{P}^n$, in the number theoretic case. It is a fundamental fact that:

The height $T_{f,D}$ depends only on the rational equivalence class of D mod $O(1)$.

This is sometimes called the **first main theorem**, although it is simple to prove. Then the height can be characterized by the following conditions.

Let $f: \mathbf{C} \to X$ be a holomorphic map into a projective variety X. To each Cartier divisor D on X one can associate a function

$$T_{f,D}: \mathbf{R}_{\geqq 1} \to \mathbf{R},$$

well defined mod $O(1)$, *depending only on the rational equivalence class of D, and uniquely determined by the following properties.*

H 1. *The map $D \mapsto T_{f,D}$ is a homomorphism* mod $O(1)$.

H 2. *If E is very ample and $\psi: X \to \mathbf{P}^n$ is an imbedding into projective space, such that $E = f^{-1}$(hyperplane), then*

$$T_{f,E} = T_{\psi \circ f} + O(1)$$

where $T_{\psi \circ f}$ denotes the Cartan–Nevanlinna expression for the height of a map into projective space.

In addition, the height satisfies the further properties:

Vojta's Dictionary from [Vo 87]	
Nevanlinna Theory	*Roth's Theorem*
$f: \mathbf{C} \to \mathbf{C}$, non-constant	$\{b\} \subseteq k$, infinite
r	b
θ	$v \in S$
$\lvert f(re^{i\theta})\rvert$	$\lVert b \rVert_v$, $v \in S$
$\operatorname{ord}_z f$	$\operatorname{ord}_v f$, $v \notin S$
$\log \dfrac{r}{\lvert z \rvert}$	$\log N_v$
Characteristic function	Logarithmic height
$T(r) = \int_0^{2\pi} \log^+ \lvert f(re^{i\theta})\rvert \dfrac{d\theta}{2\pi} + N(\infty, r)$	$h(b) = \dfrac{1}{[k:\mathbf{Q}]} \sum_v \log^+ \lVert b \rVert_v$
Proximity function	
$m(a, r) = \int_0^{2\pi} \log^+ \left\lvert \dfrac{1}{f(re^{i\theta}) - a} \right\rvert \dfrac{d\theta}{2\pi}$	$m(a, b) = \dfrac{1}{[k:\mathbf{Q}]} \sum_{v \in S} \log^+ \left\lVert \dfrac{1}{b - a} \right\rVert_v$
Counting function	
$N(a, r) = \sum_{\lvert w \rvert < r} \log \dfrac{r}{\lvert w \rvert}$	$N(a, b) = \dfrac{1}{[k:\mathbf{Q}]} \sum_{v \notin S} \log^+ \left\lVert \dfrac{1}{b - a} \right\rVert_v$
First Main Theorem	Property of heights
$N(a, r) + m(a, r) = T(r) + O(1)$	$N(a, b) + m(a, b) = h(b) + O(1)$
Second Main Theorem	Conjectured refinement of Roth
$\sum_{i=1}^{m} m(a_i, r) \leqq 2T(r) - N_1(r) + O(r \log T(r))$ //	$\sum_{i=1}^{m} m(a_i, b) \leqq 2h(b) + O(\log h(b))$
Defect	
$\delta(a) = \liminf_{r \to \infty} \dfrac{m(a, r)}{T(r)}$	$\delta(a) = \liminf_{b} \dfrac{m(a, b)}{h(b)}$
Defect Relation	Roth's theorem
$\sum_{a \in C} \delta(a) \leqq 2$	$\sum_{a \in k} \delta(a) \leqq 2$
Jensen's formula	Artin–Whaples Product Formula
$\log \lvert c_\lambda \rvert = \int_0^{2\pi} \log \lvert f(re^{i\theta})\rvert \dfrac{d\theta}{2\pi} + N(\infty, r) - N(0, r)$	$\sum_v \log \lVert b \rVert_v = 0$

The Dictionary in the One-Dimensional Case.

H 3. *For any Cartier divisor D and ample E we have*

$$T_{f,D} = O(T_{f,E}).$$

H 4. *If D is effective and $f(\mathbf{C}) \not\subset D$, then $T_{f,D} \geqq -O(1)$.*

H 5. *The association $(f, D) \mapsto T_{f,D}$ is functorial in (X, D). In other words, if $\psi: X \to Y$ is a morphism of varieties, and $D = \psi^{-1}(D')$, where D' is a divisor on Y, then*

$$T_{f,D} = T_{\psi \circ f, D'} + O(1).$$

H 6. *Let $\psi: X \to \mathbf{P}^m$ be a morphism, and suppose $D = \psi^{-1}(H)$ where H is a hyperplane. Then*

$$T_{f,D} = T_{\psi \circ f} + O(1)$$

where $T_{\psi \circ f}$ is the Cartan–Nevanlinna height.

Let X be a projective non-singular variety over $\mathbf{C}$ and let $f: C \to X$ be holomorphic. In the neighborhood of a point of $\mathbf{C}$, let w be a complex coordinate, and let $z_1, \ldots, z_n$ be complex coordinates in a neighborhood of the image of this point in X. We suppose that $f(0) = (0, \ldots, 0)$. Write

$$f(w) = w^e(g_1(w), \ldots, g_n(w)) \qquad \text{so} \qquad z_i = w^e g_i(w)$$

such that $g_1, \ldots, g_n$ do not all vanish at 0. We define the **ramification order** of f at the point to be $e - 1$. Thus for each point of $\mathbf{C}$ we have assigned an integer $\geqq 0$, defining the **ramification divisor** of f. We can then define $N_{f,\mathrm{Ram}}$ in a way similar to $N_{f,D}$.

Finally, let D be a divisor on X. We say that D has **simple normal crossings** if we can write

$$D = \sum D_j$$

as a sum of irreducible components D_j which are non-singular and have normal crossings (defined at the end of the preceding section). What would be the **second main theorem** of Nevanlinna theory concerns such divisors, but is only a conjecture today.

Conjecture 5.1 *Let X be a projective non-singular variety over $\mathbf{C}$. Let D be a divisor on X with simple normal crossings. There exists a proper Zariski closed subset Z_D having the following property. Let*

$$f: \mathbf{C} \to X$$

be a holomorphic map such that $f(\mathbf{C}) \not\subset Z_D$. Let K be the canonical class and let E be an ample divisor. Then

$$m_{f,D}(r) + T_{f,K}(r) + N_{f,\mathrm{Ram}}(r) \leqq O(\log r + \log^+ T_{f,E}(r))$$

for $r \to \infty$ outside a set of finite Lebesgue measure.

Only a linear case of this conjecture is known today when X has dimension greater than 1, and we now turn to this case. Note that on projective space $\mathbf{P}^n$, the canonical class is given by

$$K = -(n+1)H \qquad \text{where } H \text{ is a hyperplane.}$$

As to the ramification, suppose we deal with a non-constant holomorphic map

$$f: \mathbf{C} \to \mathbf{P}^n$$

into projective space, represented by $(f_0, \ldots, f_n)$, where the functions f_i are entire without common zeros. We have the **Wronskian**

$$W = W(f_0, \ldots, f_n) = \det\left(\left(\frac{d}{dz}\right)^i f_j\right)$$

with $i, j = 0, \ldots, n$. The zeros of the Wronskian define a discrete set of points on $\mathbf{C}$, which are the ramification points of f. We define:

$$\begin{aligned} N_W(r, 0) &= \sum_{\substack{a \in \mathbf{D}(r) \\ a \neq 0}} \mathrm{ord}_a(W) \log \left|\frac{r}{a}\right| + \mathrm{ord}_0(W) \log r. \\ &= N_{f,\mathrm{Ram}}(r) \qquad \text{(the \textbf{ramification counting function}).} \end{aligned}$$

Then we have the **main theorem** in this case.

Theorem 5.2. *Let $D = \sum D_j$ be a divisor on $\mathbf{P}^n$ such that the irreducible components are hyperplanes in general position. There exists a finite union of hyperplanes Z_D having the following property. For every holomorphic map*

$$f = (f_0, \ldots, f_n): \mathbf{C} \to \mathbf{P}^n$$

represented by entire functions f_i without common zeros, and such that $f(\mathbf{C}) \not\subset Z_D$ the inequality holds:

$$m_{f,D}(r) + T_{f,K}(r) + N_{f,\mathrm{Ram}}(r) \leqq O(\log r + \log^+ T_{f,E}(r))$$

for $r \to \infty$ outside a set of finite Lebesgue measure.

Nevanlinna proved the theorem when $n = 1$. [Nev 25], [Nev 70]. In higher dimension, Cartan proved the theorem under the assumption that $f(\mathbf{C})$ is not contained in any hyperplane [Car 29], [Car 33], see also [La 87]. Vojta showed that the existence of the finite union of hyperplanes Z_D sufficed [Vo 89c].

For the case when D has components of higher degree, or for varieties other than $\mathbf{P}^n$, the situation today is in flux. Siu has made an attempt to get the ramification term and the main theorem by using meromorphic connections [Siu 87], but results are very partial, and his error term is not very good.

Let us return to the general case, and suppose that the canonical class is ample, or pseudo ample. Then the error term on the right-hand side is of a lower order of growth than the left-hand side $T_{f,K}$ if the map f is not algebraic. Thus the main theorem implies the conjecture that if the variety X is pseudo canonical, then there exists a proper Zariski closed subset Z of X such that the image of every non-constant holomorphic map of $\mathbf{C}$ into X is contained in Z. But the Nevanlinna type inequality in the main theorem would give a quantitative estimate. In particular, even if every holomorphic map $\mathbf{C} \to X$ is constant, one could rephrase the estimate in terms of maps of discs into X. We shall do so in a variation given later as Theorem 5.4.

Let us now deal with the error term, that is the term on the right-hand side of the inequality in the main theorem. We need to normalize the height so that it has certain smoothness properties. That means we have to pick the Weil functions to have such properties. Let us deal with maps into $\mathbf{P}^1$, which constitute the original Nevanlinna case. Given w, $w' \in \mathbf{C}$ we define

$$\|w, w'\|^2 = \frac{|w - w'|^2}{(1 + |w|^2)(1 + |w'|^2)},$$

and similarly if a or $a' = \infty$. Then define more precisely

$$m_f(a, r) = \int_0^{2\pi} - \log\|f(re^{i\theta}), a\| \frac{d\theta}{2\pi} + \log\|f(0), a\|.$$

assuming $f(0) \neq a, \infty$, and then define

$$T_f(r) = m_f(a, r) + N_f(a, r).$$

This value is independent of a.

In analogy with number theory, let ψ be a positive (weakly) increasing function of a real variable such that

$$\int_e^{\infty} \frac{1}{u\psi(u)} du = b_0(\psi)$$

is finite. For any positive increasing function F of class C^1 such that $r \mapsto rF'(r)$ is positive increasing, and for $r, c > 0$ we define the **error function**

$$S(F, c, \psi, r) = \log F(r) + \log \psi(F(r)) + \log \psi(crF(r)\psi(F(r))).$$

We let $r_1(F)$ be the smallest number $\geqq 1$ such that $F(r_1) \geqq 1$, and we let $b_1(F)$ be the smallest number $\geqq 1$ such that

$$b_1 rF'(r) \geqq e \quad \text{for} \quad r \geqq 1.$$

Theorem 5.3 (Absolute case). *Let*

$$\gamma_f = \frac{|f'|^2}{(1 + |f|^2)^2}.$$

Then for $r \geqq r_1(T_f)$ outside a set of measure $\leqq 2b_0(\psi)$, and for all $b_1 \geqq b_1(T_f)$ we have

$$-2T_f(r) + N_{f,\mathrm{Ram}}(r) \leqq \tfrac{1}{2}S(T_f, b_1, \psi, r) - \tfrac{1}{2}\log \gamma_f(0).$$

(Relative case). *Let $a_1, \ldots, a_q$ be distinct points of $\mathbf{P}^1$. Suppose that $f(0) \neq 0, \infty, a_j$ for all j, and $f'(0) \neq 0$. Let*

$$s = \tfrac{1}{3} \min_{i \neq j} \|a_i, a_j\| \quad \text{and} \quad b_2 = \frac{1}{s^{2(q-1)}}.$$

Then

$$-2T_f + \sum m_f(a_j, r) + N_{f,\mathrm{Ram}}(r) \leqq \tfrac{1}{2}S(B_q T_f^2, b_1, \psi, r) + b$$

where

$$B_2 = 12q^2 + q^3 \log 4 \quad \text{and} \quad b = \tfrac{1}{2}\log b_2 - \tfrac{1}{2}\log \gamma_f(0) + 1.$$

This formulation results from the work of Ahlfors [Ah 41], Lang [La 88], [La 90a, b], and Wong [Wo 89]. The absolute case with the precise error term is due to Lang. The relative case with the precise error term is due to Wong, except for the use of the general Khintchine type function ψ, which I suggested. It is important to note the difference between the appearance of T_f in the error term in the absolute case, and T_f^2 (dating back to Ahlfors) in the relative case. A more structural description will be given in the higher dimensional context of Theorem 5.5 below.

For suitable function ψ, one sees that the error term on the right is of the form

$$(1+\varepsilon)\log T_f(r) + O_\varepsilon(1) \qquad \text{for every} \quad \varepsilon > 0.$$

In analogy with number theory (see Chapter IX, §2) I raised two questions in analysis:

(a) Is this the best possible error term for "almost all" meromorphic functions, in a suitable sense of "almost all"?
(b) What is the best possible error term for each one of the classical functions such as $\wp$, θ, Γ, J, ζ?

I would define the **type** of a meromorphic function f to be a function ψ such that the error term has the form

$$\log \psi(T_f) + O(1).$$

The problem is to determine best possible types for the classical functions.

If instead of $\mathbf{P}^1$ we take maps of $\mathbf{C}$ into a curve of genus 1, that is, a complex torus of dimension 1, then one gets the same inequality except that the term corresponding to the canonical class is 0.

If one considers a map into a curve of genus $\geqq 2$, then the canonical class is ample, and the inequality gives a contradiction to the existence of such a map. But one can restrict attention to a map of a disc into the Riemann surface, and one can thus get a measure of hyperbolicity. We give one example of such a result.

We need first a differential geometric definition of the height which often gives greater insight into its behavior, following Ahlfors–Shimizu. Specifically, let Y be a complex manifold (not assumed compact!), and let

$$f\colon \mathbf{D}(R) \to Y$$

be a non-constant holomorphic map. If η is a (1, 1)-form on Y then we define the **height** for $r < R$ by

$$T_{f,\eta}(r) = \int_0^r \frac{dt}{t} \int_{\mathbf{D}(t)} f^*\eta.$$

The integral converges if $df(0) \neq 0$. We write

$$f^*\eta = \gamma_f \Phi \qquad \text{where} \quad \Phi = \frac{\sqrt{-1}}{2\pi}\, dz \wedge d\bar{z}.$$

Recall that

$$\operatorname{Ric} f^*\eta = dd^c \log \gamma_f.$$

One can define the order of vanishing at a given point of $\mathbf{D}(R)$ for the derivative of f, whence a ramification divisor. Actually, there exists a holomorphic function Δ on $\mathbf{D}(R)$, and a positive C^∞ function h such that

$$\gamma_f = |\Delta|^2 h,$$

so we can define the ramification counting function $N_{f,\mathrm{Ram}}(r) = N_\Delta(r, 0)$.

The following theorem stems from Griffiths–King [GrK 73] and Vojta [Vo 87], Theorem 5.7.2, with the improvement on the error term stemming from [La 90]. It gives a quantitative measure of hyperbolicity, in the context of differential geometry and Griffiths functions.

Theorem 5.4. *Let Y be a complex manifold with a positive $(1, 1)$-from η. Let $f: \mathbf{D}(R) \to Y$ be a holomorphic map. Suppose there is a constant $B > 0$ such that*

$$Bf^*\eta \leqq \operatorname{Ric} f^*\eta.$$

Assume $df(0) \neq 0$. Let $b_1 = b_1(T_{f,\eta})$. Then for $r < R$ we have

$$BT_{f,\eta}(r) + N_{f,\mathrm{Ram}}(r) \leqq \tfrac{1}{2}S(T_{f,\eta}, b_1, \psi, r) - \tfrac{1}{2}\log \gamma_f(0)$$

for $r \geqq r_1(T_{f,\eta})$ outside a set of measure $\leqq 2b_0(\psi)$.

Note that the theorem is formulated for a manifold which is not necessarily compact, and that the map f is defined on a disc. Also no assumption is made on a compactification or normal crossings. Furthermore, every holomorphic map of $\mathbf{C}$ into Y is constant because Y is hyperbolic, so to get a non-empty estimate in the main inequality, one has to use a formulation involving a map from a disc, not a map from $\mathbf{C}$ into Y.

We shall now state a higher dimensional version because it exhibits still another feature of the error term. With a Nevanlinna type error term, such a version was given by Carlson–Griffiths [CaG 72]. The improvement on the error term then went through [La 88b], Wong [Wo 89], and [La 90a], [Cher 90]. We need to make some definitions.

On $\mathbf{C}^n$ we consider the euclidean norm $\|z\|$ of a point $z = (z_1, \ldots, z_n)$. We define the differential forms

$$\omega(z) = dd^c \log\|z\|^2 \qquad \text{and} \qquad \sigma(z) = d^c \log\|z\|^2 \wedge \omega^{n-1}.$$

Let X be a projective non-singular variety over $\mathbf{C}$. Let D be a divisor on X. Let $\dim X = n$ and let

$$f: \mathbf{C}^n \to X$$

be a holomorphic map which is **non-degenerate**, in the sense that f is locally a holomorphic isomorphism at some point. Let L_D be a line bundle having a meromorphic section s whose divisor is D. Let ρ be a hermitian metric on L_D and let λ_D be the Weil function given by

$$\lambda_D(P) = -\log|s(P)|_\rho = \lambda_{\rho,s}(P).$$

Let $\mathbf{S}_r$ be the sphere of radius r centered at the origin, and define the **proximity function**

$$m_{f,\rho}(r) = \int_{\mathbf{S}(r)} (\lambda_D \circ f)\sigma + \log|s \circ f(0)|_\rho.$$

For simplicity we have assumed $f(0) \notin D$ and $df(0)$ is an isomorphism. Recall the **Chern form**

$$c_1(\rho) = -dd^c \log|s|_\rho^2,$$

Let η be a $(1,1)$-form on X. Define the **height**

$$T_{f,\eta}(r) = \int_0^r \frac{dt}{t} \int_{\mathbf{B}(t)} f^*\eta \wedge \omega^{n-1},$$

where $\mathbf{B}(r)$ is the ball of radius r. If $\eta = c_1(\rho)$ then we write $T_{f,\rho}$ instead of $T_{f,\eta}$. Then $T_{f,\rho}$ is independent of ρ, mod $O(1)$.

Let Ω be a volume form on X (i.e. a positive (n,n)-form). Then Ω defines a metric κ on the canonical line bundle L_K, and essentially by definition,

$$c_1(\kappa) = \operatorname{Ric}\Omega.$$

The height $T_{f,\kappa} = T_{f,\operatorname{Ric}\Omega}$ is one choice of height $T_{f,K}$ associated with the canonical class K. We can write

$$f^*\Omega = |\Delta|^2 h\Phi, \qquad \text{where} \quad \Phi = \prod \frac{\sqrt{-1}}{2\pi} dz_i \wedge d\bar{z}_i$$

and Δ is a holomorphic function on $\mathbf{C}^n$, while h is C^∞ and > 0. Then $\Delta = 0$ defines the **ramification divisor** of f, denoted by Z. Define the **counting function**

$$N_{f,\mathrm{Ram}}(r) = \int_0^r \frac{dt}{t} \int_{Z \cap \mathbf{B}(r)} \omega^{n-1}.$$

We say that D has **simple normal crossings** if $D = \sum D_j$ is a formal sum of non-singular irreducible divisors, and locally at each point of X there exist complex coordinates $z_1, \ldots, z_n$ such that in a neighborhood of this point, D is defined by $z_1 \ldots z_k = 0$ with some $k \leqq n$.

When $n = 1$, the property of D having simple normal crossings is equivalent to the property that D consists of distinct points, taken with multiplicity 1. The maximal value of k which can occur will be called the **complexity** of D. Finally, in higher dimension n, we suppose that $r \mapsto F(r)$ and $r \mapsto r^{2n-1}F'(r)$ are positive increasing functions of r, and we define the **error function**

$$S(F, c, \psi, r) = \log F(r) + \log \psi(F(r)) + \log \psi(cr^{2n-1}F(r)\psi(F(r))).$$

We let $b_1(F)$ be the smallest number $\geqq 1$ such that $b_1 r^{2n-1}F'(r) \geqq e$ for all $r \geqq 1$. The definition of $r_1(F)$ is the same as for $n = 1$. Then the analogue of Theorem 5.3 in higher dimension runs as follows. We let:

$D = \sum D_j$ be a divisor on X with simple normal crossings;

$\rho_j =$ metric on L_{D_j};

$\Omega =$ volume form on X;

$\eta =$ closed positive (1, 1)-form such that $\Omega \leqq \eta^n/n!$ and also $c_1(\rho_j) \leqq \eta$ for all j.

$\gamma_f =$ the function such that $f^*\Omega = \gamma_f \Phi$.

For a function α define the **height transform**

$$F_\alpha(r) = \int_0^r \frac{dt}{t^{2n-1}} \int_{B(r)} \alpha\Phi.$$

Theorem 5.5. *Suppose that $f(0) \notin D$ and $0 \notin \mathrm{Ram}_f$. Suppose that D has complexity k. Then*

$$\sum m_{f,\rho_j}(r) + T_{f,\mathrm{Ric}\,\Omega}(r) + N_{f,\mathrm{Ram}}(r) \leqq \frac{n}{2} S(BT_{f,\eta}^{1+k/n}, b_1, \psi, r) - \tfrac{1}{2}\log \gamma_f(0) + 1$$

for all $r \geqq r_1(F_{\gamma_f^{1/n}})$ outside a set of measure $\leqq 2b_0(\psi)$, and some constant $B = B(D, \eta, \Omega)$ which can be given explicitly, via the choice of sections s_j and the metrics ρ_j.

The above theorem stems from the work of Ahlfors, Wong and Lang as in the one-dimensional case. I want to emphasize the exponent $1 + k/n$, which applies for all $k = 0, \ldots, n$. The case $k = 0$ is when there is no divisor, and with such good error term is due to Lang. Thus the value distribution of the map f would be determined in the error term on the right-hand side by the local behavior of the divisor at its singular points.

No such good result is known in the number theoretic case, but the analytic theory suggests what may be the ultimate answer in that case. This is the reason for our having stated the theorem in higher dimension, since the structure of $1 + k/n$ did not appear in dimension 0.

It should be emphasized that Theorems 5.4 and 5.5 can be formulated and proved for normal coverings Y of $\mathbf{C}$ or $\mathbf{C}^n$, as the case may be. Then the degree $[Y:\mathbf{C}]$ or $[Y:\mathbf{C}^n]$ appears in the error term. Stoll [St 81], following Griffiths–King [GrK 72], obtains factors and a dependence on the degree which do not properly exhibit the conjectured structure. Extending the proof of Theorem 5.5, William Cherry showed that the degree occurs only as a factor, as follows.

Let

$$p: Y \to \mathbf{C}^n$$

be a possibly ramified covering, and assume that Y is normal. Let $[Y:\mathbf{C}^n]$ be the degree of the covering. For all the objects (σ, ω, Φ, etc.) defined on $\mathbf{C}^n$, put a subscript Y to denote their pull back to Y by p. For instance

$$\sigma_Y = p^*\sigma, \qquad \omega_Y = p^*\omega, \qquad \Phi_Y = p^*\Phi.$$

We denote by $Y\langle 0\rangle$ the set of points $y \in Y$ such that $p(y) = 0$. Let $f: Y \to X$ be a non-degenerate holomorphic map. As before, we define γ_f by

$$f^*\Omega = \gamma_f \Phi_Y.$$

We suppose that η is a closed positive (1, 1)-form on X such that

$$\Omega = \frac{1}{n!}\eta^n.$$

The height $T_{f,\eta}$ is defined as for maps of $\mathbf{C}^n$ into X, except that the integral over $\mathbf{B}(t)$ is replaced by an integral over $Y(t)$, inverse image of $\mathbf{B}(t)$ under p.

Theorem 5.6 ([Cher 90], [Cher 91]). *Assume p, f unramified above* 0. *Then*

$$T_{f,\mathrm{Ric}\,\Omega}(r) + N_{f,\mathrm{Ram}}(r) - N_{p,\mathrm{Ram}}(r)$$

$$\leqq [Y:\mathbf{C}^n]\frac{n}{2}S(T_{f,\eta}/[Y:\mathbf{C}^n], \psi, r) - \frac{1}{2}\sum_{y \in Y\langle 0\rangle} \log \gamma_f(y)$$

for all $r \geqq r_1(T_{f,\eta}/(n-1)![Y:\mathbf{C}^n])$ *outside a set of measure* $\leqq 2b_0(\psi)$.

When there is a divisor D, Cherry gets a similar error term. The height can be normalized right away as in number theory, dividing by the degree, to make everything look better. In any case, we note that the term $N_{p,\mathrm{Ram}}$ occurs with coefficient 1, in a very simple way.

CHAPTER IX

Weil Functions, Integral Points and Diophantine Approximations

The height can be decomposed as a sum of local functions for each absolute value v. These functions are intersection multiplicities at the finite v, and essentially Green's functions in one form or another at the infinite v. Whereas a height is associated with a divisor class, those local functions are associated with a divisor, and measure the distance of a point to the divisor in some fashion. They are normalized to be continuous outside the support of the divisor, and to have a logarithmic singularity on the divisor, so they tend to infinity on the divisor.

There are many uses for those functions. They give the natural tool to express results in diophantine approximations, and they play a role analogous to the proximity function in Nevanlinna theory, as Vojta observed. We shall run systematically through their various aspects.

Proofs for the foundational results of §1 can be found in [La 83]. References for other proofs will be given as the need arises.

Whereas previously we have concentrated on diophantine questions involving rational points, we now come to integral points and conditions under which their heights are bounded and there is only a finite number of such points. Again we meet curves, subvarieties of abelian varieties, hyperbolic conditions, but in the context of non-compact varieties, notably affine varieties.

Vojta actually integrated the theories of rational points and integral points by subsuming them under a general formalism transposed from Nevanlinna theory. We shall state Vojta's most general conjectures, and we shall indicate his proof of Faltings' theorem along lines which had been used before only in the context of diophantine approximations and integral points. He showed for the first time how one can globalize and sheafify this approach to obtain results on rational points in the case of genus > 1. Thus Vojta provided an entirely new approach and proof for

Mordell's conjecture, which does not pass through the Shafarevich conjecture and accompanying l-adic representations. In addition, Vojta by casting his approach in the context of Arakelov theory also shows how to use the recently developed higher dimensional theory of Gillet–Soulé and Bismut–Vasserot for a specific application. Vojta used the higher dimension because even though one starts with a curve, he applies that theory to the product of the curve with itself a certain number of times, at least equal to 2, but even higher to get more precise results. Thus we behold the grand unification of algebraic geometry, analysis and PDE, diophantine approximations, Nevanlinna theory, and classical diophantine problems about rational and integral points.

Following Vojta's extension of diophantine approximation methods, Faltings then succeeded in applying this method to prove my two conjectures: finiteness of integral points on affine open subsets of abelian varieties, and finiteness of rational points on a subvariety of an abelian variety which does not contain translations of abelian subvarieties of dimension > 0. We describe briefly these results.

Actually, there are two major aspects of diophantine approximation methods: the one as above, relying on the Thue–Siegel–Schneider–Roth–Schmidt method; the other relying on diophantine approximations on toruses, especially Baker's method and its extensions.

This chapter, and the preceding chapter, give two examples of a general principle whereby diophantine properties of varieties result from their behavior at the completion of the ground field at one absolute value, in the present cases taken to be archimedean. In Chapter VIII, we saw that under one imbedding, we could determine an exceptional set, namely the Zariski closure of the union of all non-constant images of $\mathbf{C}$ into the variety by holomorphic maps. Conjecturally, this special set does not depend on the imbedding of the ground field into the complex numbers and has an algebraic characterization. The conjecture that the complement of this exceptional set is Mordellic shows how a qualitative diophantine property is determined by the behavior of the variety at one archimedean place. By the way, the non-archimedean analogue of this property remains to be worked out. In the present chapter, we consider quantitative diophantine properties, namely bounds on the height, and we shall see in §7 how certain inequalities obtained at one imbedding of the ground field into $\mathbf{C}$ give rise to bounds on the heights of rational points. These inequalities concern lower bounds for linear combinations of logarithms (ordinary or abelian) with integer or algebraic coefficients. The optimal conjectures are still far from being proved, but sufficient results following a method originated by Baker are known already to yield important diophantine consequences. One of these culminated with the Masser–Wustholz theorem, to be described in Theorem 7.3.

Some of the methods of diophantine approximation arose originally in the theory of transcendental numbers and algebraic independence. I have not gone into this subject, and I have only extracted those aspects of diophantine approximations which are directly relevant (as far as one can see today) to diophantine questions. Over three decades I wrote several surveys where I went further into the theory of transcendental numbers in connection with diophantine analysis, and which some readers might find useful: [La 60b], [La 65b], [La 71] and [La 74].

IX, §1. WEIL FUNCTIONS AND HEIGHTS

Let F be a field with a proper absolute value v, which we assume extended to the algebraic closure F^a. As usual, $v(x) = -\log|x|_v$. We let X be a projective variety defined over F.

Let V be a Zariski open subset of X, defined over F. Let B be a subset of $V(F^a)$. We say that B is **affine-bounded** if there exists a coordinatized affine open subset U of V with coordinates $(x_1, \ldots, x_n)$ and a constant $\gamma > 0$ such that for all $x \in B$ we have $\max|x_i|_v \leqq \gamma$. We say that B is **bounded** if it is contained in the finite union of affine bounded subsets. We note that X assumed projective implies that $X(F^a)$ is bounded.

Let $\alpha: V(F^a) \to \mathbf{R}$ be a real valued function. We define α to be **bounded from above** in the usual sense. We say that α is **locally bounded from above** if α is bounded from above on every bounded subset of $V(F^a)$. We define **locally bounded from below** and **locally bounded** in a similar way.

Let D be a Cartier divisor on X. By a **Weil function** associated with D we mean a function

$$\lambda_D = \lambda_{D,v}: X(F^a) - \operatorname{supp}(D) \to \mathbf{R}$$

having the following property. If D is represented by a pair (U, φ) then there exists a locally bounded continuous function

$$\alpha: U(F^a) \to \mathbf{R}$$

such that for all points $P \in U - \operatorname{supp}(D)$ we have

$$\lambda_D(P) = v \circ \varphi(P) + \alpha(P).$$

The continuity of α is v-continuity, not Zariski topology continuity. (Note: Néron [Ne 65] in his exposition and extension of Weil's work called Weil functions **quasi functions**. But there is nothing "quasi" about these functions, and I found his terminology misleading.)

The association

$$D \mapsto \lambda_D$$

is a homomorphism mod $O(1)$.

Let F_v be the completion as usual, and $\mathbf{C}_v$ the completion of the algebraic closure of F_v. Then we may have defined a Weil function on the base change of X to $\mathbf{C}_v$, and this Weil function then defines a Weil function on X itself. In practice, F may be a number field, in which case F_v is locally compact, and if we have a Weil function when X is a variety over F_v, then a continuous function on $X(F_v)$ is necessarily locally bounded. However, we also want to consider Weil functions on $X(F_v^{\mathrm{a}})$. These could of course be viewed as a compatible family of Weil functions on the set of points of X in finite extensions of F_v. In the function field case, however, when F is a function field of one variable, say, over some constant field which is not finite, then F_v is not locally compact. Thus we made a general definition which applies to all cases.

If λ_D and λ'_D are Weil functions associated with the same divisor then their difference $\lambda_D - \lambda'_D$ is continuous and bounded on $X(F^{\mathrm{a}})$. Thus two Weil functions differ by $O(1)$.

Weil functions behave functorially in the following sense. Let

$$f: X' \to X$$

be a morphism defined over F. Let D be a Cartier divisor on X. Assume that $f(X')$ is not contained in the support of D. Then $\lambda_D \circ f$ is a Weil function on X', associated with f^*D.

Weil functions preserve positivity in the following sense. Assume that X is projective, or merely that $X(F^{\mathrm{a}})$ is bounded. Let D be an effective Cartier divisor. Then there exists a constant $\gamma > 0$ such that $\lambda_D(P) \geqq -\gamma$ for all $P \in X(F^{\mathrm{a}})$. Furthermore, let $D_j = D + E_j$ $(j = 1, \ldots, m)$ be Cartier divisors, with D, E_j effective for all j, and such that the supports of $E_1, \ldots, E_m$ have no point in common. Then

$$\lambda_D = \inf_j \lambda_{D_j} + O(1).$$

Example 1. One way to construct a Weil function is as follows. Let $\mathscr{L}$ be a metrized line sheaf with a meromorphic section s whose divisor is D. Then the function

$$\lambda_{D,v}(P) = -\log|s(P)|_v \qquad \text{for} \quad P \in X(F^{\mathrm{a}}) - \mathrm{supp}(D)$$

is a Weil function.

Example 2. Let R be a discrete valuation ring with quotient field F and valuation v. Let $Y = \mathrm{spec}(R)$ and let

$$X \to Y$$

be a flat morphism. Assume that X is regular, and that the generic fiber is a complete non-singular variety X_F. Let D be a divisor on X and let

D_F be its restriction to the generic fiber. For each point $P \in X(F^a)$ and P not lying in D_F, let D be represented by a rational function φ on a Zariski open neighborhood of P in X. Then $v(\varphi(P))$ is independent of the choice of φ, and the function

$$P \mapsto v(\varphi(P))$$

is a Weil function associated with D_F. We call this choice of Weil function the one arising from **intersection theory**, because $v(\varphi(P))$ may be viewed as an intersection number of D and E_P, where E_P is the Zariski closure of P in X. This example applies to, and in fact stemmed originally from, Néron models for abelian varieties.

Example 3. Let X be a non-singular complete curve over the complex numbers. Let $\mathbf{g}_D$ be the Green's function associated with an effective divisor D. Our normalizations are such that for the ordinary absolute value v, the function

$$\tfrac{1}{2}\mathbf{g}_D$$

is a Weil function associated with D.

Example 4. Let A be an abelian variety over the complex numbers, so we have an analytic isomorphism

$$p\colon \mathbf{C}^n/\Lambda \to A(\mathbf{C}),$$

where Λ is a lattice in $\mathbf{C}^n$. To each divisor D on $A(\mathbf{C})$ there is associated a normalized theta function F_D on $\mathbf{C}^n$ whose divisor is $p^{-1}(D)$, and which is uniquely determined up to a constant factor. By definition, a **normalized theta function** is a meromorphic function on $\mathbf{C}^n$ satisfying the condition

$$F(z+u) = F(z)\exp\left[\pi H(z,u) + \frac{\pi}{2}H(u,u) + 2\pi\sqrt{-1}K(u)\right]$$

$$\text{for} \quad z \in \mathbf{C}^n, \quad u \in \Lambda,$$

where H is a hermitian form called the **Riemann form**, and $K(u)$ is real valued. (Cf. my *Introduction to Algebraic and Abelian Functions* for the basic properties.) Then the function λ_D defined by

$$\lambda_D(z) = -\log|F_D(z)| + \frac{\pi}{2}H(z,z)$$

is a Weil function whose divisor is D. In fact, this function is normalized in such a way that it satisfies an additional property under translation by

$a \in A$, namely,

$$\lambda_{D_a}(z) = \lambda_D(z - a) + c(a),$$

where D_a is the translation of D by a, and $c(a)$ is a constant depending on a and D but independent of z. As such, the function λ_D is called a **Néron function**. See the last section of [Ne 65].

Suppose that instead of one absolute value v we have a proper set of absolute values, satisfying the product formula, and suppose that X is a projective variety defined over a finite extension of the base field $\mathbf{F}$. Then we can pick a Weil function $\lambda_{D,v}$ for each v. We want to take the sum. To do this, it is necessary to make the choice such that certain uniformity conditions are satisfied. One can do this a priori to get:

Theorem 1.1. *Assume that the set of absolute values satisfies the product formula. Let D be a Cartier divisor on X. Then there exists a choice of Weil functions $\lambda_{D,v}$ for each v such that, if we put*

$$h_\lambda(P) = \frac{1}{[F:\mathbf{F}]} \sum_v [F_v : \mathbf{F}_v] \lambda_{D,v}(P)$$

for $P \in X(F)$, $P \notin \operatorname{supp}(D)$, then h_λ is a height associated with the divisor class of D.

The choice of Weil functions can be made following Example 1, as follows. Say D is effective. Let $\mathscr{L}$ be a line sheaf with a section s whose divisor (s) is D. For each v select a norm on the v-adic extension such that

$$|at|_v = |a|_v |t|_v \qquad \text{for} \quad a \in F_v, \quad t \in \mathscr{L}$$

and such that for each t, $|t|_v = 1$ for all but a finite number of v. Then the choice of $\lambda_{D,v}$ as in Example 1 will work for Theorem 1.1.

Suppose that the set of absolute values satisfying the product formula comes from the discrete valuations of a Dedekind ring, together with a finite number of other valuations. Suppose in addition that X_F is the generic fiber of a morphism

$$X \to Y$$

where $Y = \operatorname{spec}(\mathfrak{o})$ such that X is regular, proper and flat over Y as in Example 2. The regularity assumption is a crucial one, since in general it involves a resolution of singularities. Fix a divisor on X. Then for all the discrete valuations of $\mathfrak{o}$ we have a Weil function as in Example 2, defined by the intersection theory. For the remaining finite number, we may choose any Weil function. Then this set of Weil functions can be used in Theorem 1.1, to take their sum and obtain the height associated with the divisor class. The arbitrary choice at a finite number of v simply contributes to the term $O(1)$, but in the present context, without further normalizations, we do not expect to achieve more.

One basic idea of diophantine approximation theory is to determine, in some sense, how close a point can be to, say, a divisor D. To measure this closeness, we introduce the proximity function. We work under the standard situation of a field $\mathbf{F}$ with a proper set of absolute values satisfying the product formula. Let X be a projective non-singular variety defined over $\mathbf{F}^{\mathrm{a}}$. Let S be a finite set of absolute values of $\mathbf{F}$, and for each finite extension F let S_F be the extension of S to F. If $\mathbf{F}$ contains a set of archimedean absolute values S_∞, we assume that $S \supset S_\infty$. Suppose X is defined over F. Let D be a divisor on X. Define the **proximity function**

$$m_{D,S}(P) = \frac{1}{[F:\mathbf{F}]} \sum_{v \in S_F} [F_v : \mathbf{F}_v] \lambda_{D,v}(P)$$

for $P \in X(F)$ and $P \notin \operatorname{supp}(D)$. If we replace F by a finite extension F' and S_F by its extension $S_{F'}$ on F', then the right-hand side is unchanged. So we can legitimately omit F from the notation on the left-hand side, and the definition of $m_{D,S}(P)$ applies to algebraic points P over $\mathbf{F}$. As Vojta pointed out, this proximity function is the analogue of the proximity function in Nevanlinna theory, and from known results in Nevanlinna theory, Vojta then conjectured diophantine inequalities in the number theoretic case. We shall deal with these in §4.

With respect to the finite S we define the analogue of the **counting function** in Nevanlinna theory to be

$$N_{D,S}(P) = h_D(P) - m_{D,S}(P) \qquad \text{for} \quad P \in X(\mathbf{F}^{\mathrm{a}}).$$

Having chosen the Weil functions $\lambda_{D,v}$ suitably to give a decomposition of the height into a sum of these Weil functions over all v with suitable multiplicities, we may also write the counting function in the form

$$N_{D,S}(P) = \frac{1}{[F:\mathbf{F}]} \sum_{v \notin S} [F_v : \mathbf{F}_v] \lambda_{D,v}(P)$$

for points P lying in $X(F)$.

We now return to an arbitrary fixed proper absolute value v, and look into the possibility of normalizing Weil functions more precisely than up to bounded functions. We want them normalized up to an additive constant. So we let Γ be the group of **constant functions** on $X(F^{\mathrm{a}})$.

Observe that if φ is a rational function on X, then φ determines a Weil function

$$\lambda_\varphi(P) = -\log|\varphi(P)|_v.$$

If X is complete, then the divisor of a rational function determines the function up to a non-zero multiplicative constant, so the Weil function defined above is determined up to an additive constant. The normalizations of Weil functions as in the next two theorems are due to Néron [Ne 65].

Theorem 1.2. *Let A be an abelian variety defined over F. To each divisor D on A there exists a Weil function λ_D associated with D, and uniquely determined up to an additive constant by the following properties.*

(1) *The association $D \mapsto \lambda_D$ is a homomorphism* mod Γ.
(2) *If $D = (\varphi)$ is principal, then $\lambda_D = \lambda_\varphi +$* constant.
(3) *Let $a \in A(F^a)$. Let T_a be translation by a, and put $D_a = T_a(D)$. Then there exists a constant $\gamma_{a,D}$ such that*

$$\lambda_{D_a} = \lambda_D \circ T_{-a} + \gamma_{a,D}.$$

Functions normalized as in Theorem 1.2 are called **Néron functions.** They satisfy the additional property:

(4) *Let $f: B \to A$ be a homomorphism of abelian varieties over F. Then*

$$\lambda_{f^*D} = \lambda_D \circ f \bmod \Gamma.$$

On arbitrary varieties, one cannot get such a general characterization without a further assumption.

Theorem 1.3. *Let X be a projective non-singular variety defined over F. To each divisor D algebraically equivalent to 0 on X one can associate a Weil function unique mod constant functions, satisfying the following conditions.*

(1) *The association $D \mapsto \lambda_D$ is a homomorphism* mod Γ.
(2) *If $D = (\varphi)$ is principal, then $\lambda_D = \lambda_\varphi$* mod Γ.
(3) *If $f: X' \to X$ is a morphism defined over F, and D is algebraically equivalent to 0 on X, such that f^*D is defined, then*

$$\lambda_{f^*D} = \lambda_D \circ f \bmod \Gamma.$$

Again, the Weil functions as in Theorem 1.3 are called **Néron functions.**

Having normalized the Néron functions up to additive constants, we can get rid of these constants if we evaluate these functions by additivity on 0-cycles of degree 0 on A or X as the case may be. We then obtain a bilinear pairing between divisors (algebraically equivalent to 0 on an arbitrary variety) and 0-cycles of degree 0. This pairing is called the **Néron pairing** or **Néron symbol.** As with heights, relations between divisors are reflected in relations between the Néron functions or the Néron symbol. We refer to [Ne 65] or [La 83] for a list of such relations.

In addition, the theorems concerning algebraic families of heights which we gave in Chapter III, §2 extend mutatis mutandis to **algebraic families of Néron functions**. See [La 83], Chapter 12.

IX, §2. THE THEOREMS OF ROTH AND SCHMIDT

Let α be an algebraic number. **Roth's theorem** states:

Given ε, one has the inequality

$$\left|\alpha - \frac{p}{q}\right| \geqq \frac{1}{q^{2+\varepsilon}}$$

for all but a finite number of fractions p/q in lowest form, with $q > 0$.

The inequality can be rewritten

$$-\log\left|\alpha - \frac{p}{q}\right| - 2\log q \leqq \varepsilon \log q$$

for all but a finite number of fractions p/q. If a fraction p/q is close to α, then p, q have the same order of magnitude, so instead of $\log q$ in the above inequality, we can use the height and rewrite the inequality as

$$-\log\left|\alpha - \frac{p}{q}\right| - 2h(p/q) \leqq \varepsilon \log q.$$

More generally, let α be any real irrational number. Following [La 66a], [La 66c] we define a **type** for α to be a positive increasing function ψ such that

$$-\log|\alpha - \beta| - 2h(\beta) \leqq \log \psi(h(\beta))$$

for all rational $\beta \in \mathbf{Q}$. A theorem of Khintchine states that Lebesgue almost all numbers α have type ψ if

$$\sum_{q=1}^{\infty} \frac{1}{q\psi(q)} < \infty.$$

A basic question is whether Khintchine's principle applies to algebraic numbers, although possibly some additional restrictions on the function might be needed. Roth's theorem can be formulated as saying that an algebraic number has type $\leqq q^{\varepsilon}$ for every $\varepsilon > 0$. In the sixties, I conjectured that this could be improved to having a type along the Khintchine line, say with $(\log q)^{1+\varepsilon}$. See also Bryuno [Bry 64] and [RDM 62]. Then the Roth-type inequality could be written

$$-\log\left|\alpha - \frac{p}{q}\right| - 2\log q \leqq (1+\varepsilon)\log\log q$$

for all but a finite number of fractions p/q. However, except for quadratic numbers, which all have bounded type (trivial exercise), there is no example of an algebraic number about which one knows that it is or is not of type $(\log q)^k$ for some number $k > 1$. It becomes a problem to determine the type for each algebraic number, and for the classical numbers. For instance, it follows from Adams' work [Ad 66], [Ad 67] that e has type

$$\psi(q) = \frac{C \log q}{\log\log q}$$

with a suitable constant C, which is much better than the "probability" type and goes beyond Khintchine's principle: the sum $\sum 1/q\psi(q)$ diverges.

In light of Vojta's analogy of Nevanlinna theory and the theory of heights, it occurred to me to transpose my conjecture from number theory to Nevanlinna theory, thus giving rise to the error terms which have been stated in Chapter VIII, §5, in terms of a function ψ analogous to the Khintchine function.

In [La 60a] I pointed out that Roth's theorem could be axiomatized to fit the general pattern of height theory, as follows. Let $\mathbf{F}$ be a field with a family of proper absolute values satisfying the product formula. Let F be a finite extension of $\mathbf{F}$. Let S_∞ be a finite set of absolute values of F containing all the archimedean absolute values if any, but not empty, at any rate. Let R_S denote the subset of elements $x \in F$ such that

$$|x|_v \leqq 1 \qquad \text{for all} \quad v \in S.$$

Then R_S is a ring, called the ring of S-**integers**. Let $R_\infty = R_{S_\infty}$. One needs to assume an additional property, essentially a weak form of a Riemann–Roch theorem, which is true in the number field case and the function field case. This form of Riemann–Roch guarantees the existence of many functions having sufficiently large absolute values at those $v \in S_\infty$. The point is that one needs to solve linear equations and one needs to bound the solutions as a function of the height of the coefficients. Riemann–Roch is precisely the tool which accomplishes this for us. Under these hypotheses, one can formulate Roth's theorem as follows. See also [La 83].

For each $v \in S$ let α_v be algebraic over F, and assume v extended to $F(\alpha_v)$. Given ε, the elements $\beta \in F$ satisfying the condition

$$\frac{1}{[F:\mathbf{F}]} \sum_{v \in S} \log \max(1, \|\alpha_v - \beta\|_v) - 2h(\beta) \leqq \varepsilon h(\beta)$$

have bounded height.

Thus the analogy between algebraic numbers and algebraic functions held also in this case. For an account of the method of proof, from Thue–Siegel to Vojta, see §6.

To go further, let F be a number field, and let S be a finite set of absolute values containing the archimedean set S_∞. Let $\mathfrak{o}_{F,S}$ be the ring of integers of F, localized at all primes not in S. For

$$x = (x_0, \ldots, x_N) \quad \text{with} \quad x_i \in \mathfrak{o}_{F,S} \quad \text{so} \quad x \in \mathfrak{o}_{F,S}^{N+1},$$

define the **size**

$$\operatorname{size}(x) = \max_{v \in S, i} \|x_i\|_v.$$

Also linear forms $L_0, \ldots, L_M$ are said to be in **general position** if $M \leqq N$ and the forms are linearly independent, or $M > N$ and any $N + 1$ of them are linearly independent.

A higher dimensional version of Roth's theorem was proved by Schmidt [Schm 70], [Schm 80]. As Vojta remarked, this theorem was analogous to Cartan's theorem in Nevanlinna theory, and Vojta improved the statement of Schmidt's theorem to the following.

Theorem 2.1. *Let N be a positive integer. Let $\mathbf{L}$ be a finite set of linear forms in $N + 1$ variables with coefficients in $\mathbf{Q}^{\mathrm{a}}$. There exists a finite union Z of proper linear subspaces of $\mathbf{Q}^{\mathrm{a}(N+1)}$ having the following property. Given a number field F, a finite set of absolute values S containing S_∞, and for each $v \in S$ given linear forms $L_{v,0}, \ldots, L_{v,M} \in \mathbf{L}$ with $M \geqq N$, we have for every $\varepsilon > 0$*

$$\prod_{v \in S} \prod_{i=0}^{M} \|L_{v,i}(x)\|_v \gg \operatorname{size}(x)^{M-N-\varepsilon}$$

for all but a finite number of $x \in \mathfrak{o}_{F,S}^{N+1}$ lying outside Z.

In Schmidt's version, the exceptional set Z depends on ε, F and S, but Vojta succeeded in eliminating this dependence [Vo 89c]. The finite set of exceptional points lying outside Z still depends on ε, F and S, however.

The above statement reflects the way inequalities have been written on affine space. However, it is also useful to rewrite these inequalities in terms of heights, and following Vojta, in a way which makes the formal analogy with Nevanlinna theory clearer. For this purpose, if L is a linear form on projective space $\mathbf{P}^N$ and H is the hyperplane defined by $L = 0$, then we can define a **Weil function** associated with H by the formula

$$\lambda_{H,v}(P) = -\log \frac{|L(P)|_v}{\max_i |x_i|_v}$$

for any point $P \in \mathbf{P}^N(F)$ with coordinates $P = (x_0, \ldots, x_N)$ and $x_i \in F$. Theorem 2.1 can then be formulated with Weil functions as follows.

Theorem 2.2. *Let $H_0, \ldots, H_M$ be hyperplanes in general position in $\mathbf{P}^N$ over $\mathbf{Q}^{\mathrm{a}}$. There exists a set Z equal to a finite union of hyperplanes having the following property. Given a number field F, a set of absolute values S, and $\varepsilon > 0$, we have*

$$\sum_{i=0}^{M} m_{H_i,S}(P) - (N+1)h(P) \leqq \varepsilon h(P)$$

except for a finite number of points in $\mathbf{P}^N(F)$ outside Z.

For a discussion of conjectures concerning similar inequalities when the points P are allowed to vary over all algebraic points, see [Vo 89c]. Under such less restrictive conditions, a term must be added on the right-hand side involving the discriminant, of the form $f(N)\, d(P)$ for some function $f(N)$. Vojta discusses which functions can reasonably occur, for instance $f(N) = N$, which would result from his general conjectures, which we state in §4.

Since already in Roth's theorem one does not know how to improve the type from q^{ε} to a power of $\log q$ or better, a fortiori no such result is known for the higher dimensional Schmidt case. But as we remarked in the Nevanlinna case, the good error term with the complexity of the divisor suggests the ultimate answer in this higher dimensional case.

Finally it is appropriate to mention here the direction given by Osgood [Os 81], [Os 85] for diophantine approximations and Nevanlinna theory, having to do with differential fields which provide still another context besides the number field case, function field case, or holomorphic Nevanlinna case.

IX, §3. INTEGRAL POINTS

Let F be a field with a proper set of absolute values, and let S be a finite set of these absolute values containing all the archimedean ones if such exists. We let $R_{F,S}$ be the subring of F consisting of those elements $x \in F$ such that $|x|_v \leqq 1$ for $v \notin S$.

Let V be an affine variety defined over F. We let $F[V]$ be the global ring of functions on V. Then $F[V]$ is finitely generated, i.e. we can write

$$F[V] = F[x_1, \ldots, x_n].$$

The function field of V is the quotient field $F(V) = F(x_1, \ldots, x_n)$. We call $(x_1, \ldots, x_n)$ a set of **affine coordinates** on V.

Let R be a subring of some field containing F, and let I be a subset of points of V rational over the quotient field of R. We say that I is **R-integralizable**, or **R-integral**, if there exists a set of affine coordinates such that $x_i(P) \in R$ for all i and all $P \in I$. If $R = R_{F,S}$ then we also say that I is **S-integralizable**, or **S-integral**.

A set of points in $V(F)$ is S-integral if and only if there exists a set of affine coordinates such that the values $x_i(P)$ have bounded denominators for all i and all $P \in I$. By **bounded denominator**, we mean that there exists $b \neq 0$ in $R_{F,S}$ such that $bx_i(P) \in R_{F,S}$ for all i, $P \in I$.

Let V be the complement of the hyperplane at infinity in projective space $\mathbf{P}^n$, and let D be this hyperplane. A subset I of $Y(F)$ is S-integral if and only if there exists a Weil function $\lambda_{D,v}$ for each $v \notin S$, such that

$$\lambda_{D,v}(P) \leqq c_v \qquad \text{for all} \quad v \notin S, \quad \text{all } P \in I.$$

This follows immediately from the definitions. For instance, if $(x_1, \ldots, x_n)$ is a set of affine coordinates integralizing the points in I, we can take the Weil function to be

$$\lambda_{D,v} = \log \max\bigl(1, |x_1(P)|_v, \ldots, |x_n(P)|_v\bigr).$$

In [Vo 87] Vojta works with possibly non-ample effective divisors D, so with non-affine open sets V, for instance certain moduli varieties. For this purpose, he defines the notion of **(S, D)-integrality** or **(S, D)-integralizable** set of rational points by using the condition stated above in terms of the Weil functions, applicable in this more general case.

One basic theorem about integral points concerns curves.

Theorem 3.1. *Let F be a field finitely generated over $\mathbf{Q}$ and let R be a subring finitely generated over $\mathbf{Z}$. Let V be an affine curve defined over F and let X be its projective completion. If the genus of X is $\geqq 1$, or if the genus of X is 0, but there are at least three points in the complement of V in X, then every set of R-integralizable points on V is finite.*

When F is a number field and R is the ring of integers, the theorem is due to Siegel [Sie 29]. After the work of Mahler [Mah 33] for curves of genus 1 over $\mathbf{Q}$, the theorem was extended to the more general rings in

[La 60a]. In light of Faltings' theorem, only the cases of genus 1 and genus 0 are relevant in the qualitative statement we have given. However, even in higher genus, bounds on the heights of integral points may be of a different type than bounds for the height of rational points, so quantitative forms of the theorem are of interest independently of Faltings' theorem.

Siegel's method uses Roth's theorem (in whatever weaker form it was available at the time). The method is of sufficient interest so we shall describe it briefly in the form given in [La 60a]. Let us consider first a curve defined over a finite extension of a field $\mathbf{F}$ with a proper set of absolute values satisfying the product formula, so we have heights, and the decomposition of heights into a sum of Weil functions. We let X be the complete non-singular curve, defined over F, and we let $\varphi \in F(X)$ be a non-constant function, which will define integrability for us. That is, we consider the set of points in $X(F)$ such that $\varphi(P) \in R_{F,S}$. We may call these the **φ-integral points**, and we want to show that they have bounded height. This is accomplished by putting together a geometric formulation of Roth's theorem, together with a lifting procedure using coverings of the curve. We state both steps as propositions. The first proposition applies to a curve of any genus $\geqq 0$.

Proposition 3.2. *Let X be a projective non-singular curve defined over F, and $\varphi \in F(X)$ not constant. Let r be the largest of the multiplicities of the poles of φ. Let κ be a number > 2 and C a number > 0. Let S be a finite set of absolute values. Then the set of points $P \in X(F)$ which are not among the poles of φ, and are such that*

$$\frac{1}{[F:\mathbf{F}]} \sum_{v \in S} \log \max(1, \|\varphi(P)\|_v) \geqq \kappa r h(P) - C$$

have bounded height. (The height h is taken with respect to the given projective imbedding.)

The above proposition is merely a version of Roth's theorem. The next proposition shows what it implies for curves of higher genus.

Proposition 3.3. *Suppose that X has genus $\geqq 1$. Then given $\varepsilon > 0$ the set of points $P \in X(F)$ such that*

$$\log|\varphi(P)|_v \geqq \varepsilon h(P)$$

has bounded height.

Note that when we have the factor ε on the right-hand side, the sum becomes irrelevant, since the estimate applies to each term. The inequality for curves of genus > 0 is reduced to the inequality for curves in general by means of the method described in the next proposition.

Proposition 3.4. *Suppose that X has genus $\geqq 1$. Let m be an integer > 0, unequal to the characteristic of F. Let J be the Jacobian of X over F, and assume that $J(F)/mJ(F)$ is finite. Let I be an infinite set of points in $X(F)$. Then there exist an unramified covering $f: X' \to X$ defined over F, an infinite set of rational points $I' \subset X'(F)$, such that f induces an injection of I' into I, and a projective imbedding of X' such that*

$$h \circ f = m^2 h' + o(h') \qquad \text{for} \quad h' \to \infty,$$

as functions on $X'(F^{\mathrm{a}})$. The heights h and h' refer to the heights on X and X' respectively, in their projective imbeddings.

The unramified covering $f: X' \to X$ is obtained from the Jacobian. Indeed, suppose infinitely many points $P \in I$ lie in the same coset of $J(F)/mJ(F)$. Then there is a point $P_0 \in J(F)$ such that all the points $P \in I$ in that coset can be written in the form

$$P = mQ + P_0.$$

We restrict the covering $J \to J$ given by $x \mapsto mx + P_0$ to the curve to obtain Proposition 3.4, using the functorial properties of the height, and especially the quadraticity. Thus we have a form of descent by coverings.

We use Proposition 3.2 in combination with Proposition 3.4, applied to the function $\varphi \circ f$. Since the covering of Proposition 3.4 is unramified, the zeros of φ and of $\varphi \circ f$ have the same multiplicities, so Proposition 3.3 follows at once.

As an application to φ-integral points P, we have

$$\frac{1}{[F : \mathbf{F}]} \sum_{v \in S} \log \max(1, \|\varphi(P)\|_v) = h(\varphi(P)) \geqq \varepsilon h(P)$$

for some $\varepsilon > 0$. Applying Proposition 3.3 shows that the height of integral points is bounded.

We have emphasized the method of proof because variations and substantial extensions occur systematically in the theory. As a first example, consider the equation (called the **unit equation**)

$$(*) \qquad a_1 u_1 + a_2 u_2 = 1$$

with $a_1, a_2 \in R$ (where R is finitely generated over $\mathbf{Z}$) and u_1, u_2 are to lie in a finitely generated multiplicative group Γ. Then Γ/Γ^m is finite.

Writing

$$u_i = w_i^m b_i \qquad \text{with} \quad w_i \in \Gamma,$$

we see that infinitely many solutions of the equation (*) in Γ give rise to infinitely many solutions of the equation

$$(**) \qquad a_1 b_1 w_1^m + a_2 b_2 w_2^m = 1,$$

and for $m \geqq 3$ the new equation (**) has genus $\geqq 1$ so we can apply Theorem 3.1 to see that (*) has only finitely many solutions with $u_i \in \Gamma$. The case of genus 0 in Theorem 3.1 is reduced to the case of genus $\geqq 1$ and Proposition 3.3 by taking similar ramified coverings. However, this method is inefficient for the unit equation, and results showing a much tighter structure are conjecturable, as we shall do in §7. Originally, the equation $x_1 + x_2 = x_3$ in relatively prime integers divisible by only a finite number of primes over **Z** was considered by Mahler, as an application of his p-adic extension of the Thue–Siegel theorem (pre-Roth version) [Mah 33], Folgerung 2. It was considered as a "unit equation" (for units of a finitely generated ring) explicitly in [La 60a].

In higher dimension, I conjectured [La 60a] and Faltings proved [Fa 90]:

Theorem 3.5 (Faltings). *Let A be an abelian variety defined over a finitely generated field over the rational numbers. Let V be an affine open subset of A. Let R be a finitely generated ring over* **Z** *contained in some finitely generated field F. Then every subset of R-integral points in $V(F)$ is finite.*

Note also that Theorem 3.5 follows from Vojta's conjectures which will be mentioned in §4, and give a general framework for this type of finiteness.

Faltings proves his theorem by going through the higher dimensional analogue of Proposition 3.3. Before we state his inequality, note that for each absolute value v and subvariety Z of A one can define a v-adic distance $d_v(P, Z)$ in a natural way.

Theorem 3.6 (Faltings [Fa 90]). *Let Z be a subvariety of A over a number field F. Fix an absolute value v on F. Given $\varepsilon > 0$, there is only a finite number of rational points $P \in A(F) - Z$ such that*

$$\operatorname{dist}_v(P, Z) < \frac{1}{H(P)^{\varepsilon}},$$

where $H(P) = \exp h_E(P)$, and h_E is the height with respect to any ample divisor E.

If Z is a divisor D, in terms of Weil functions and a finite number of absolute values S, we can rewrite this inequality in the equivalent form

$$m_{S,D}(P) \leqq \varepsilon h_E(P)$$

for all but a finite number of points $P \in A(F) - Z$. For comments on the proof, see §6. Faltings' inequality fits the general type of Vojta inequality stated below in Conjecture 4.1, because the canonical class on an abelian variety is 0, and for rational points the term with the discriminant $d(P)$ does not appear. On abelian varieties I had conjectured actually a stronger version of such an inequality [La 74] which we shall discuss in §7.

Of course, there are also the relative cases of finiteness, both in dimension 1 and higher dimension. Let k be an algebraically closed field of characteristic 0, and let F be a function field over k. Let R be a finitely generated subring of F over k, so R is the affine ring of some affine variety over k. In [La 60a] I proved:

Theorem 3.7. *Let V be an affine curve over F, of genus $\geqq 1$, or of genus 0 but with at least three points at infinity in its projective completion. Then every set of R-integralizable points in $V(F)$ has bounded height.*

And in higher dimension, I conjectured the analogue:

Let V be an affine open subset of an abelian variety defined over the function field F. Then a set of R-integral points in $V(F)$ has bounded height, and so is finite modulo the F/k-trace.

This conjecture was proved by Parshin [Par 86] using his hyperbolicity method under the additional assumption that the hyperplane at infinity does not contain the translation of an abelian subvariety of dimension $\geqq 1$. In the direction of differential equations, see also Osgood [Os 81], [Os 85].

Also in higher dimension, the number theoretic analogue of Borel's theorem that the complement of $2n + 1$ hyperplanes in general position in $\mathbf{P}^n$ is hyperbolic was proved by Ru and Wong [RuW 90], namely:

Theorem 3.8. *Let F be a number field and let $H_1, \ldots, H_q$ be hyperplanes in general position in $\mathbf{P}^n$. Let S be a finite set of absolute values and let*

$$D = \sum H_i.$$

Then for every integer $1 \leqq r \leqq n$ the set of (S, D)-integralizable points in $\mathbf{P}^n(F) - D$ is contained in a finite union of linear subspaces of

$\mathbf{P}^n(F)$ of dimension $r-1$, provided that $q > 2n - r + 1$. In particular, if $q \geqq 2n + 1$ the set of (S, D)-integralizable points of $\mathbf{P}^n(F) - D$ is finite.

The proof extends the proof of Schmidt's subspace theorem, by refining the approximation method using certain weights for the approximation functions. The use of these weights is related to Vojta's extension of Schmidt's theorem as stated in Theorem 2.1, but the relation has not yet been made explicitly.

IX, §4. VOJTA'S CONJECTURES

These conjectures provide a general framework for a large number of previous results or conjectures, which are seen as special cases. The framework covers both rational points and integral points. We work over number fields.

Conjecture 4.1. *Let X be a projective non-singular variety over a number field F. Let S be a finite set of absolute values containing the archimedean ones, and let K be the canonical class on X. Let D be a divisor with simple normal crossings on X. Let r be a given positive integer and $\varepsilon > 0$. Let E be a pseudo-ample divisor on X. There exists a proper Zariski closed subset Z of X (depending on the above data), such that*

$$m_{S,D}(P) + h_K(P) \leqq d(P) + \varepsilon h_E(P) + O_\varepsilon(1)$$

for all points $P \in X(\mathbf{Q}^a)$ not lying in Z for which $[F(P): F] \leqq r$.

Several comments need to be made concerning the extent to which certain hypotheses are needed in this conjecture. Vojta has raised the possibility that these hypotheses can be weakened as follows.

(a) In his improvement of Schmidt's theorem, Vojta showed how the exceptional set does not depend on ε. To what extent is there such independence in the more general case at hand? From my point of view, the error term should anyhow be of the form

$$d(P) + (1 + \varepsilon) \log h_E(P) + O_\varepsilon(1),$$

as for the error terms in Nevanlinna theory, Chapter VIII, §5, even with a Khintchine-type function.

(b) A restriction was made for the algebraic points to have bounded degree. In current applications, the estimate of the conjecture suffices to imply numerous other conjectures concerning rational points in $X(F)$, as in [Vo 87]. Indeed, the various proofs of implication rely on the construction of coverings which lift rational points to points of bounded degree. Thus the stronger property that the inequality should hold without the restriction of bounded degree would exhibit a phenomenon which has not been directly encountered in the applications. When I wrote up whatever results were known around 1960, I was careful to separate the parts of the proofs which, on the one hand, imply that certain sets of points have bounded height; and on the other hand, show that certain sets of bounded height satisfy certain finiteness conditions. The two parts are quite separate. Dealing with all algebraic points in $X(F^{a})$ makes this separation quite clear. The entire basic theory which makes the heights functorial with respect to relations among divisor classes goes through for $X(F^{a})$. The question Vojta raises in his inequalities is whether the more refined estimates for the canonical height also hold uniformly. For instance, to what extent does the term $O_{\varepsilon}(1)$ depend on the degree of the points. A current result of William Cherry in the analogous case of Nevanlinna theory shows that the analogue of Vojta's conjecture is true, with essentially best possible error terms with a Khintchine type function. See [Cher 90] and [Cher 91]. Cf. Chapter VIII, §1, Theorem 5.6. The fact that the degree occurs only as a factor without any extraneous constant term provides evidence that *Vojta's conjecture should follow a similar pattern and be valid uniformly for all algebraic points.*

We note that Vojta proved that his conjecture implied my conjecture concerning the finiteness of integral points on affine open subsets of abelian varieties [Vo 87] Chapter 4, §2. He obtains this from the following corollary of Conjecture 4.1 (so the corollary is itself conjectural).

Corollary 4.2. *Let X be a non-singular projective variety defined over a number field F, and let D be a divisor with simple normal crossings. Let K be the canonical class and assume that $K + D$ is pseudo ample. Let S be a finite set of places. Then an (S, D)-integralizable set of points in $X(F)$ is not Zariski dense in X.*

The reader will note the persistent hypothesis that the divisor D in the statements of theorems and conjectures has simple normal crossings. Sometimes one wants to apply an estimate on heights with respect to a divisor which comes up naturally but does not satisfy that hypothesis. Vojta shows in several cases how his theorem applies to a blow up of the divisor which does have normal crossings. In each case, the estimate applied to the blow up gives the expected estimates on the heights.

A formulation for the inequality in Vojta's Conjecture 4.1 for rational points on an abelian variety was already given in [La 74] p. 783 and [La 64], in the context of diophantine approximation on toruses. I did not consider algebraic points with the corresponding estimate of the logarithmic discriminant. But the approach by considering linear combinations of logarithms (ordinary or abelian) and their properties of diophantine approximation led to the conjectured better error term $O(\log h_E(P))$ rather than $\varepsilon h_E(P)$. See §7. We now consider a second conjecture of Vojta having to do with coverings.

Conjecture 4.3. *Let X, X' be projective non-singular varieties over a number field F. Let D be a divisor with simple normal crossings on X. Let E be a pseudo-ample divisor on X. Let $\varepsilon > 0$. Let*

$$f: X \to X'$$

be a finite surjective morphism. Let S be a finite set of absolute values. Then there exists a proper Zariski closed subset Z of X depending on the previous data, such that

$$m_{S,D}(P) + h_K(P) \leqq d(P) + \varepsilon h_E(P) + O_\varepsilon(1)$$

for all points $P \in X(F^a) - \mathrm{supp}(Z)$ for which $f(P) \in X'(F)$.

Conjecture 4.3 and also Conjecture 4.1 are applied to coverings, and both contain the discriminant term on the right-hand side. Hence the conjectures must be consistent with taking finite coverings, which may be ramified. To show this consistency and other matters, Vojta compares the discriminant term in coverings as follows.

Theorem 4.4. *Let $f: X \to X'$ be a generically finite surjective morphism of projective non-singular varieties over a number field F. Let S be a finite set of absolute values. Let Δ be the ramification divisor of f. Then for all $P \in X(F^a) - \mathrm{supp}(\Delta)$ we have*

$$d(P) - d(f(P)) \leqq N_{\Delta,S}(P) + O(1).$$

We defined $N_{\Delta,S}$ previously, and briefly $N_{\Delta,S} = h_\Delta - m_{\Delta,S}$. Vojta's Theorem 4.4 is a generalization to the ramified case of a classical theorem of Chevalley–Weil which we give as a corollary. See [ChW 32], [We 35], and [La 83].

Corollary 4.5 (Chevalley–Weil). *Let $f: X \to X'$ be an unramified finite covering of projective non-singular varieties over a number field F. Then for every pair of points $P \in X(F^a)$ and $Q = f(P)$, the relative discriminant of $F(P)$ over $F(Q)$ divides a fixed integer d.*

By using ramified coverings, Vojta has shown that the case of Conjecture 4.3 when $D = 0$, implies the general case with a divisor D. Similarly, if $\dim X = 1$ so X is a curve, then the case of Conjecture 4.1 with

$D = 0$ implies the general case with a divisor D. See [Vo 87], Proposition 5.4.1.

Technical remark. Actually, the Chevalley–Weil theorem was proved for normal varieties, in the unramified case. The non-singularity is a convenient assumption when there are singularities in the ramified case.

Since there is only a finite number of number fields of bounded degree and bounded discriminant, one obtains:

Corollary 4.6. *Let $f: X \to X'$ be as in the previous corollary. Then $X(F)$ is finite for all number fields F if and only if $X'(F)$ is finite for all number fields F.*

Example. Let Φ_n be the Fermat curve of degree n, and let $X(N)$ be the modular curve over $\mathbf{Q}$ of level N. Over the complex numbers, we have $X(N)(\mathbf{C}) \approx \Gamma(N)\backslash\mathfrak{H}$, where $\mathfrak{H}$ is the upper half plane. Then there is a correspondence

$$\begin{array}{ccc} \Phi_n & & X(2n) \\ & \searrow \quad \swarrow & \\ & \mathbf{P}^1 & \end{array}$$

such that the liftings of Φ_n and $X(2n)$ over each other are unramified. Cf. Kubert–Lang [KuL 75].

IX, §5. CONNECTION WITH HYPERBOLICITY

We have already remarked that Parshin used a hyperbolic method to prove part of the function field conjectures on rational and integral points on subvarieties of abelian varieties.

Let V be an *affine* variety over a number field contained in the complex numbers. Since it is not known if Kobayashi hyperbolicity for $V(\mathbf{C})$ is equivalent to Brody hyperbolicity for $V(\mathbf{C})$, there is some problem today about being sure what form the transposition of my conjecture to the affine case takes for affine varieties, whereby a finiteness condition on integral points is equivalent to some hyperbolicity condition. I formulated one possibility as follows in [La 87].

Conjecture 5.1. *If $V(\mathbf{C})$ is hyperbolically imbedded in a projective closure, then V has only a finite number of integral points in every finitely generated ring over $\mathbf{Z}$.*

Lacking the equivalence of the hyperbolicity conditions, I would formulate the converse only more weakly, that the diophantine condition implies that $V(\mathbf{C})$ is Brody hyperbolic. The problem is to what extent does one need additional restrictions near the boundary where the Kobayashi distance may degenerate (perhaps none as in Faltings' Theorem 3.5).

Roughly speaking, taking out a sufficiently large divisor from a projective variety leaves a hyperbolic variety, of which I expect that it has only a finite number of integral points as above. The question is what does "sufficiently large" mean. A theorem of Griffiths [Gri 71] asserts that one can always take out a divisor such that the remaining variety has a bounded domain as covering space, which is one of the strongest forms of hyperbolicity. Classically, if we take out three points from $\mathbf{P}^1$, then the remaining open set is Brody hyperbolic (Picard's theorem). In higher dimension the complement of $2n + 1$ hyperplanes in general position in $\mathbf{P}^n$ is Kobayashi hyperbolic. For examples of Bloch, Fujimoto, Green in this direction, see [La 87], Chapter VII, §2. A general idea is that one can take out several irreducible divisors of low degree, even degree 1 which means hyperplanes, or one can take out a divisor of high degree. Today, there is no systematic theory giving conditions for hyperbolicity in the non-compact case, which is less developed than the compact case. For our purposes here, the problem is to prove that such hyperbolic non-compact varieties have only a finite number of integral points. In that line, Vojta proved [Vo 87], Theorem 2.4.1.

Theorem 5.2. *Let X be a projective non-singular variety over a number field F. Let r be the rank of the Mordell–Weil group $A'(F)$, where A' is the Picard variety of X and let ρ be the rank of the Néron–Severi group* NS(X). *Let D be a divisor on X consisting of at least*

$$\dim X + \rho + r + 1$$

components. Let S be a finite set of absolute values of F. Then every (S, D)-integralizable set of points in $X(F)$ is not Zariski dense.

Note: "Components" in NS(X) and D are meant to be F-irreducible components, not necessarily geometrically irreducible. Vojta's improvement of Schmidt's theorem also lies in this direction. Furthermore, Vojta has given a quantitative form to estimates for the heights of integral points, under the strongest possible form of hyperbolicity, by using hyperbolic (1, 1)-forms as follows.

Conjecture 5.3 (The (1, 1)**-form conjecture** [Vo 87], Chapter 5, §7). *Let X be a projective non-singular variety over a number field F contained in* $\mathbf{C}$. *Let D be a divisor with normal crossings on X and let $V = X - D$. Assume that there exists a positive* (1, 1)*-form ω on* $V(\mathbf{C})$ *which is* *strongly hyperbolic, and in fact, that there exists a constant $B > 0$ such that, if $f: \mathbf{D} \to V(\mathbf{C})$ is a non-constant holomorphic map, then*

$$\operatorname{Ric} f^*\omega \geqq Bf^*\omega.$$

Also assume that $\omega \geqq c_1(\rho)$ for some metric ρ on a line sheaf $\mathscr{L}$ on X. Let E be pseudo ample on X. Let S be a finite set of absolute values. Let I be a set of (S, D)-integralizable points of bounded degree over F. Then for all points $P \in I$ we have

$$h_{\mathscr{L}}(P) \leqq \frac{1}{B} d(P) + \varepsilon h_E(P) + O_\varepsilon(1).$$

Remark 1. In this conjecture, there is no Zariski closed subset acting as an exceptional set.

Remark 2. If the set I is contained in the rational points $X(F)$, then the conjecture implies that I is finite.

Remark 3. In light of the analogous result in Nevanlinna theory which motivated the conjecture, but for which the assumption that D has normal crossings turned out to be superfluous, I expect it to be equally superfluous in the present arithmetic case. Also the error term εh_E should be replaced by $O(\log h_E)$ or better.

Remark 4. The question whether the restriction that the points should have bounded degree applies as well to the present case.

Vojta applies the (1, 1)-form conjecture to deduce several number theoretic applications. We mention two of them. First he proves in [Vo 87]:

Conjecture 5.3 *implies the Shafaverich conjecture.*

Specifically, in [Vo 87], Chapter 5, §7, by applying the conjecture to the moduli space and its boundary divisor, he proves that Conjecture 5.3 gives somewhat more uniformity, and implies:

Corollary 5.4. *Given a finite set of places S; positive integers n, r; and $\varepsilon > 0$, there exists a number $C = C(S, n, r, \varepsilon)$ such that for every semi-stable principally polarized abelian variety A of dimension n and good reduction outside S, over a number field F of degree $\leqq r$, we have*

$$h_{\mathrm{Fal}}(A) \leqq \left(\frac{n}{2} + \varepsilon\right) d(F) + C.$$

But this corollary is itself conjectural. Note that the moduli space is not affine, so Conjecture 5.3 is applied in a rather delicate case, and the notion of (S, D)-integralizability is used in a rather strong way, as distinguished from the notion of integral points on affine varieties.

Second, Vojta shows in [Vo 88] how the (1, 1)-form conjecture implies a bound on $(\omega_{X/Y}^2)$ in Arakelov theory, and also for the height, similar to those in the function field case, and similar to those which we recalled in Chapter VII.

IX, §6. FROM THUE–SIEGEL TO VOJTA AND FALTINGS

A basic approximation method which started with Thue–Siegel went through a number of developments due to Schneider, Dyson, Gelfond, Roth, Viola, Schmidt, and culminated with the recent work of Vojta who combined all the aspects of previous work into an Arakelov context, thus expanding enormously the domain of applicability of this method. Vojta's current program is still in progress, but something can be said to give an idea of this program, both in the results achieved and its prospects. See [Vo 90a], [Vo 90b], [Vo 90c]. Faltings boosted the method further by proving a higher dimensional result [Fa 90].

We begin by a few words concerning intermediate results before Roth's theorem.

In a relatively early version of determining the best approximations of algebraic numbers by rational numbers, one had the Thue–Siegel–Dyson–Gelfond result:

Given $\varepsilon > 0$ and an algebraic number α of degree n over $\mathbf{Q}$, there are only finitely many rational numbers p/q $(p, q \in \mathbf{Z}, q > 0)$ such that

$$\left|\alpha - \frac{p}{q}\right| \leqq \frac{1}{q^{\sqrt{2n+\varepsilon}}}.$$

Of course this was short of the conjectured result ultimately proved by Roth, but it sufficed to prove the finiteness of integral points on curves as discussed in §3. The method of proof used two approximations $\beta_1 = p_1/q_1$ and $\beta_2 = p_2/q_2$ such that β_1 and β_2 have large heights, and also such that the quotient of the heights $h(\beta_2)/h(\beta_1)$ is large. If there are infinitely many solutions to the above inequality, then such β_1, β_2 can be found. However, the logic starting from such numbers β_1 and β_2 is such that the proof is not effective, since we don't have an effective starting point for the existence of β_1, β_2. One then shows that there exists a polynomial $G(T_1, T_2)$ with integer coefficients which are not too large, such that G vanishes of high order at (α, α), that is

$$D_1^{i_1} D_2^{i_2} G(\alpha, \alpha) = 0$$

for i_1, i_2 satisfying certain linear conditions, namely

$$\frac{i_1}{d_1} + \frac{i_2}{d_2} < t$$

with suitable positive integers d_1, d_2 and a fairly large t. Then in a crucial lemma, one shows that in fact, some derivative does not vanish at (β_1, β_2), that is

$$\xi = D^{j_1} D^{j_2} G(\beta_1, \beta_2) \neq 0$$

with j_1, j_2 not too large. Such a lemma was provided especially by Dyson and Gelfond. Since ξ is a rational number, one can then estimate the height of ξ from above to be small because many derivatives of G vanish at (α, α); on the other hand, this height has a lower bound depending only on the original degree of G and the heights of β_1, β_2. Solutions of linear equations and linear inequalities in the above construction can be found to yield a contradiction at this point.

Schneider [Schn 36] showed that instead of using two approximating numbers, by using arbitrarily many, say $\beta_1, \ldots, \beta_m$, and constructing a polynomial $G(T_1, \ldots, T_m)$ in m variables, then one would obtain the desired approximation with $q^{2+\varepsilon}$. He showed how the appearance of the $2 + \varepsilon$ is due to a combinatorial probabilistic estimate in the course of the proof having nothing to do with algebraic numbers. However, Schneider did not see how to complete the second step in the proof where one needs that a suitable derivative

$$D_1^{j_1} \ldots D_m^{j_m} G(\beta_1, \ldots, \beta_m)$$

does not vanish, and so he obtained only a partial result. Roth's achievement was to see very clearly through the entire situation, and to prove this second step by means of a classical Wronskian method which he saw how to adjust to yield the non-vanishing of the derivative.

The Wronskian method, however, was not well suited for extensions to algebraic geometric contexts. Viola [Vio 85] proved a non-vanishing result at the level of the Dyson–Gelfond lemma, by using methods of algebraic geometry having to do with the analysis of singularities. Although that method did not yield the $q^{2+\varepsilon}$, it did yield a fruitful alternative approach to one of the key steps in the proof of that theorem. That approach was vastly expanded by Esnault–Viehweg [EV 84] to get a result in arbitrarily many variables, yielding the full Roth theorem. Esnault–Viehweg work on products of projective spaces. However, as they point out, their result does not yet give what is needed for Schmidt's theorem.

In *Diophantine Geometry* ([La 62], see also [La 83]) I axiomatized the Roth proof so that it applied also to the function field case. I specifically pointed out that the construction of the polynomial $G(T_1, \ldots, T_m)$ depends on solving linear equations which amounted to a weak form of the Riemann–Roch theorem. Mordell in his 1964 review of that book commented: "The reader might prefer to read [Roth's proof] which requires only a knowledge of elementary algebra and then he need not be troubled with axioms which are very weak forms of the Riemann–Roch theorem." On the other hand, readers might prefer to read proofs which do use Riemann–Roch theorems, not only my adjustments of Roth's own proof but Vojta's subsequent contributions which jazz things up even more:

(a) By making the analogy with Nevanlinna theory.
(b) By globalizing and sheafifying not only on the projective line but on curves of arbitrary genus over, say, the ring of integers of a number field.

In this case, by following a pattern stemming from the previous, weaker versions of Roth's theorem, among many other patterns, Vojta needs a Hirzebruch–Grothendieck–Riemann–Roch along the most substantial lines developed in Arakelov theory by Gillet–Soulé in relative dimension > 1. Using such Riemann–Roch theorem, and other tools from algebraic geometry, analysis, and diophantine approximations, he was able to give an entirely new proof of Mordell's conjecture—Faltings' theorem. So different people prefer different things at different times.

Furthermore, Vojta developed his method first in the function field case, and then translated the method into the number field case, thus showing once again the effectiveness of the analogy. We describe Vojta's method at greater length, to show the connections not only with algebraic geometry, but with analysis and Arakelov theory, including partial differential equations. See [Vo 89b] for the function field case, and [Vo 90a] for the number field case.

In Roth's theorem, the choice of β_1, β_2 amounts to a choice of point on the product $\mathbf{A}^1 \times \mathbf{A}^1$ of the affine line with itself, or if you wish, the product $\mathbf{P}^1 \times \mathbf{P}^1$ of the projective line with itself. We now consider a projective non-singular curve C of genus $g \geqq 2$ defined over a number field F, and we consider the product $C \times C$. We let $P = (P_1, P_2)$ be a point in $C(F) \times C(F)$. For $i = 1, 2$ let x_i be a local coordinate on C in a neighborhood of P_i. Then (x_1, x_2) are coordinates on $C \times C$ at P. We suppose $x_i(P_i) = 0$. Let s be a section of a line sheaf $\mathscr{L}$ on $C \times C$, defined in a neighborhood of P by a formal power series

$$f(x_1, x_2) = \sum_{i_1, i_2 \leqq 0} a_{i_1 i_2} x_1^{i_1} x_2^{i_2}.$$

Let d_1, d_2 be positive integers. We define the **index** of s at P relative to d_1, d_2 to be the largest real number $t = t(s, P, d_1, d_2)$ such that

$$a_{i_1 i_2} = 0$$

for all pairs (i_1, i_2) of natural numbers satisfying

$$\frac{i_1}{d_1} + \frac{i_2}{d_2} < t.$$

If D is an effective divisor on $C \times C$, then we define the **index of** D to be the index of a section s on $\mathcal{O}_{C \times C}(D)$ for which $(s) = D$. The above definitions are independent of the choice of local parameters, and globalize the notion of index stemming from Dyson–Gelfond and Schneider–Roth.

As usual, we let $Y = \text{spec}(\mathfrak{o}_F)$ *where* $\mathfrak{o}_F$ *is the ring of integers of* F. *We let*

$$\pi: X \to Y$$

be a regular semi-stable family of curves over Y *with generic fiber* $X_F = C$. *In addition, we assume that all double points occurring on the geometric fibers of* π *are rational over the residue class fields, and that the tangents for both branches of the fibers (in the complete local ring) are also rational over the residue field.*

The semistability and rationality assumptions can always be realized over a finite extension of a given base. The main steps of Vojta's proof then run as follows.

(1) For certain values of a parameter r, find a divisor D_r on $C \times C$ which is ample if r is sufficiently large.

(2) Show that there exists a thickening of D_r to a divisor on $X \times_Y X$, that is, find an Arakelov divisor Z_r on $X \times_Y X$ which restricts to D_r on the generic fiber and satisfies certain bounds at infinity, such that the line sheaf $\mathcal{O}(mZ_r)$ with suitably large m has a section s_r satisfying certain bounds at the infinite places.

(3) Choose rational points P_1, P_2 on C such that $h(P_1)$ is sufficiently large, and $h(P_2)/h(P_1)$ is sufficiently large; also such that certain sphere packing conditions are satisfied.

(4) Prove a lower bound for the index of the section s_r of Step 2 at the point $P = (P_1, P_2)$, when r is close to $h(P_2)/h(P_1)$.

(5) Get a contradiction between this lower bound and an upper bound obtained by a suitably globalized version of Dyson's lemma.

The difficulties which arose previously exist in even stronger form in the present globalized context. The first one to be surmounted was Vojta's globalization of Dyson's lemma [Vo 89a], motivated by Viola's proof.

A much more elaborate difficulty came from proving the existence of the section s_r in step (2). The existence of a section in the standard context of algebraic geometry (without metrics on line sheaves) is given by the Hirzebruch–Grothendieck–Riemann–Roch theorem, which in Vojta's case is applied to $X \times_Y X$, having relative dimension 2. Whereas in the function field case, Hirzebruch–Grothendieck–Riemann–Roch sufficed, the number field case required an Arakelov version since bounds at infinity have to be taken into account. This required estimating eigenvalues of Laplacians and applying such bounds to the geometric situation of metrized line sheaves. Major results in this direction were obtained by Gillet–Soulé [GiS 88] and Bismut–Vasserot [BiV 89]. Vojta complemented these results to fit his situation. They could not be applied directly since certain positivity conditions were assumed previously on the metrics, so Vojta had to provide additional work to deal with certain metrics which are not positive. See Chapter VI, §3.

Vojta's work is in progress in at least two directions: First, he develops the use of points $P_1, \ldots, P_m$ instead of P_1, P_2 to get a better measure of approximation corresponding to Roth's theorem instead of Thue–Siegel–Dyson's theorem. The second direction is to extend the result to include the proximity term $m_{D,S}(P)$, where D is a divisor consisting of distinct points on C, and forward along the lines of Vojta's conjectures stated in §4.

On the other hand, Faltings [Fa 90] influenced by Vojta's paper, restructured and extended the diophantine approximation method. He succeeded in proving thereby that a subvariety of an abelian variety, not containing the translation of an abelian subvariety of dimension > 0, has only a finite number of rational points in any number field F, or any finitely generated field over the rationals. Faltings eliminated the use of the Gillet–Soulé Riemann–Roch theorem in higher dimension to obtain the desired section in step (2). Instead he uses a globalized version of a lemma of Siegel, solving integrally equations with integer coefficients (or algebraic integers), and giving suitable bounds for the heights of the solutions in terms of the height of the coefficients. Secondly, instead of using something like Dyson's lemma, Faltings uses a new method of algebraic geometry to show that some suitable derivative does not vanish in step (4). Finally, Faltings works with a sequence of points $P_1, \ldots, P_m$, not just two points, such that the ratios of successive heights is large.

At the moment of writing, the situation is in flux, so it is not clear what use will be made in the future of Vojta's or Faltings versions of the general method globalizing the Roth–Schmidt theorems. In any case, Faltings' result was the first time that a variety of dimension > 1 was proved to be Mordellic, except of course for the product of curves, finite unramified covers, and finite unramified quotients of the above.

A simplification of Vojta's proof also eliminating the use of Arakelov

Gillet–Soulé theory (but still using Riemann–Roch) was given by Bombieri [Bom 90]. In addition, this simplification also shows how to use Roth's lemma instead of the version following Viola.

IX, §7. DIOPHANTINE APPROXIMATION ON TORUSES

We have used the word "torus" in two senses: one sense is that of complex torus, and so the group of complex points of an abelian variety; the other sense is that of linear torus, that is, a group variety isomorphic to a product of multiplicative groups over an algebraically closed field. We have already seen analogies between these two cases, notably in the formulation of results or conjectures describing the intersection of subvarieties of semiabelian varieties with finitely generated subgroups. We shall go more deeply into this question here, following [La 64] and [La 74].

Let us start with the linear case. In Proposition 3.2, we gave a geometric formulation of Roth's theorem, serviceable to study all integral points. But as we have also seen, we also want to consider the more special situation when we restrict our attention to a finitely generated subgroup of the multiplicative group, so let us start with a conjecture from [La 64]. We let $G = \mathbf{G}_m$ be the multiplicative group.

Conjecture 7.1. *Let F be a number field and let Γ be a finitely generated subgroup of the multiplicative group $G(F)$. Let φ be a non-constant rational function on G defined over F, and let m be the maximum of the multiplicities of the zeros on G (so distinct from 0 or ∞). Let r be the rank of Γ. Then given ε, the height of points P in Γ which are bounded away from 0 and ∞ and satisfy the inequality*

$$(1) \qquad |\varphi(P)| < \frac{1}{h(P)^{rm+\varepsilon}}$$

is bounded.

We shall transform inequality (1). The function field $F(G)$ is generated by a single function, and φ is just a rational function. A point $P \in G(F)$ is represented by an element of F, say β, and if $\varphi(\beta)$ is small in absolute value, then there is some root $\alpha \neq 0$ of the rational function φ such that β is close to α. If β is close to α, then its distance from any other zero or pole of φ is bounded away from 0 (approximately by the distance of α itself from another zero or pole). The multiplicity of α in a factorization of φ is at most m. The worst possible case is that in which this multiplicity is m. In that case, $|\varphi(\beta)|$ is approximately equal to $|\alpha - \beta|^m$, up to a constant factor. Hence our inequality amounts to

$$|\alpha - \beta| \ll \frac{1}{h(\beta)^{r+\varepsilon}}.$$

This approximation can be transferred to the additive group via the logarithm. Let $\beta_1, \ldots, \beta_r$ be free generators of Γ modulo its torsion group, i.e. modulo the group of roots of unity. Then we can write

$$\beta = \beta_1^{q_1} \ldots \beta_r^{q_r} \zeta$$

for some torsion element ζ. Define

$$q = q(\beta) = \max |q_i|.$$

As a function on Γ, it is easy to see that $q \gg\ll h$. Furthermore, $|\alpha - \beta|$ is of the same order of magnitude as $|\log \alpha - \log \beta|$. Since ζ ranges over a finite set, in dealing with solutions of inequality 7.1(1) we may assume that we have always the same ζ. Let $u_0 = \log(\alpha\zeta)$. Then we may rewrite our inequality in the form

$$|u_0 - q_1 \log \beta_1 - \cdots - q_r \log \beta_r + q_{r+1} 2\pi i| < \frac{1}{q^{r+\varepsilon}}$$

where the log is one fixed value of the logarithm.

We have therefore transferred our diophantine approximation on the multiplicative group over a number field into an inhomogeneous approximation on the additive group. The period $2\pi i$ of the exponential function contributes one term to the sum on the left, and gives rise to $r + 1$ free choices of the coefficients $q_1, \ldots, q_{r+1}$. A standard application of Dirichlet's box principle shows that r cannot be replaced by a smaller exponent on the right-hand side. In fact, given real numbers $\xi_1, \ldots, \xi_r$ and an integer $q > 0$ there exist integers $q_1, \ldots, q_r$ not all 0 such that

$$|q_1 \xi_1 + \cdots + q_r \xi_r| \ll \frac{1}{q^{r-1}}, \qquad \text{and} \qquad |q_i| \leqq q.$$

Hence the exponent rm in the conjecture cannot be improved upon.

Khintchine's theorem already mentioned in §2, and suitable generalizations by Schmidt [Schm 60], imply that for Lebesgue almost all sets $\xi_1, \ldots, \xi_r, \xi_{r+1}$ of real numbers, the solutions of the inequality

$$|\xi_{r+1} + q_1 \xi_1 + \cdots + q_r \xi_r| \ll \frac{1}{q^{r-1+\varepsilon}} \qquad \text{with} \quad q = \max |q_i|,$$

are finite in number. Thus the conjecture essentially asserts that the logarithms of algebraic numbers behave like "almost all numbers". Dirichlet's box principle shows that the conjecture gives the best possible exponent.

Gelfond [Gel 60] had obtained some inequalities for linear combinations of logs of two algebraic numbers, but Baker made a breakthrough when he extended such inequalities to arbitrarily many numbers, from [Ba 66] to [Ba 74]. Feldman made a key improvement in Baker's method. The full conjecture stated above is still unproved. At this time, one has the

Baker–Feldman inequality. *For $q_i \in \mathbf{Z}$ not all zero,*

$$|q_1 \log \beta_1 + \cdots + q_r \log \beta_r| > \frac{c'}{q^c}$$

except for a finite number of $(q_1, \ldots, q_r)$.

Here c, c' depend on the heights of $\beta_1, \ldots, \beta_r$. An explicit dependence can be given, for which we refer to other expositions. Besides those of Baker himself, see especially Wustholz [Wu 85], [Wu 88], who reconsidered linear forms of logarithms in light of more recent insights using certain techniques of algebraic geometry.

The same sort of arguments as above apply to elliptic curves or abelian varieties. To avoid more complicated exponents, let us consider an elliptic curve A defined over a number field F, imbedded in the complex numbers.

Conjecture 7.2. *Let Γ be a finitely generated subgroup of $A(F)$. Let φ be a non-constant rational function on A defined over F. Let m be the maximum of the multiplicities of the zeros on A. Let r be the rank of Γ. Then given ε, there is only a finite number of points $P \in \Gamma$ satisfying the inequality*

$$|\varphi(P)| < \frac{1}{h(P)^{(1/2)m(r+1)+\varepsilon}}. \tag{2}$$

Indeed, we parametrize the complex points $A(\mathbf{C})$ by an exponential map

$$\exp\colon \mathbf{C} \to A(\mathbf{C})$$

whose kernel is a lattice, the period lattice, in $\mathbf{C}$. The rational function φ then becomes a meromorphic function Φ of a variable $z \in \mathbf{C}$. Let $\{P_1, \ldots, P_r\}$ be a basis of Γ mod torsion, and let

$$u_j = \log P_j \qquad \text{meaning that} \qquad P_j = \exp(u_j).$$

Let us assume for the moment for concreteness that our exponential map is given explicitly by the Weierstrass functions $(\wp, \wp')$. Let $\{\omega_1, \omega_2\}$ be a basis for the period lattice. Let

$$P = q_1 P_1 + \cdots + q_r P_r + Q$$

where $Q \in A(F)_{\text{tor}}$ is a torsion point. If a sequence of points P approaches a zero of φ, then the logarithms $\log P$ approach a point $u_0 = \log(P_0 - Q)$. Of course P_0 may be algebraic, not necessarily in $A(F)$. Hence there exist integers q_{r+1}, q_{r+2} such that

$$q_1 u_1 + \cdots + q_r u_r + q_{r+1}\omega_1 + q_{r+2}\omega_2$$

is close to u_0. But by the quadraticity of the Néron–Tate height, and the fact that on a finite dimensional vector space all norms are equivalent, we have

$$h(P) \gg\ll q^2 \qquad \text{where} \quad q = \max |q_i|.$$

Therefore the inequality (2) in Conjecture 7.2 amounts to the inequality given among the elliptic logarithms by

$$|-u_0 + q_1 u_1 + \cdots + q_r u_r + q_{r+1}\omega_1 + q_{r+2}\omega_2| < \frac{1}{q^{r+1+\varepsilon}}.$$

Thus our conjecture asserts that from the point of view of diophantine approximations, the logs of algebraic points on the elliptic curve behave again like "almost all numbers".

Actually, the periods ω_1, ω_2 are linearly independent over the reals. This would seem to indicate that the exponent $r + 1$ should be replaced by r, just as for the multiplicative group. This also raises the following question. Given a non-zero period ω, are there infinitely many real values of t such that $\wp(t\omega)$ are algebraic and linearly independent over the integers? The extent to which r can be further lowered depends on the existence of such values of t.

Similarly, considerations apply to an abelian variety A over F, with F contained in $\mathbf{C}$. The exponential map is a complex analytic homomorphism

$$\exp\colon \operatorname{Lie}(A)_{\mathbf{C}} \to A(\mathbf{C})$$

whose kernel is the period lattice Λ. If we identify $\operatorname{Lie}(A)_{\mathbf{C}}$ with $\mathbf{C}^n$, then we **normalize the exponential map** so that the tangent linear map at the origin is algebraic. We write

$$P = \exp u \qquad \text{and} \qquad u = \log P$$

for $u \in \mathbf{C}^n$. The log here is sometimes called the **abelian logarithm**. We obtain similar conjectured diophantine inequalities for linear combinations of logarithms of algebraic points of $A(F)$, using a basis for the Mordell–Weil group. The exponent of q on the right-hand side is determined by the probabilistic model and Dirichlet's box principle.

From this point of view, Faltings' inequality in Theorem 3.6 should be replaced by the inequality

$$\operatorname{dist}_v(P, Z) < \frac{1}{h(P)^c}$$

where c is an exponent reflecting the complexity of the singularities of Z and the rank of the Mordell–Weil group. Taking the log, and supposing $Z = D$ is a divisor represented by a rational function φ on a Zariski open set, this inequality would also read

$$-\log|\varphi(P)| < c \log h(P), \tag{3}$$

which was already conjectured in [La 74], p. 783. Such an inequality is in line with the error terms found for the analytic Nevanlinna theory in Chapter VIII, §5. When D has simple normal crossings, then the constant c will reflect both the complexity of the singularities and the rank of the Mordell–Weil group, according to the probabilistic model and Dirichlet's box principle.

Even for elliptic curves, and even in the case of complex multiplication, as of today, the analogue of the Baker–Feldman inequality for elliptic logarithms is not known. Only worse estimates are known, albeit non-trivial ones, but we shall not give an account of these partial results, which get constantly improved. The best known result at this time (getting close) is due to Hirata–Kohno [Hir 90]. I have tried however to give references which will help the reader get acquainted with currently known methods. I have emphasized the conjectures, which personally I find more satisfactory to describe the theoretical framework behind the mass of partial results available today in the direction of diophantine inequalities for the height.

I would like to bring up one more possible application of methods of diophantine approximations to abelian varieties. Let A be an abelian variety defined over a number field F. Let $A[m]$ denote the subgroup of points of order m in $A(\mathbf{Q}^a)$. The problem is to give a lower bound for the degree

$$[F(A[m]) : F].$$

The first results (other than in the case of complex multiplication) were due to Serre, for elliptic curves. In this case, let G denote the Galois group of $F(A_{\mathrm{tor}})$ over F. Then the l-adic representations of G on $T_l(A)$ make G an open subgroup of the product $\prod \operatorname{Aut} T_l(A)$. In [La 75] I showed how techniques of diophantine approximation used in transcendence theory could also be applied to give some lower bound for the degrees of $F(A[p^n])$ for a fixed prime p and $n \to \infty$. However, the bound so far obtained falls very short of Serre's theorem even when m is a prime power.

The results of Masser–Wustholz [MaW 91] are partly concerned with bounds for the size of Galois invariant subgroups of A_{tor}, and will now be discussed. I am much indebted to Masser and Wustholz for making available preliminary copies of their manuscript as well as for their guidance, so that I could report on their results in this book.

We return to the considerations of Chapter IV, §6 having to do with the proof of Masser–Wustholz's theorem. Let A be an abelian variety defined over a number field F of degree d over the rationals. We let H be a non-degenerate Riemann form (positive definite) on $\mathbf{C}^n$ with respect to Λ. Denoting $\operatorname{Lie}(A)$ the Lie algebra of A, we let $\operatorname{Lie}(A)_{\mathbf{C}}$ be its extension to the complex numbers. Then we identify

$$\mathbf{C}^n = \operatorname{Lie}(A)_{\mathbf{C}}$$

with its exponential map defined earlier in this section. The Riemann form has a unique normalized theta function (up to a constant factor) associated with it (cf. Chapter IX, §1, Example 4), and the zeros of this theta function define a divisor on A whose class modulo algebraic equivalence is a polarization, called the **associated polarization**. We let m be the degree of this polarization.

On the other hand, let B be an abelian subvariety of A. We define its **degree with respect to the Riemann form** (or associated polarization) by the formula

$$(\deg_H B)^2 = (\dim B)!^2 \, |\det \operatorname{Im} H_{\Lambda(B)}|,$$

where $\Lambda(B) = \operatorname{Lie}(B)_{\mathbf{C}} \cap \Lambda$ and $H_{\Lambda(B)}$ is the restriction of the Riemann form to $\Lambda(B)$. If the polarization associated with H is very ample, then this degree is essentially a normalization of the projective degree in the corresponding projective imbedding. The main result linking Theorem 6.1 of Chapter IV and the methods used in the theory of diophantine approximation is the following [MaW 91].

Theorem 7.3. *Let $w \in \operatorname{Lie}(A)_{\mathbf{C}} \cap \Lambda$ and let G_w be the smallest abelian subvariety defined over F such that $\mathbf{C}w \subset \operatorname{Lie}(G_w)_{\mathbf{C}}$. Then*

$$\deg_H G_w \leqq c_1(m, n, d) \max\bigl(1, h_{\mathrm{Fal}}^{\mathrm{st}}(A), H(w, w)\bigr)^{k_1(n)}$$

where c_1 and k_1 are constants depending only on (m, n, d) and n respectively.

The proofs allow the writing down of explicit values for c_1 and k_1, although these are by no means the best possible conjectured values. For instance, the proofs yield a value for $k_1(n)$ involving $n!$, but a much lower value is expected to hold.

I shall now indicate briefly how Theorem 7.3 is used to imply Theorem 6.1 of Chapter IV. Suppose that A is simple for simplicity. Suppose given an isogeny

$$\alpha: A \to B.$$

We select one suitably reduced period ω_1 of A in Λ. In the case of an elliptic curve, we would pick τ in a fundamental domain. We select a basis of periods $\{\eta_1, \ldots, \eta_{2n}\}$ of B also suitably reduced. We apply Theorem 7.1 to the abelian variety $A \times B^{2n}$ and to the single period

$$w = (\omega_1, \eta_1, \ldots, \eta_{2n}).$$

Let G_w be the smallest abelian subvariety defined over F as given by the theorem. There exists a positive integer r and a factor B of B^r such that if we let

$$G'_w = G_w \cap (A \times B^r),$$

$$\Gamma = \text{projection of } G'_w \text{ on } A \times B,$$

then Γ is the graph of the isogeny we are looking for in Theorem 6.1. When A is not simple, the situation is considerably more complicated.

In transcendental terms, we consider the complex representation ρ of α, so that we may write $\rho(\alpha)\omega_1$ as a linear combination

$$\rho(\alpha)\omega_1 = \sum_{j=1}^{2n} b_j \eta_j$$

with integers b_j. In particular, this is a linear relation in abelian logarithms. Roughly speaking Baker's method then yields another relation with suitably smaller coefficients bounded in terms of various heights. To show the analogy I shall state the simple essentially analogous version on the multiplicative group.

Theorem 7.4. *Let $\alpha_1, \ldots, \alpha_n$ be algebraic numbers of degree $\leqq d$. Let R be the $\mathbf{Z}$-submodule of $\mathbf{Z}^n$ consisting of all relations among* $\log \alpha_1$,

...,$\log \alpha_n$*; that is, all vectors* $(b_1, \ldots, b_n) \in \mathbf{Z}^n$ *such that*

$$b_1 \log \alpha_1 + \cdots + b_n \log \alpha_n = 0.$$

Then there exists a basis of R *over* $\mathbf{Z}$ *such that for all elements* $(b_1, \ldots, b_n)$ *in this basis, we have the estimate*

$$|b_j| \leqq c_2(n, d) \max_i \left(1, h(\alpha_i)\right)^{n-1} \qquad \text{for} \quad j = 1, \ldots, n.$$

The similarity of this statement with Theorem 7.3 is evident, except for the presence of the algebraic subgroup G_w. To make the analogy exact, one must actually consider the subspace $R_{\mathbf{Q}}$ generated by the relations over $\mathbf{Q}$, and use an algebraic subgroup of G^n where G is the multiplicative group. Instead of the height of basis elements one must use the height of this subspace in the Grassmanian, essentially as in Schmidt's papers.

In the present context, and also the context of Chapter III, §1 among others, we want to find a basis of a finitely generated abelian group with the smallest norm, for a suitable norm on this group. See for instance [La 60], [La 78], Chapter IV, §5, [La 83a], Chapter V, §7, Theorem 7.6, [Wa 80], and Masser [Mass 88], §2 and §4. I shall give here very simple proofs for relevant statements which already show what is involved to get minimal norms. The method stems from Stark, who first applied it to the group of units in a number field, and is simpler than Baker's method. The question arises whether Baker's method is really necessary for the Masser–Wustholz result.

We shall deal with a finitely generated abelian group Γ written additively. We let $|\ |$ be a seminorm on Γ, i.e. $|\ |$ satisfies:

$$\begin{aligned} &|P| \geqq 0 \qquad \text{for all} \quad P \in \Gamma; \\ &|nP| = |n||P| \qquad \text{for} \quad n \in \mathbf{Z}; \\ &|P + Q| \leqq |P| + |Q| \qquad \text{for} \quad P, Q \in \Gamma. \end{aligned}$$

We let Γ_0 the **kernel** of $|\ |$, that is the subgroup of elements P such that $|P| = 0$. We assume that there exists $\delta > 0$ such that if $|P| \neq 0$ then $|P| \geqq \delta$. We let r be the rank of Γ/Γ_0, and assume $r \geqq 1$.

Lemma 7.5. *Let* $P_1, \ldots, P_n$ *be generators of* Γ. *Assume* $n > r$. *Then there exists a relation*

$$\sum m_i P_i \in \Gamma_0$$

with integers m_i *not all* 0 *satisfying*

$$|m_i|^{1/r} \leqq \delta^{-1} \sum_{j \neq i} |P_j|.$$

Proof. Passing to the factor group Γ/Γ_0, we may assume without loss of generality that $|\ |$ is a norm on Γ. After renumbering the generators, say $\{P_1, \ldots, P_r\}$ is a maximal family of linearly independent elements among $P_1, \ldots, P_n$. Consider say $P = P_{r+1}$. Then $\langle P_1, \ldots, P_{r+1} \rangle$ has rank r, and the **Z**-module of linear relations $(m_1, \ldots, m_{r+1})$ such that

$$m_1 P_1 + \cdots + m_{r+1} P_{r+1} = 0$$

is free of rank 1 over **Z**, generated by a vector such that $(m_1, \ldots, m_{r+1}) = 1$. Let $N = -m_{r+1}$, so we write the relation in the form

$$NP = m_1 P_1 + \cdots + m_r P_r.$$

Rather than use Dirichlet's box principle as in [La 78] and [La 83a], I use a variation from Waldschmidt [Wa 80]. We are going to show that

$$|N| \leqq c_0 \qquad \text{where} \qquad c_0 = (r/\delta)^r |P_1| \ldots |P_r|.$$

This will yield Lemma 7.5 by using the inequality

$$(|P|_1 \ldots |P_r|)^{1/r} \leqq \frac{1}{r}(|P_1| + \cdots + |P_r|).$$

We define

$$c_j = \delta/r|P_j| \qquad \text{for} \quad j = 1, \ldots, r.$$

Then

$$c_0 \ldots c_r = 1.$$

By Minkowski's theorem, there exist integers $q, s_1, \ldots, s_r$ such that

$$|qm_j/N - s_j| < c_j \text{ for } j = 1, \ldots, r \qquad \text{and} \qquad |q| \leqq c_0.$$

(See [Schm 80], p. 33, Theorem 2C. The inequality is strict for $j = 1, \ldots, r$.) Let

$$n_j = qm_j - Ns_j \qquad \text{so that} \qquad |n_j| < Nc_j.$$

Let $Q = qP - s_1 P_1 - \cdots - s_r P_r$. Then

$$NQ = n_1 P_1 + \cdots + n_r P_r$$

whence

$$|Q| \leqq \frac{1}{N} \sum |n_j P_j| < \frac{1}{N} \sum N c_j |P_j| \leqq \delta.$$

Hence $Q = O$, and $qP = s_1P_1 + \cdots + s_rP_r$. Hence the $(r+1)$-tuple $(q, s_1, \ldots, s_r)$ is a non-zero multiple of $(N, m_1, \ldots, m_r)$ and $|N| \leqq |q| \leqq c_0$. This proves Lemma 7.5.

Remark. Using the Dirichlet box principle as in [La 78] (corrected in [La 83a], Chapter 5, Theorem 7.6) is a simpler technique, but leads to a slightly less elegant inequality

$$[|m_i|^{1/r}] \leqq \delta^{-1} \sum_{j \neq i} |P_j|.$$

Lemma 7.6. *Let $P_1, \ldots, P_r \in \Gamma$ be linearly independent mod Γ_0. Then there exists a basis $\{P_1', \ldots, P_r'\}$ of Γ/Γ_0 such that*

$$|P_i'| \leqq \sum_{j=1}^{i} |P_j|.$$

Proof. The standard arguments of algebra give the desired estimated. I reproduce these arguments. Without loss of generality we can assume $\Gamma_0 = 0$, by working in the factor group Γ/Γ_0. Let $\Gamma_1 = \langle P_1, \ldots, P_r \rangle$. Let N be the index $(\Gamma : \Gamma_1)$. Then $N\Gamma$ has finite index in Γ_1. Let $n_{j,j}$ be the smallest positive integer such that there exist integers $n_{j,1}, \ldots, n_{j,j-1}$ satisfy

$$n_{j,1}P_1 + \cdots + n_{j,j}P_j = NP_j' \qquad \text{with some} \quad P_j' \in \Gamma.$$

Without loss of generality we may assume $0 \leqq n_{j,k} \leqq N - 1$. Then the elements $P_1', \ldots, P_r'$ form the desired basis.

Examples. The most important for us are the multiplicative group and abelian varieties. If Γ is a finitely generated multiplicative group of elements in a number field K, then a point P in Γ is a non-zero element α, and we define as in [La 64]

$$|P| = h(\alpha) \quad \text{where } h \text{ is the absolute height.}$$

On an abelian variety over a number field, we let Γ be a finitely generated group of algebraic points, and we let

$$|P| = \hat{h}(P)^{1/2}$$

where $\hat{h}$ is the Néron–Tate height associated with an even ample divisor class.

Next we have an example having to do with estimating relations. Let $\Gamma = \langle P_1, \ldots, P_n \rangle$ as above, and let R be the $\mathbf{Z}$-module of relations

$$M = (m_1, \ldots, m_n) \mod \Gamma_0$$

with $m_i \in \mathbf{Z}$, that is, n-tuples of integers satisfying $\sum m_i P_i \in \Gamma_0$. We put the sup norm on R, denoted by $\|M\| = \max|m_i|$.

Proposition 7.7. *There exists a* **Z**-*basis for* R *such that all elements* M *in the basis satisfy the bound*

$$\|M\| \leqq rB^r \qquad \textit{where} \quad B = \delta^{-1} \sum_{i=1}^{r} |P_i|.$$

Proof. By Proposition 7.5, for each $j = r + 1, \ldots, n$ there exists a relation between $P_1, \ldots, P_r, P_j$ satisfying the bound

$$\|M_j\|^{1/r} \leqq B.$$

The proof is concluded by applying Lemma 7.6.

In the applications one needs minima for the heights of algebraic numbers of bounded degree not equal to roots of unity, and one needs minimal heights for the non-torsion points on abelian varieties over number fields. For elliptic curves and abelian varieties see [Mass 84], and for the multiplicative group see the best known result in Dobrowolski [Do 79], in the direction of a conjecture of Lehmer.

Lehmer's conjecture. *Let* h *denote the absolute height on* $\mathbf{Q}^{\mathrm{a}}$. *There exists a constant* $c > 0$ *such that for all algebraic numbers* α *not equal to* 0 *or to a root of unity, we have*

$$h(\alpha) \geqq c/[\mathbf{Q}(\alpha) : \mathbf{Q}].$$

Taking into account the weaker results proved in this direction, Theorem 7.4 is a special case of Proposition 7.7.

CHAPTER X

Existence of (Many) Rational Points

In most of the book, we have dealt with cases when the main idea was to show the existence of few rational points. Roughly speaking this situation prevailed when the canonical class is ample. There is an opposite situation, when minus the canonical class (also called the **anti-** or **co-canonical class**) is ample, and one expects lots of rational points if there is one. Then one can propose ways of counting them asymptotically, when ordered by ascending height.

To me at the moment, one striking aspect of this direction is that it connects with an older idea of Artin that when the number of variables n is greater than the degree d (say of a homogeneous polynomial), then one expects the polynomial equation to have non-trivial solutions in various contexts. The $n > d$ condition appears today as precisely the condition which insures an ample co-canonical class. How many solutions may be measured in various ways. One possibility is to determine the extent to which a variety is **unirational**, i.e. the rational image of a projective space or is generically fibered by linear group varieties as in **Sp 2**, **Sp 3** describing the special set of Chapter I, §3. Artin's conjectures (and those cases when they have been proved) remain to be looked at in this context. Other ways may be given via a zeta function.

To some extent the present chapter may be viewed as dealing with the case when the special set defined in Chapter I, §3 is the whole variety, or conjecturally when a variety is not pseudo canonical. To the three possibilities canonical, very canonical and pseudo canonical can now be added the prefix anti, with the problem of determining the structure of the variety under this opposite set of conditions.

Conditions for the existence of a global rational point involve not only a global hypothesis such as the ampleness of the anti-canonical class, but also the determination of conditions under which the Hasse principle holds: if there exist (non-singular) rational points locally in every completion, then there exists a global (non-singular) rational point. We shall deal with some cases, but results in this direction still appear to me fragmentary, except for those pertaining to linear algebraic groups. For various reasons I have chosen to end the book before going fully into that direction. Aside from some reasons given at the end of §2, it seems to me that this direction would fit better in an encyclopedia volume dealing wholly with linear algebraic groups.

I am much indebted to Manin and Colliot-Thélène for valuable suggestions concerning this chapter.

X, §1. FORMS IN MANY VARIABLES

Artin called a field F **quasi-algebraically closed** if every form (homogeneous polynomial) of degree d in n variables with $n > d$ and coefficients in F has a non-trivial zero in F. More generally, F is said to be $\boldsymbol{C_i}$ if every form of degree d in n variables with $n > d^i$ has a non-trivial zero in F. Thus C_1 is the same as quasi-algebraically closed. Instead of considering homogeneous polynomials, one may also consider polynomials without constant terms. Such polynomials f always have the **trivial zero**

$$f(0, \dots, 0) = 0.$$

A field satisfying the analogous properties for such polynomials will be called **strongly quasi-algebraically closed**, resp. **strongly** C_i. In practice, whenever a field has been proved to be C_i, it has also been proved to be strongly C_i.

If a field is C_i then every finite extension is C_i. Also in practice if F is C_i then the rational function field $F(t)$ in one variable is C_{i+1}. Cf. [La 51].

The situation with C_i fields provided to my knowledge the first example whereby diophantine problems over a finite extension are reduced to a lower field by what became later known as **restriction of scalars**. Indeed, if we pick a basis $\{\alpha_1, \dots, \alpha_m\}$ for a finite extension F' of F, then a system of r polynomial equations in n unknowns over F' is equivalent to a system of $r[F' : F]$ equations in $n[F' : F]$ unknowns over F itself, by writing a variable T over F' as a linear combination

$$T = t_1\alpha_1 + \cdots + t_m\alpha_m$$

with variables $t_1, \dots, t_m$ in F.

Furthermore, supposing F' separable over F for simplicity, let the **norm form** be

$$N(t_1, \dots, t_m) = \prod_\sigma (t_1\alpha_1^\sigma + \cdots + t_m\alpha_m^\sigma),$$

where the product is taken over all conjugates σ of F' over F. Then $N(t_1,\ldots,t_m)$ is a form in m variables, of degree m, with coefficients in F, and having only the trivial zero in F.

Suppose F has a discrete valuation, with ring of integers R and prime element π. Let $f(t_1,\ldots,t_m)$ be a form of degree m in m variables in $R[t_1,\ldots,t_m]$ having only the trivial zero mod π. Let

$$t^{(1)} = (t_1^{(1)},\ldots,t_m^{(1)}),\ \ldots,t^{(m)} = (t_1^{(m)},\ldots,t_m^{(m)})$$

be m independent sets of m independent variables. Let

$$g(t^{(1)},\ldots,t^{(m)}) = f(t^{(1)}) + \pi f(t^{(2)}) + \cdots + \pi^{m-1}f(t^{(m)}).$$

Then g is a form of degree m in m^2 variables with coefficients in F and having only the trivial zero. So from this point of view, the conditions $d = n$ and $d^2 = n$ form natural boundaries for the property that if the number of variables is sufficiently large compared to the degree then there exists a non-trivial zero.

In the thirties, Tsen proved that a function field F in one variable over an algebraically closed constant field has no non-trivial division algebra of finite dimension over F. Artin noted that his method of proof implied something stronger, which in Artin's terminology states that such a field is quasi-algebraically closed. In light of Wedderburn's theorem that finite fields admit no non-trivial division algebras over them, Artin conjectured that a finite field is C_1. This was proved by Chevalley [Che 35].

The diophantine property of being C_1 implies the non-existence of division algebras as above: if E is a finite extension of F and $N(t_1,\ldots,t_n)$ is the norm form, then one applies the C_1 property to the equation

$$N(t_1,\ldots,t_n) = at_{n+1}^n \qquad \text{with} \quad a \in F, \quad a \neq 0$$

to show that every element of F is a norm from E, whence the non-existence of the division algebras. There was no reason to believe the converse, and Ax found an example of a field F such that every finite extension of F is cyclic, but F is not C_1 [Ax 66]. In cohomological terms, we have the implication:

If F is C_1 then $H^2(G_F, F^{a*}) = 0$

but not the converse.

Artin also conjectured that certain fields obtained by adjoining certain roots of unity are C_1, both globally and locally. Locally, let k be a p-adic field (finite extension of $\mathbf{Q}_p$) or a power series field in one variable over a finite constant field of characteristic p. Let F be the maximal unramified extension of k. One may characterize F as the field obtained by adjoining to k all n-th roots of unity with n prime to p. Then I proved Artin's conjecture that F is quasi-algebraically closed. More generally [La 51]:

Theorem 1.1. *Let F be a field complete under a discrete valuation with perfect residue class field. Then:*

(a) *The maximal unramified extension of F is C_1.*
(b) *If the residue class field is algebraically closed, then F is C_1. In particular, the field of formal power series over an algebraically closed constant field is C_1.*
(c) *The field of convergent power series over* **C** (say) *is C_1.*

Looking back with today's perspective, I realize that it is not known whether the algebraic set defined over such fields as in Theorem 1.1 (or global fields which are C_1 conjecturally) by a homogeneous polynomial with $n > d$ contains a rational curve, and what is the dimension of the largest unirational subvariety. Looking at these old questions from this point of view would give them new life.

Artin conjectured that a p-adic field and the power series in one variable over a finite field are C_2. I proved the case of power series in [La 51], but the conjecture for p-adic fields turned out to be false by examples of Terjanian [Ter 66], [Ter 77]. On the other hand, Ax–Kochen proved [AxK 65]:

Theorem 1.2. *For each positive integer d there exists a finite set of primes $S(d)$, such that if $p \notin S(d)$ then every polynomial $f \in \mathbf{Q}_p[T_1, \ldots, T_n]$ of degree d in n variables with $n > d^2$ has a non-trivial zero in $\mathbf{Q}_p$.*

A special case of a conjecture of Kato–Kuzumaki [KatK 86] states that a modification of the C_2 property holds, asserting the existence of a 0-cycle of degree 1 rational over a p-adic field if $n > d^2$. This is proved for forms of prime degree. (Note that a more general conjecture over arbitrary fields has been shown to be false by Merkuriev.)

When $d = 3$, Demjanov and Lewis proved that a cubic form in 10 variables over a p-adic field has a non-trivial zero [Lew 52].

Global results are still fragmentary. Davenport worked on the problem of the existence of non-trivial zeros of cubic forms over the rationals, and got the number of variables down to 16 [Dav 63]. Then Heath–Brown got it down to 10 for non-singular forms [H–B 83], and Hooley got it down to 9 if the form is non-singular and has a non-trivial zero in every p-adic field (see Hasse's principle below) [Ho 88]. It is still expected that a cubic form in 10 variables over **Q** has a non-trivial zero in **Q**.

For any degree, over number fields, Peck [Pe 49] proved:

Theorem 1.3. *Over a totally imaginary number field F, a form of degree d in n variables has a non-trivial zero in F if n is sufficiently large with respect to d. ["Totally imaginary" means F has no imbedding in the real numbers.]*

For forms of odd degree, Birch eliminated the restriction that the field be totally imaginary [Bir 57]. These theorems establish some special cases of **Hasse's principle**, which states for a variety, or an algebraic set X:

If $X(F_v)$ is not empty for each absolute value v on F, then $X(F)$ is not empty. In other words, if X has a rational point in every completion of F, then X has a rational point in F.

The problem is to determine which X satisfy the Hasse principle, and if it is not satisfied, what are the obstructions for its satisfaction. Experience (starting with Birch [Bir 61]) shows that it may be more useful to deal with the **non-singular Hasse principle**, where one assumes the existence of a simple rational point in each completion F_v, and one then concludes the existence of a simple global rational point in F.

The above mentioned authors use refinements and variations of the **Hardy–Littlewood circle method**. Roughly speaking, the circle method applies to complete intersections. Birch [Bir 61] showed

Theorem 1.4. *Let F be a number field and let X be the hypersurface defined over F by a form of degree d in n variables. If n is sufficiently large with respect to d and with respect to the dimension of the set of singular points, then the non-singular Hasse principle holds for X.*

Thus Birch's method applies for forms of even or odd degree, and contains the previous results of Peck and himself as mentioned above. In addition, the non-singular Hasse principle of [Bir 61] worked with fewer variables than in [Bir 57]. The method also gives an estimate on the number of rational points of bounded height, which is important in the context of §4. See Conjecture 4.3.

For an exposition of the circle method, see Davenport [Da 62], Schmidt [Schm 84], and Vaughan [Vau 81]. Adelic versions are given at a more sophisticated level by Lachaud [Lac 82], Danset [Dan 85], and Patterson [Pat 85].

Greenleaf [Grlf 65] proved my conjecture:

Theorem 1.5. *Let f be a polynomial of degree d in n variables with zero constant term and $n > d$ over a number field F. Then f has a non-trivial zero in all but a finite number of v-adic completions F_v.*

Note that in all results of the above type, no hypothesis of geometric irreducibility is made. The role of such irreducibility and of non-singularity is not completely cleared up. For instance, when I conjectured Greenleaf's theorem [La 60b], I had already observed:

Remark 1.6. *A variety over a number field has a rational point in all but a finite number of completions.*

Greenleaf reduced Theorem 1.5 to the above remark. Remember that for us, a variety is geometrically irreducible. For the proof of the remark, cutting the variety with sufficiently general hyperplane sections we are reduced to the case of dimension 1, that is, the case of curves. Then by Weil's Riemann hypothesis in function fields, the curve has a non-singular point mod $\mathfrak{p}$ for all but a finite number of primes $\mathfrak{p}$, whence a $\mathfrak{p}$-adic point by variants of Hensel's lemma.

Over a complete field, the existence of one simple rational point implies the existence of a whole neighborhood in the topology defined by the field, again by using a refinement method, e.g. Newton's approximation method. If for instance the variety is defined by one equation

$$f(x, y) = f(x_1, \dots, x_n, y) = 0,$$

with a rational point (x, y) such that $D_2 f(x, y) \neq 0$, then for all values $\bar{x}$ close to x, there exists a value $\bar{y}$ close to y such that $f(\bar{x}, \bar{y}) = 0$. We shall go deeper into this phenomenon in §3.

Remark 1.7. *The property for a complete non-singular variety to have a rational point is a birational invariant* (cf. [La 54]). Indeed, let X be a complete variety over a field k with a simple point $P \in X(k)$. Then there exists a k-valued place of the function field $k(X)$ lying above P, and this place induces a point on every other complete variety birationally equivalent to X over k, that is, having the same function field.

From a totally different method, see also [CTSal 89], Colliot-Thélène, Sansuc and Swinnerton-Dyer proved a sharp result along the statements of this section, namely:

Theorem 1.8 ([CTSSD 87]). *Let k be a totally imaginary number field. Then two quadratic forms over k with at least 9 variables have a non-trivial common zero over k.*

Up to technical details, for an arbitrary number field k, the obstructions arising from the real places are the only ones. The result of Theorem 1.8 is the analogue of Meyer's classical result over $\mathbf{Q}$ that a quadratic form in 5 variables which has a non-trivial zero in all real completions has a global non-trivial zero (Hasse over number fields).

Igusa has introduced a zeta function in connection with the problem of finding a zero for a form with many variables [Ig 78], but it has not so far led to the sharp conjectured results, for instance for cubic forms.

Finally we mention Artin's global conjecture that the field $\mathbf{Q}(\boldsymbol{\mu})$ obtained by adjoining all roots of unity to the rationals is C_1. For non-singular cubic surfaces, and more generally non-singular projective rational varieties, Kanevsky [Kan 85] establishes a connection with the question whether the Manin obstruction to the Hasse principle is the only one (see his Theorem 3). We shall discuss the Manin obstruction in §2. One of Kanevsky's results is as follows.

Theorem 1.9. *Let X_F be a non-singular cubic surface over a number field F. Assume that for every finite extension E of F the Manin obstruction to the Hasse principle for X_E is the only one. Then X_F has a rational point in an abelian extension of F.*

X, §2. THE BRAUER GROUP OF A VARIETY AND MANIN'S OBSTRUCTION

All known counterexamples to Hasse's principle for a variety X are accounted for by an obstruction defined by Manin [Man 70], where he shows that a generalization of the Brauer group, the so-called Brauer–Grothendieck group, gives a general obstruction which:

when applied to abelian varieties leads to the Shafarevich–Tate group;
when applied to rational varieties leads to the above mentioned obstruction.

For many results obtained in this direction and similar ones, as well as bibliographies, I refer to the appendix of Manin's book [Ma 74], Second Edition 1986; to the survey by Colliot-Thélène [C-T 86]; to the survey by Manin–Tsfasman [MaT 86]; and to the extensive survey [CTKS 85].

In this section I shall go into greater detail into the Brauer group and the Manin obstruction. I am much indebted to Colliot-Thélène for his guidance in writing this section, both for explaining theorems and for indicating the literature.

We shall use the cohomology of groups, in particular over p-adic fields and number fields. I recommend Shatz [Shat 72] as a reference for proofs of basic or local facts, and Artin–Tate [AT 68] for global facts.

Let F be a field of characteristic 0. By definition, the **Brauer group** $\mathrm{Br}(F)$ is

$$\mathrm{Br}(F) = H^2(G_F, \mathbf{G}_m(F^a)),$$

where $\mathbf{G}_m$ is the multiplicative group, and $\mathbf{G}_m(F^a)$ is the multiplicative group of the algebraic closure of F.

Local remarks

Suppose first that $F = F_v$ is complete under a discrete valuation v. Let:

F_v^{nr} be the maximal unramified extension of F_v;

$G^{\mathrm{nr}} = \mathrm{Gal}(F_v^{\mathrm{nr}}/F_v)$ be the Galois group of this maximal unramified extension;

I = inertia group = $\mathrm{Gal}(F_v^{\mathrm{a}}/F_v^{\mathrm{nr}})$.

Thus we have a tower of fields:

$$G_{F_v}\left\{\begin{array}{l} F_v^a \\ \Big| \Big\} I \\ F_v^{\mathrm{nr}} \\ \Big| \Big\} G^{\mathrm{nr}} \\ F_v \end{array}\right.$$

The inflation-restriction exact sequence of Galois cohomology reads:

$$(1)\quad 0 \to H^2(G^{\mathrm{nr}}, \mathbf{G}_m(F_v^{\mathrm{nr}})) \xrightarrow{\mathrm{inf}} H^2(G_{F_v}, \mathbf{G}_m(F_v^{\mathrm{a}})) \xrightarrow{\mathrm{res}} H^2(I, \mathbf{G}_m(F_v^{\mathrm{a}})) = 0.$$

The term 0 on the right has various justifications, one of them being Theorem 1.1(a). Thus the inflation gives an isomorphism.

In addition, the order at the valuation gives a homomorphism

$$\mathrm{ord}_v \colon \mathbf{G}_m(F_v^{\mathrm{nr}}) \to \mathbf{Z},$$

whence an induced homomorphism on the cohomology

$$(2)\qquad H^2(\mathrm{ord}_v)\colon H^2(G^{\mathrm{nr}}, \mathbf{G}_m(F_v^{\mathrm{nr}})) \to H^2(G^{\mathrm{nr}}, \mathbf{Z}).$$

From the short exact sequence $0 \to \mathbf{Z} \to \mathbf{Q} \to \mathbf{Q}/\mathbf{Z} \to 0$ one gets a natural isomorphism (the inverse of the coboundary)

$$H^2(G^{\mathrm{nr}}, \mathbf{Z}) \xrightarrow{\approx} H^1(G^{\mathrm{nr}}, \mathbf{Q}/\mathbf{Z}).$$

Hence we can view (2) also as a homomorphism of $H^2(G^{\mathrm{nr}}, \mathbf{G}_m(F_v^{\mathrm{nr}}))$ into $H^1(G^{\mathrm{nr}}, \mathbf{Q}/\mathbf{Z})$, which is just the group of characters of G^{nr} with values in $\mathbf{Q}/\mathbf{Z}$. Composing the inverse isomorphism of the inflation-restriction sequence in (1) and the homomorphism of (2), we obtain a homomorphism which we denote by

$$\partial_v \colon \mathrm{Br}(F_v) \to H^1(G^{\mathrm{nr}}, \mathbf{Q}/\mathbf{Z}).$$

We define an element $b \in \mathrm{Br}(F_v)$ to be **unramified** if $\partial_v b = 0$, in other words the image of this element under ∂_v is 0.

The global case

Next let X be a variety, non-singular in codimension 1, and defined over a field of characteristic 0. We let $F = k(X)$ be its function field. We define the **Brauer group of** X to be a certain subgroup of $\mathrm{Br}(F)$ as follows. For each discrete valuation v of $k(X)$ given by a point of X of codimension 1, or equivalently a subvariety of codimension 1, we can apply the local remarks to the completion F_v. Corresponding to the inclusion $F \subset F_v$ we have a natural homomorphism

$$\mathrm{Br}(F) \to \mathrm{Br}(F_v).$$

If $b \in \mathrm{Br}(F)$ then the image of b in $\mathrm{Br}(F_v)$ is unramified for all but a finite number of v. We say that b is **unramified** if its image is unramified for all v. The unramified elements of $\mathrm{Br}(F)$ form a subgroup which we denote by $\mathrm{Br}(X)$, and which we call the **Brauer group of** X.

For X projective non-singular, the group $\mathrm{Br}(X)$ *is a birational invariant*, in the sense that if Y is another projective non-singular variety over k with the same function field F, then $\mathrm{Br}(X)$ and $\mathrm{Br}(Y)$ are the same subgroup of $\mathrm{Br}(F)$. We denote this group by $\mathrm{Br}^{\mathrm{nr}}(F)$, and call it the **unramified Brauer group** or **Brauer–Grothendieck group**.

In addition, for each rational point $P \in X(k)$ we have a **specialization homomorphism**

$$\mathrm{Br}(X) \to \mathrm{Br}(k) \qquad \text{denoted by} \qquad b \mapsto b(P).$$

In other words, each rational point on X allows us to specialize an element of the unramified Brauer group of the function field to an element of the Brauer group of the constant field k. For proofs of the above two properties, see Grothendieck [Grot 68].

We are now ready to define the Manin obstruction to the Hasse principle. Let X be a projective non-singular variety over a number field k. We recall some facts about Brauer groups over k. We use a fact essentially from local class field theory that there is a natural injection, the **invariant** for each absolute value v on k,

$$\mathrm{inv}_v \colon \mathrm{Br}(k_v) \hookrightarrow \mathbf{Q}/\mathbf{Z},$$

which is an isomorphism for v finite, trivial if $k_v = \mathbf{C}$, and maps $\mathrm{Br}(k_v)$ on $\{0, 1/2\}$ if v is real. If $b_0 \in \mathrm{Br}(k)$, then a classical theorem of Albert–

Brauer–Hasse–Noether states that

$$\sum_v \mathrm{inv}_v(b_0) = 0.$$

For proofs fitting the present context, see [Shat 72] and [AT 68]. If there exists a rational point $P \in X(k)$, then for all $b \in \mathrm{Br}(X)$ we have

$$\sum_v \mathrm{inv}_v\, b(P) = 0.$$

In particular:

Assume that $X(k_v)$ is not empty for all v. If for all elements $\{P_v\}$ in $\prod_v X(k_v)$ there exists $b \in \mathrm{Br}(X)$ such that

$$\sum_v \mathrm{inv}_v\, b(P_v) \neq 0,$$

then $X(k)$ is empty.

The existence of elements $b \in \mathrm{Br}(X)$ having the above property will be called the **Manin obstruction** to the Hasse principle. We shall say that there is **no Manin obstruction to the Hasse principle for** X if there exists a family

$$\{P_v\} \in \prod_v X(k_v)$$

such that for all $b \in \mathrm{Br}(X)$ we have

$$\sum_v \mathrm{inv}_v\, b(P_v) = 0.$$

Theorem 2.1 (Manin [Man 74], consequence of Chapter VI, §41, p. 228, Theorem 41.24). *Let X be a curve of genus 1 over a number field k. Assume that $\text{Ш}(J(X))$ is finite. Then the Manin obstruction for X to the Hasse principle is the only one, in other words, if there is no Manin obstruction, then there exists a global rational point $P \in X(k)$.*

Actually, the existence of the Manin obstruction for curves of genus 1 is hard to verify, but is easier for varieties which are birationally equivalent to projective space over the algebraic closure k^{a}. Examples of Manin obstructions are given in Iskovskih [Isk 71], for instance the surface defined by the affine equation

$$y^2 + z^2 = (3 - x^2)(x^2 - 2),$$

which has non-singular rational points in $\mathbf{Q}_v$ for all v, but such that no projective desingularization of this surface has a rational point in $\mathbf{Q}$. For generalizations of this example and further discussions, see [CTCS 80].

In addition, one finds in [CKS 85] an example, due to Cassels–Guy, of a diagonal surface which has a p-adic point for all p but no global rational point, namely:

$$5x^3 + 9y^3 + 10z^3 + 12t^3 = 0.$$

[CKS 85] gives an algorithm such that if Manin's obstruction to the Hasse principle is the only one, then one can decide whether there is a rational point on a diagonal cubic surface over $\mathbf{Q}$.

As Colliot-Thélène has pointed out to me, it is unlikely that the Manin obstruction is the only one for all projective non-singular varieties X. Indeed, over a number field k, if that were the case, then for all non-singular complete intersections in $\mathbf{P}_k^n$ of dimension $\geqq 3$, the Hasse principle would be true, since for such X,

$$\mathrm{Br}(X)/\mathrm{Br}(k) = 0.$$

This is not expected, because for instance the Hasse principle would then hold for a non-singular hypersurface

$$f(x_0, \ldots, x_4) = 0$$

of any degree. But for high degree a general such hypersurface is conjectured to be hyperbolic, and thus has a tendency not to have rational points at all, whereas the presence of local rational points is not expected to be so rare. An actual example is lacking at this time, however.

Colliot-Thélène has proposed a more likely conjecture, namely:

The Manin obstruction formulated for 0*-cycles of degree* 1*, rational over a number field* k*, is the only obstruction to the existence of a global rational* 0*-cycle of degree* 1.

There have been some results of Saito [Sai 89] concerning curves. Let X_F be a complete non-singular curve over a number field F, and let X over spec$(\mathfrak{o}_F)$ be a regular proper model of X_F. Thus X is a regular 2-dimensional scheme. For such a scheme, one can define the **Brauer group** $\mathrm{Br}(X)$ exactly as we did for a non-singular variety. The constant field never played an essential role. Saito proves:

Theorem 2.2. *Assume that* $\mathrm{Br}(X)$ *is finite, that* X_F *has a* 0*-cycle of degree* 1 *rational over each completion* F_v *of* F*, and that the Manin obstruction for a* 0*-cycle vanishes. Then there exists a* 0*-cycle on* X_F *rational over* F *and of degree* 1.

Note that the hypothesis on the finiteness of the Brauer group is essentially equivalent to the finiteness of the Shafarevich–Tate group of the

Jacobian of X_F; and when the curve has genus 1, then one gets a theorem of Manin (see Theorem 2.1).

By methods of algebraic K-theory, initiated by Spencer Bloch, Salberger [Sal 88] showed that for a non-singular surface (projective variety of dimension 2) over a number field, fibered by conics over the projective line, the Manin obstruction for the existence of a 0-cycle of degree 1 is the only one. Thus in recent years, the theory of such 0-cycles has developed in various directions.

There is another interpretation of the Manin obstruction. Again we take X projective non-singular over k. For convenience we write X^{a} for the base change $X_{k^{\mathrm{a}}}$, so X^{a} is the extension of X to the algebraic closure of k. Then there is a long exact sequence (see [CTS 87] p. 466):

$$0 \to \mathrm{Pic}(X) \to \mathrm{Pic}(X^{\mathrm{a}})^{G_k} \to \mathrm{Br}(k) \to$$
$$\mathrm{Ker}\big(\mathrm{Br}(X) \to \mathrm{Br}(X^{\mathrm{a}})\big) \to H^1(G_k, \mathrm{Pic}\, X^{\mathrm{a}}) \to H^3\big(G_k, \mathbf{G}_m(k^{\mathrm{a}})\big).$$

To shorten the notation we have written $\mathrm{Pic}\, X^{\mathrm{a}}$ instead of $(\mathrm{Pic}\, X^{\mathrm{a}})(k^{\mathrm{a}})$. If k is a number field, then by a theorem of Tate ([AT 68], Chapter 7, Theorem 14) we have

$$H^3\big(G_k, \mathbf{G}_m(k^{\mathrm{a}})\big) = 0.$$

Hence modulo a constant part which is irrelevant for the Manin obstruction, we see that $\mathrm{Ker}\big(\mathrm{Br}(X) \to \mathrm{Br}(X^{\mathrm{a}})\big)$ is the same as $H^1(G_k, \mathrm{Pic}\, X^{\mathrm{a}})$, and the Manin obstruction can be interpreted in terms of the Picard group. For a discussion of this interpretation see [Man 70b] and [Man 74]. It would be desirable to have a book giving systematically the general properties of the Brauer group, as well as its applications to unirational varieties.

To a large (if not exclusive) extent, this book has been concerned with complete varieties, although on occasions we have seen that non-complete varieties play an essential role, as in the Néron model of an abelian variety. Given the size of the book and mostly my incompetence, I do not expand the book to include the diophantine theory of linear algebraic groups or group varieties in general. However, I conclude this section by emphasizing the existence of this theory. The theory of Ш (the Shafarevich–Tate group), the Brauer group, the Manin obstruction, extend to linear group varieties. That the Hasse principle is valid for principal homogeneous spaces of semisimple simply connected linear group varieties is due principally to Kneser and Harder. See for instance [Kn 69]. A bibliography of several basic papers of Kneser and Harder is given by Sansuc [San 81], who develops these theories systematically, including the theory of the Manin obstruction. I cite one of his results, his Corollary 8.7. (An exceptional case need not be mentioned because of a more recent result of Chernoussov.)

Theorem 2.3. *Let G be a linear group variety over a number field F. Let V be a principal homogeneous F-space of G, and let X be a non-singular completion of V over F. Then the Manin obstruction for V and for X is the only obstruction to the Hasse principle for V and X respectively.*

Actually, the proof for V goes through the proof for the completion X, using some commutative diagrams.

The descent method of Chapter III, §4 extends to other varieties. Colliot-Thélène and Sansuc's idea is that one may replace the use of isogenies (which are principal homogeneous spaces under finite abelian groups) by the use of principal homogeneous spaces under tori. This method can then be applied to reduce the study of rational points on a variety over F which becomes rational over F^{a} to the study of rational points on auxiliary varieties for which the Manin obstruction vanishes. The first examples date back to Châtelet. A general theory was developed by Colliot-Thélène and Sansuc [CTS 87], and many special cases have been studied. Theorem 1.8 fits among them.

The theory of homogeneous spaces for linear groups and the Galois cohomology with coefficients in the (linear) group of automorphisms of a variety are also used in an essential way to study the problems which arise concerning projective non-singular varieties which do not necessarily have a rational point, and become isomorphic to projective space, or are k-birationally equivalent to principal homogeneous spaces under linear algebraic group varieties. For such varieties, one raises the question whether they satisfy Hasse's principle. By definition, a **Severi–Brauer** variety over a field k is a variety which becomes isomorphic to a projective space over a finite extension of k. In his thesis [Chat 44], Châtelet generalized to such varieties what was known before on conics, and in particular:

Theorem 2.4 (Châtelet). *Let X be a Severi–Brauer variety over a number field. Then X satisfies Hasse's principle. If X is a Severi–Brauer variety over a finite field, then X is isomorphic to projective space over this field.*

As Colliot-Thélène pointed out, in light of Châtelet's basic results, the terminology would have been more appropriate to call the varieties in question **Severi–Châtelet**. For a survey of Châtelet's contributions in this direction, I refer to [C-T 88].

Example (*The Châtelet surface*). An example exhibiting non-trivial diophantine properties from the present point of view is the **Châtelet surface** $V = V_{a,P}$ defined over a field k by the affine equation

$$y^2 - az^2 = P(x),$$

with $a \in k$, $a \neq 0$, and $P \in k[x]$ is a polynomial without multiple roots. For simplicity we assume that k has characteristic 0. Thus $V_{a,P}$ is an affine non-singular surface. In [CTSSD 87] the reader will find a simple description of a projective non-singular completion of V, which we denote by $X_{a,P}$. There is a morphism

$$f\colon X_{a,P} \to \mathbf{P}^1$$

projecting on the x-line, whose generic fiber is a curve of genus 0. For degree $P = 3$ or 4, a result of Iskovskih can be formulated as saying that $-K_X$ is pseudo very ample, and the image of the birational imbedding into $\mathbf{P}^4$ can be described explicitly, cf. [Izk 72]. In [CTSSD 87] one will find a theory of the Châtelet surface *when P has degree* $\leqq 4$, containing in particular the next two theorems.

Theorem 2.5. *If there exists a k-rational point on $V_{a,P}$ then $V_{a,P}$ is k-unirational.*

The above fact is complemented over number fields by the following result.

Theorem 2.6. *Suppose k is a number field. Then:*

(a) *The Manin obstruction to the Hasse principle for $X_{a,P}$ is the only one.*

(b) *If P is irreducible over k then* $\mathrm{Br}(X)/\mathrm{Br}(k) = 0$, *and the Hasse principle is true for $V_{a,P}$ and $X_{a,P}$.*

When $k = \mathbf{Q}$, Colliot-Thélène and Sansuc [CTS 82] have shown that the non-singular Hasse principle also holds for $X_{a,P}$ when P is an irreducible polynomial of any degree, provided a conjecture of Schinzel is true:

an irreducible polynomial with relatively prime integer coefficients, positive leading coefficient, and such that $P(\mathbf{Z})$ has no common prime factor, represents infinitely many primes.

This conjecture is a generalization of the conjecture that there are infinitely many primes of the form $n^2 + 1$. However, today when the degree of P is at least 8, say, and when the Châtelet surface over $\mathbf{Q}$ has solutions in $\mathbf{Q}$, one does not know if there exist such solutions for infinitely many rational values of x. See also [C-T 86] and [San 85] for a more extensive survey of results in this direction and a more extensive bibliography.

In line with Chapter I, **3.9**, **3.10** and **3.11**, we now have the following possibility:

2.7. *Let X be a projective non-singular variety over a number field k. Assume that $-K_X$ is pseudo ample. Is the Manin obstruction the only obstruction to the Hasse principle? (Perhaps only for cycles of degree* 1, *following Colliot-Thélène's conjecture.)*

The question also arises to what extent one would need $-K_X$ pseudo ample, in other words is there a natural condition on K_X which is sufficient or necessary so that the Hasse principle holds? How far back must one go to get such a condition, e.g. is the hypothesis that $-K_X$ contains some effective divisor sufficient? Because of Chapter I, 3.11, the hypothesis that $-K_X$ is pseudo ample would not cover the case of Manin's Theorem 2.1 concerning elliptic curves, but the weaker hypothesis would. I am lacking examples or counterexamples to make a coherent general conjecture.

X, §3. LOCAL SPECIALIZATION PRINCIPLE

Let $R_0 \subset R$ be a subring of an integral ring. We say that R_0 is **relatively algebraically closed in** R if given a non-zero polynomial $P(T) \in R_0[T]$ in one variable and coefficients in R_0, if α is a root of P in R then in fact $\alpha \in R_0$. Examples of such a pair first arose as follows.

Let k be a p-adic field or a power series field in one variable over a finite constant field of characteristic p. Let F_0 be the maximal unramified extension of k. We can characterize F_0 also as the field obtained by adjoining to k all n-th roots of unity with n prime to p. Let F be the completion of F_0. Then F_0 is relatively algebraically closed in F.

Let R_0 be the ring of convergent power series inside the ring of formal power series. Then R_0 is relatively algebraically closed in R.

I formulated the following principle for such a pair of rings (R_0, R), which are local rings such that R is the formal completion of R_0.

Local specialization principle. *Let*

$$f_j(T_1, \ldots, T_n) = 0 \qquad (j = 1, \ldots, r)$$

be a finite family of polynomial equations with coefficients in R_0. Suppose this family has a solution $(x_1, \ldots, x_n)$ with $x_i \in R$. Then these equations have a solution with $\bar{x}_i \in R_0$, such that $\bar{x}_i$ lies arbitrarily close to x_i in the local ring topology of R.

I proved this result in the case when the rings arise from absolute values in the context of complete fields, including the two cases mentioned above [La 51], namely:

> *Let k be a field complete under an absolute value. Let k_0 be a dense subfield, which is relatively algebraically closed in k, and such that k is separable over k_0. Then the above formulated specialization principle applies to the pair k_0 in k, where instead of the local ring topology, we use the topology defined by the absolute value.*

In particular, I proved the specialization principle for convergent power series in one variable inside the formal power series. My conjecture [La 54] that the result would also hold for power series in several variables was proved by M. Artin [Art 68]. Other cases of the specialization principle have been proved by Bosch [Bos 81], Robba [Rob 81a, b] and van den Dries [vanD 81].

In working with fields rather than the rings, or in a one-variable situation, to specialize the solution from R to R_0, I used the Newton approximation method, I believe for the first time in the context of such diophantine problems. Such a refined version for finding a zero from an approximate zero was needed because a solution of a system of equations might be singular modulo high powers of the maximal ideal, and the usual version of Hensel's lemma could not be applied.

A finite system of polynomial equations over a complete local ring amounts to an infinite system of equations in the residue class field (under mild conditions on the local ring). The typical case is that of power series, whereby a polynomial equation over the power series amounts to infinitely many equations among the infinitely many coefficients of those power series, or equivalently, amounts to a projective system of equations in these coefficients, obtained by truncating the equations modulo a power of the maximal ideal. This procedure which arose first in [La 51] was schematized and functorized by M. Greenberg, thus giving rise to the **Greenberg functor** [Grbg 61], [Grbg 63].

X, §4. ANTI-CANONICAL VARIETIES AND RATIONAL POINTS

As noted already in Chapter I, §3, if W is an algebraic set in $\mathbf{P}^{n-1}$ defined as a complete intersection by a system of homogeneous equations

$$f_j(T_1, \ldots, T_n) = 0, \qquad j = 1, \ldots, r,$$

of degree d_j, then the anti-canonical sheaf is $\mathcal{O}(n-d)$ where $d = \sum d_j$. Thus $-K_W$ is ample if and only if $n > d$. The theory of quasi-algebraic closure, to the extent it exists, developed long before people became conscious of the interpretation of the inequality $n > d$ in terms of the canonical sheaf. Contrary to the expectation of few, or only a finite number of rational points when the canonical class is ample, one expects either none or many rational points, and even possibly the unirationality of the algebraic set, when minus the canonical class is ample. I know only very few results in this direction, aside from Theorem 12.11 of Manin's book [Man 74]. One expects certain asymptotic estimates for the number of rational points of bounded height, extending and similar to Schanuel's theorem for projective space. Lower bounds in certain cases were obtained by Schmidt [Schm 85] by the circle method. We mostly rely on some conjectures to give an idea of what may go on.

Let X be a projective non-singular variety. We say that X is **anti-canonical** if $-K_X$ is ample. Such a variety is also called a **Fano variety**. For such a variety one expects an abundance of rational points if there is at least one, and we shall discuss these varieties.

Let X be a projective non-singular variety over a number field F. Let $c \in \operatorname{Pic}(X, F)$ be a divisor class, and let h_c be one of the height functions (defined mod $O(1)$) associated with c. Define:

$$N(X, c, B) = \text{number of points } P \in X(F) \text{ such that } h_c(P) \leqq \log B.$$

Following Arakelov [Ara 74b] and Faltings [Fa 84c], Theorem 8, Batyrev, Franke, Manin and Tschinkel [BaM 90] and [FrMT 89] have also considered a zeta function in the context of counting points. Let U be a Zariski open subset of a projective variety X over the number field F. Let H_c be the exponential height,

$$H_c = \exp h_c.$$

The **zeta function** is defined by

$$\zeta_U(c, s) = \sum_{P \in U(F)} H_c(P)^{-s}.$$

Of course one must normalize the height, as follows. Let $\mathscr{L}$ be a line sheaf corresponding to the class c. Let s be a rational section defined and non-zero at a given rational point P. Then

$$h_c(P) = \sum_v - \log\|s(P)\|_v,$$

where $\| \ \|_v$ is a norm on $\mathscr{L}_v$ such that $\|as\|_v = \|a\|_v\|s\|_v$ for $a \in F$, and $\|1\|_v = 1$ for almost all v. (These are the usual conditions.) In one case, having to do with generalized flag manifolds, Franke observed that this zeta function can be identified with a Langlands–Eisenstein series, which would thereby bring in the whole machinery of zeta functions of automorphic forms into diophantine analysis via this route. In general, Batyrev–Manin make some conjectures for which we need various definitions. Let

$$\beta_U(c) = \inf\{\sigma \in \mathbf{R} \text{ such that } \zeta_U(c, s) \text{ converges for } \mathrm{Re}(s) > \sigma\}.$$

Then

$$N(U, c, B) = O(B^{\beta+\varepsilon}) \qquad \text{for all} \quad \varepsilon > 0,$$
$$N(U, c, B) \neq O(B^{\beta-\varepsilon}) \qquad \text{for all} \quad \varepsilon > 0.$$

Furthermore:

The function $c \mapsto \beta_U(c)$ depends only on the class of c in the Néron–Severi group NS(X), *and β_U extends to a continuous function on the positive cone* $\mathbf{RNS}_+(X)$ *of ample elements in the Néron–Severi group.*

In addition, Batyrev–Manin define another function α as follows. Let

$\alpha(c) = \inf\{t \in \mathbf{Q}$ such that for some $n > 0$, $n(tc + K_X)$ is linearly effective, that is, the linear equivalence class contains an effective divisor$\}$.

Then again, $\alpha(c)$ depends only on the class of c in NS(X). Furthermore, if $\alpha(c) < 0$ for some ample class c, then X is pseudo canonical. Thirdly, as for the function β, the function α extends to a continuous function on $\mathbf{RNS}_+(X)$. *In case the rank of* NS(X) $= 1$, *we have:*

$$\alpha(c) = \begin{cases} -K_X/c & \text{if } K_X \neq 0 \\ 0 \text{ on } \mathbf{RNS}_+(X) & \text{otherwise.} \end{cases}$$

The main part of Batyrev–Manin lies in the following:

Conjecture 4.1. *For every $\varepsilon > 0$ there exists a Zariski open dense set $U = U(c, \varepsilon)$ of X such that*

$$\beta_U(c) < \alpha(c) + \varepsilon.$$

Conjecture 4.2. *If X is anti-canonical, for every sufficiently large number field F and for all sufficiently small non-empty Zariski open subsets U of X we have*

$$\beta_U(c) = \alpha(c).$$

(*Sufficiently large* means containing some fixed finite extension; *sufficiently small* means contained a fixed non-empty Zariski open subset.)

In particular, suppose K_X is ample, and take c to be an ample class. Then $\alpha(c)$ is negative, so $\beta_U(c)$ is negative, so $U(F)$ is finite for sufficiently small U. This is a special case of one of my conjectures expressed in Chapter I, **3.1**, which is also implied by Vojta' conjectured quantitative height inequalities.

In addition, Batyrev conjectured [Bat 90]:

Conjecture 4.3. *Let X be anti-canonical. Then as in Conjecture* 4.2, *for sufficiently large F and for sufficiently small U (which may depend on F), there exists a number $\gamma = \gamma(X) \geqq 0$ and an integer $r \geqq 0$ such that*

$$N(X, c, B) \sim \gamma B^{\alpha(c)}(\log B)^r,$$

where r is defined by the condition

$$r + 1 = \textit{codim of the face of effective cone where } \alpha(c).c + K_X \textit{ lies.}$$

The above asymptotic expression would set Schanuel's counting of points on projective space from Chapter II, §2 in a much wider setting. For further discussion, see Franke–Manin–Tschinkel [FrMT 89]. For non-singular complete intersections, Colliot-Thélène tells me that when the number of variables is sufficiently large with respect to the degree, the circle method gives estimates of the conjectured type.

Added in extremis in proofs: The circle method attributed to Hardy–Littlewood in the text has its origins in a letter from Ramanujan to Hardy. The relevant bibliography was pointed out to me by Ram Murty, namely: R. C. VAUGHN, *The Hardy–Littlewood method*, Cambridge University Press (1981), p. 3; K. RAMACHANDRA, Srinivasan Ramanujan (the inventor of the circle method), *J. Math. Phys. Sci.* **21** (1987), No. 6, pp. 545–565; and Ram Murty's review of this paper, *Math. Reviews*, 1989, **89e–11001**; Atle SELBERG, Reflections around the Ramanujan Centenary, in his *Collected Works*, Springer-Verlag, last paper, see especially pp. 698, 701, 702, 705, 706.

Bibliography

[Ad 66] W. ADAMS, Asymptotic diophantine approximations to *e*, *Proc. Nat. Acad. Sci. USA* **55** (1966) pp. 28–31

[Ad 67] W. ADAMS, Asymptotic diophantine approximations and Hurwitz numbers, *Amer. J. Math.* **89** (1967) pp. 1083–1108

[Ara 71] S. J. ARAKELOV, Families of algebraic curves with fixed degeneracies, *Izv. Akad. Nauk SSSR Ser. Mat.* **35** (1971) pp. 1269–1293; translation; *Math. USSR Izv.* **5** (1971) pp. 1277–1302

[Ara 74a] S. J. ARAKELOV, Intersection theory of divisors on an arithmetic surface, *Izv. Akad. Nauk SSSR Ser. Mat.* **38** No. 6 (1974) pp. 1179–1192; translation: *Math. USSR Izv.* **8** No. 6 (1974) pp. 1167–1180

[Ara 74b] S. J. ARAKELOV, Theory of intersections on an arithmetic surface, *Proc. International Congress of Mathematicians*, Vancouver 1974, Vol. 1, pp. 405–408

[Art 68] M. ARTIN, On the solutions of analytic equations, *Invent. Math.* **5** (1968) pp. 277–291

[Art 86] M. ARTIN, Néron models, in *Arithmetic geometry*, eds. G. Cornell, J. Silverman, Springer-Verlag, 1986, pp. 213–230

[ArW 71] M. ARTIN and G. WINTERS, Degenerate fibers and reduction of curves, *Topology* **10** (1971) pp. 373–383

[AT 68] E. ARTIN and J. TATE, *Class field theory*, Benjamin 1968

[Ax 66] J. AX, A field of cohomological dimension 1 which is not C_1, *Bull. Amer. Math. Soc.* **71** (1966) p. 717

[Ax 72] J. AX, Some topics in differential algebraic geometry II: on the zeros of theta functions, *Amer. J. Math.* **94** (1972) pp. 1205–1213

[AxK 65] J. AX and S. KOCHEN, Diophantine problems over local fields I, II, *Amer. J. Math.* **87** (1965) pp. 605–630

[AxK 66] J. AX and S. KOCHEN, Diophantine problems over local fields III, *Ann. of Math.* **83** (1966) pp. 437–456

[Ba 66+] A. BAKER, Linear forms in the logarithms of algebraic numbers I, II, III, IV, *Mathematika* **13** (1966) pp. 204–216; **14** (1967) pp. 102–107; **14** (1967) pp. 220–228; **15** (1968) pp. 204–216

[Ba 68] A. BAKER, Contributions to the theory of Diophantine equations: I On the representation of integers by binary forms; II The Diophantine equation $y^2 = x^3 + k$, *Phil. Trans. Roy. Soc. London* **263** (1968) pp. 173–208

[Ba 71] A. BAKER, Effective methods in Diophantine problems I, II, *Proc. Symposia Pure Math.* vol. **20** AMS (1971) pp. 195–205 and vol. **24**. pp. 1–7

[Ba 72+] A. BAKER, A sharpening of the bounds for linear forms in logarithms I, II, III, *Acta Arith.* **21** (1972) pp. 117–129; **24** (1973) pp. 33–36; 27 (1974) pp. 247–252

[Ba 77] A. BAKER, The theory of linear forms in logarithms, *Transcendence theory: advances and applications*, Academic Press, London 1977, pp. 1–27

[Bat 90] V. BATYREV, The cone of effective divisors of threefolds, preprint, (1990), to appear

[BaM 88] V. BATYREV and J. MANIN, Counting points on algebraic varieties, Progress report for the Soviet–American meeting, New York, 1988

[BaM 90] V. BATYREV and J. MANIN, Sur le nombre des points rationnels de hauteur bornée des variétés algébriques, *Math. Ann.* **286** (1990) pp. 27–43

[Be 85] A. BEILINSON, Height pairing between algebraic cycles, *Current trends in arithmetical algebraic geometry, Contemporary Mathematics* Vol. **67**, ed. Kenneth Ribet, AMS 1985

[Bir 57] B. J. BIRCH, Homogeneous forms of odd degree in a large number of variables, *Mathematika* **4** (1957) pp. 102–105

[Bir 61] B. J. BIRCH, Forms in many variables, *Proc. Roy. Soc. London Ser. A* (1961) pp. 245–263

[BiV 88a] J. M. BISMUT and E. VASSEROT, Comportement asymptotique de la torsion analytique associee aux puissances d'un fibré en droites positif, *C.R. Acad. Sci. Paris* **307** (1988) pp. 779–781

[BiV 88b] J. M. BISMUT and E. VASSEROT, The asymptotics of the Ray–Singer torsion associated with high powers of a positive line bundle, Preprint IHES 1988

[Bog 80] F. BOGOMOLOV, Points of finite order on an abelian variety, *Izv. Akad. Nauk SSSR Ser. Mat.* **44** (1980); translation: *Math. USSR Izv.* **17** No. 1 (1981) pp. 55–72

[Bom 90] E. BOMBIERI, The Mordell conjecture revisited, preprint

[Bos 81] S. BOSCH, A rigid analytic version of M. Artin's theorem on analytic equations, *Math. Ann.* **255** No. 3 (1981) pp. 395–404

[BLR 90] S. BOSCH, W. LÜTKEBOHMERT and M. RAYNAUD, *Néron models*, Ergebnisse der Math. **21**, Springer-Verlag 1990

[BoMM 89] J. B. BOST, J. F. MESTRE, L. MORET-BAILLY, Sur le calcul explicite des "classes de Chern" des surfaces arithmétiques de genre 2, preprint, *École Normale Sup.* Paris 1989

[Br 83] L. Breen, Fonctions thêta et théorème du cube, *Springer Lecture Notes* **980** (1983)

[BrM 86] D. Brownawell and D. Masser, Vanishing sums in function fields, *Math. Proc. Cambridge Philos. Soc.* **100** (1986) pp. 427–434

[BruM] A. Brumer and O. McGuinness, The behavior of the Mordell–Weil group of elliptic curves, *Bull. Amer. Math. Soc.* **23** No. 2 (1990) pp. 375–382

[Bry 64] A. D. Bryuno, Continued fraction expansion of algebraic numbers, *Zh. Vichisl Mat. i Mat. Fiz.* **4** nr. 2 (1964) pp. 211–221; AMS translation *USSR Comput. Math. Phys.* **4** (1964) pp. 1–15

[CaG 72] J. Carlson and P. Griffiths, A defect relation for equidimensional holomorphic mappings between algebraic varieties, *Ann. of Math.* **95** (1972) pp. 557–584

[Car 29] H. Cartan, Sur les zéros des combinations linéaires de p fonctions holomorphes données, *C.R. Acad. Sci. Paris* **189** (1929) pp. 521–523

[Car 33] H. Cartan, same title, giving proofs of the above, *Mathematika* **7** (1933) pp. 5–31

[Cas 62a] J. W. S. Cassels, Arithmetic on curves of genus 1 (III), The Tate–Shafarevich group, *Proc. London Math. Soc.* **12** (1962) pp. 259–296

[Cas 62b] J. W. S. Cassels, Arithmetic on curves of genus 1 (IV), Proof of the Hauptvermutung, *J. reine angew. Math.* **211** (1962), pp. 95–112

[Cas 63] J. W. S. Cassels, Arithmetic on curves of genus 1 (V), Two counter-examples, *J. London Math. Soc.* **38** (1963) pp. 244–248

[Cas 64a] J. W. S. Cassels, Arithmetic on curves of genus 1 (VI), The Tate–Shafarevich group can be arbitrarily large, *J. reine angew. Math.* **214–215** (1964) pp. 65–70

[Cas 64b] J. W. S. Cassels, Arithmetic on curves of genus 1 (VII), The dual exact sequence, *J. reine angew. Math.* **216** (1964) pp. 150–158

[Cas 65] J. W. S. Cassels, Arithmetic on curves of genus 1 (VIII). On the conjectures of Birch and Swinnerton-Dyer, *J. reine angew. Math.* **217** (1965) pp. 180–189

[Chab 38] C. Chabauty, Sur les équations diophantiennes liées aux unités d'un corps de nombres algébriques fini, *Annali di Math.* **17** (1938) pp. 127–168

[Chab 41] C. Chabauty, Sur les points rationnels des variétés algébriques dont l'irrégularité est supérieure à la dimension, *C.R. Acad. Sci. Paris* **212** (1941) pp. 1022–1024

[Chab 43a] C. Chabauty, Sur le théorème fondamental de la théorie des points entiers et pseudo-entiers des courbes algébriques, *C.R. Acad. Sci. Paris* **217** (1943) pp. 336–338

[Chab 43b] C. Chabauty, Démonstration de quelques lemmes de rehaussement, *C.R. Acad. Sci. Paris* **217** (1943) pp. 413–415 [These two 1943 papers are partly incorrect. See the comments in [La 78], p. 144.]

[Chai 85] C. L. CHAI, Compactification of Siegel modular schemes, *London Math. Soc. Lecture Notes Series* **107**, 1985

[Chai 90] C. L. CHAI, A note on Manin's theorem of the kernel, *Amer. J. Math.* to appear

[ChF 90] C. L. CHAI and G. FALTINGS, *Degeneration of abelian varieties*, Ergebnisse der Math. **22**, Springer-Verlag 1990

[Chat 44] F. CHÂTELET, Variations sur un thème de Poincaré, *Ann. Sci. École Norm. Sup.* **61** (1944) pp. 249–300

[Cher 90] W. CHERRY, Nevanlinna theory for coverings, in: Topics in Nevanlinna theory, eds. S. Lang, W. Cherry. *Springer Lecture Notes* **1433** (1990) pp. 113–174

[Cher 91] W. CHERRY, The Nevanlinna error term for coverings; the generically surjective case, to appear

[Chev 35] C. CHEVALLEY, Démonstration d'une hypothèse de M. Artin, *Abh. Math. Sem. Univ. Hamburg* **11** (1935) pp. 73–75

[ChW 32] C. CHEVALLEY and A. WEIL, Un théorème d'arithmétique sur les courbes algébriques, *C.R. Acad. Sci. Paris* **195** (1932) pp. 570–572

[Cho 55] W. L. CHOW, Abelian varieties over function fields, *Trans. Amer. Math. Soc.* (1955) pp. 253–275

[Chu 85] D. and G. CHOODNOVSKY, Pade approximations and diophantine geometry, *Proc. Nat. Acad. Sci.* **82** (1985) pp. 2212–2216

[Cl 83] H. CLEMENS, Homological equivalence modulo algebraic equivalence is not finite generated, *Pub. Math. IHES* **58** (1983) pp. 19–38

[Cl 84] H. CLEMENS, Some results about Abel–Jacobi mappings, in: Topics in transcendental algebraic geometry, ed. P. Griffiths, *Ann. of Math. Stud.* **106**, Princeton University Press (1984) pp. 289–304

[ClKM 88] H. CLEMENS, J. KOLLAR and S. MORI, Higher dimensional complex geometry, *Astérisque* **166** (1988)

[CoaW 77] J. COATES and A. WILES, On the conjecture of Birch–Swinnerton-Dyer, *Invent. Math.* **39** (1977) pp. 223–251

[Col 90] R. COLEMAN, Manin's proof of the Mordell conjecture over function fields, *Enseign. Math.* to appear

[C-T 86] J. L. COLLIOT-THÉLÈNE, Arithmétique des variétés rationnelles et problèmes birationnels, *Proc. Intern. Congress Math.*, Berkeley 1986, pp. 641–653

[C-T 88] J. L. COLLIOT-THÉLÈNE, Les grands thèmes de François Châtelet, *Enseign. Math.* **34** (1988) pp. 387–405

[CTCS 80] J. L. COLLIOT-THÉLÈNE, D. CORAY and J. J. SANSUC, Descente et principe de Hasse pour certaines variétés rationnelles, *J. reine angew. Math.* **320** (1980) pp. 150–191

[CTKS 85] J. L. COLLIOT-THÉLÈNE, D. KANEVSKY, J. J. SANSUC, Arithmétique des surfaces cubiques diagonales, in: Diophantine approximation and transcendence theory, ed. G. Wüstholz, *Springer Lecture Notes* **1290** (1985) pp. 1–108

[CTSal 89] J. L. COLLIOT-THÉLÈNE and P. SALBERGER, Arithmetic on some singular cubic hypersurfaces, *Proc. London Math. Soc.* (3) **58** (1989) pp. 519

[CTS 82] J. L. COLLIOT-THÉLÈNE and J. J. SANSUC, Sur le principe de Hasse et l'approximation faible, et sur une hypothèse de Schinzel, *Acta Arith.* **41** (1982) pp. 33–53

[CTS 87] J. L. COLLIOT-THÉLÈNE and J. J. SANSUC, La descente sur les variétés rationnelles II, *Duke Math. J.* **54** (1987) pp. 375–492

[CTSSD 87] J. L. COLLIOT-THÉLÈNE, J. J. SANSUC and P. SWINNERTON-DYER, Intersections of two quadrics and Châtelet surfaces I, *J. reine angew. Math.* **373** (1987) pp. 37–107; II, *ibid.* **374** (1987) pp. 72–168

[CorH 88] M. CORNALBA and J. HARRIS, Divisor classes associated with families of stable varieties, with applications to the moduli space of curves, *Ann. Sci. École Norm. Sup.* **21** (1988) pp. 455–475

[Dan 85] R. DANSET, Méthode du cercle adélique et principe de Hasse fin, *Enseign. Math.* **31** (1985) pp. 1–66

[Dav 62] H. DAVENPORT, *Analytic methods for diophantine equations and diophantine inequalities*, Ann Arbor, Ann Arbor Publishers 1962

[Dav 63] H. DAVENPORT, Cubic forms in 16 variables, *Proc. Roy. Soc. London Ser A* **272** (1963) pp. 285–303

[Dav 65] H. DAVENPORT, On $f^3(t) - g^2(t)$, *K. Norske Vid. Selsk. Forrh.* (Trondheim) **38** (1965) pp. 86–87

[Del 70] P. DELIGNE, Les constantes des équations fonctionelles, *Séminaire Delange–Pisot–Poitou*, 11 (1969/70), No. 19 (1970) pp. 16–28

[Del 83] P. DELIGNE, Représentations l-adiques, in: Séminaire sur les pinceaux arithmétiques: La conjecture de Mordell, ed. Lucien Szpiro, *Astérisque*, **127** (1985) pp. 249–255

[DelM 69] P. DELIGNE and D. MUMFORD, The irreducibility of the space of curves of given genus, *Pub. Math. IHES* **35** (1969) pp. 75–109

[DelS 74] P. DELIGNE and J.-P. SERRE, Formes modulaires de poids 1, *Ann. Sci. École Norm. Sup.* **7** (1974) pp. 507–530

[Dem 66] V. A. DEMJANENKO, Rational points of a class of algebraic curves, *Izv. Akad. Nauk SSSR Ser. Mat.* **30** (1966) pp. 1373–1396; Transl., II. Ser., *Amer. Math. Soc.* **66** (1968) pp. 246–272

[Dem 68] V. A. DEMJANENKO, An estimate of the remainder term in Tate's formula, *Mat. Zametki* **3** (1968) pp. 271–278; translation: *Math. Notes* **3** (1968) pp. 173–177

[Dem 74] V. A. DEMJANENKO, L. Euler's conjecture, *Acta Arith.* **25** (1973–74) pp. 127–135

[DesM 78] M. DESCHAMPS and R. MÉNÉGAUD, Applications rationnelles séparables dominantes sur une variété du type général, *Bull. Soc. Math. France* **106** (1978) pp. 279–287

[Deu 41] M. DEURING, Invarianten und Normalformen elliptischer Funktionenkörper, *Math. Z.* **47** (1941) pp. 47–56

[Di 20] L. E. DICKSON, *History of the theory of numbers* II, Carnegie Institution, Washington DC 1920; reprinted by Chelsea 1966

[Do 79] E. DOBROWOLSKI, On a question of Lehmer and the number of irreducible factors of a polynomial, *Acta Arith.* **34** (1979) pp. 391–401

[El 88] N. ELKIES, On $A^4 + B^4 + C^4 = D^4$, *Math. Comp.* **51**, No. 184 (1988) pp. 825–835

[EsV 84] H. ESNAULT and E. VIEHWEG, Dyson's lemma for polynomials in several variables, *Invent. Math.* **78** (1984) pp. 445–490

[EsV 90] H. ESNAULT and E. VIEHWEG, Effective bounds for semipositive sheaves and for the height of points on curves over complex function fields, *Comput. Math.* to appear

[Fa 83a] G. FALTINGS, Arakelov's theorem for abelian varieties, *Invent. Math.* **73** (1983) pp. 337–347

[Fa 83b] G. FALTINGS, Endlichkeitssätze für abelsche Varietäten über Zahlkörpern, *Invent. Math.* **73** (1983) pp. 349–366; English translation: in Arithmetic geometry, eds. G. Cornell, J. Silverman, Springer-Verlag 1986, pp. 9–27

[Fa 84a] G. FALTINGS, Moduli spaces, in [FaW], pp. 1–33

[Fa 84b] G. FALTINGS, Heights, in [FaW], pp. 33–114

[Fa 84c] G. FALTINGS, Calculus on arithmetic surfaces, *Ann. of Math.* **119** (1984) pp. 387–424

[Fa 90] G. FALTINGS, Diophantine approximations on abelian varieties, to appear

[FaW] G. FALTINGS and G. WÜSTHOLZ, *Rational Points*, Seminar Bonn/Wuppertal 1983–1984, Vieweg 1984

[FrMT 89] J. FRANKE, J. MANIN and Y. TSCHINKEL, Rational points of bounded height on Fano varieties, *Invent. Math.* **95** (1989) pp. 421–435

[Fr 86] G. FREY, Links between stable elliptic curves and certain diophantine equations, *Annales Universitatis Saraviensis*, Series Math. 1 (1986) pp. 1–40

[Fr 87a] G. FREY, Links between elliptic curves and solutions of $A - B = C$, *J. Indian Math. Soc.* **51** (1987) pp. 117–145

[Fr 87b] G. FREY, Links between solutions of $A - B = C$ and elliptic curves, in: Number theory, Ulm 1987, eds. H. P. Schlickewei, E. Wirsing, *Springer Lecture Notes* **1380** (1989) pp. 31–62

[GiS 88a] H. GILLET and C. SOULÉ, Arithmetic intersection theory, Preprint IHES, 1988

[GiS 88b] H. GILLET and C. SOULÉ, Characteristic classes for algebraic vector bundles with hermitian metric, Preprint IHES, 1988

[GiS 88c] H. GILLET and C. SOULÉ, Analytic torsion and the arithmetic Todd genus, preprint; appeared in *Topology* **30** (1991) pp. 21–54

[GiS 88d] G. GILLET and C. SOULÉ, Amplitude arithmétique, *C.R. Acad. Sci. Paris* **307** (1988) pp. 887–890

[GiS 89] H. GILLET and C. SOULÉ, Un théorème de Riemann–Roch Grothendieck arithmétique, Preprint IHES, 1989: *C.R. Acad. Sci. Paris* **309** (1989) pp. 929–932

[Go 76] D. GOLDFELD, The class number of quadratic fields and the conjecture of Birch–Swinnerton-Dyer, *Ann. Scuola Norm. Sup. Pisa*, Serie IV, Vol. III, **4** (1976) pp. 623–663

[Go 79] D. GOLDFELD, Conjectures on elliptic curves over quadratic fields, in: Number theory, Carbondale 1979, ed. M. B. Nathanson, *Springer Lecture Notes* **751** (1979) pp. 108–118

[Gra 65] H. GRAUERT, Mordell's Vermutung über rationale Punkte auf algebraischen Kurven und Funktionenkörpern, Pub. Math. IHES **25** (1965) pp. 131–149

[Gr_M 78] M. GREEN, Holomorphic maps to complex tori, *Amer. J. Math.* **100** (1978), pp. 615–620

[Gr_W 89] W. GREEN, Heights in families of abelian varieties, *Duke Math. J.* **58** No. 3 (1989) pp. 617–632

[GrG 80] M. GREEN and P. GRIFFITHS, Two applications of algebraic geometry to entire holomorphic mappings, in: *Chern Symposium* 1979, eds. W.-Y. Hsiang, S. Kobayashi et al. Springer-Verlag 1980, pp. 41–74

[Grbg 61] M. GREENBERG, Schemata over local rings, *Ann. of Math.* **73** (1961) pp. 624–648

[Grbg 63] M. GREENBERG, Schemata over local rings II, *Ann. of Math.* **78** (1963) pp. 256–266

[Grbg 69] M. GREENBERG, *Lectures on forms in many variables*, W. A. Benjamin 1969

[Grlf 65] N. GREENLEAF, Irreducible subvarieties and rational points, *Amer. J. Math.* (1965) pp. 21–35

[Gri 71] P. GRIFFITHS, Complex analytic properties of certain Zariski open sets on algebraic varieties, *Ann. of Math.* **94** (1971) pp. 21–51

[GrK 73] P. GRIFFITHS and J. KING, Nevanlinna theory and holomorphic mappings between algebraic varieties, *Acta Math.* **130** (1973) pp. 145–220

[Gros 82] B. GROSS, *On the conjecture of Birch and Swinnerton-Dyer for elliptic curves with complex multiplication*, Conference on Fermat's Last Theorem, Birkhäuser 1982, pp. 219–236

[Gros 84] B. GROSS, Heegner points on $X_0(N)$, *Modular forms*, ed. R. Rankin, Chichester, Ellis Horwood 1984, pp. 87–106

[Gros 90] B. GROSS, *Kolyvagin's work on modular elliptic curves*, Durham conference, 1990

[GrZ 86] B. GROSS and D. ZAGIER, Heegner points and derivates of L-series, *Invent. Math.* **84** (1986) pp. 225–320

[Grot 68] A. GROTHENDIECK, Groupe de Brauer I, II, III, Dix exposés sur la cohomologie des Schemas, North-Holland, Amsterdam 1968

[Grot 72] A. GROTHENDIECK, Modèles de Néron et monodromie, Sémin. Géom. Algébrique Bois-Marie 1967–1969, SGA7 I, No. 9, *Springer Lecture Notes* **288** (1972) pp. 313–523

[Ha 77] R. HARTSHORNE, *Algebraic geometry*, Springer-Verlag 1977

[H-B 83] D. R. HEATH-BROWN, Cubic forms in ten variables, *Proc. London Math. Soc.* (3) **47** (1983) pp. 225–257

[Hi 87a] M. HINDRY, Points de torsion sur les sous-variétés de variétés abéliennes, *C.R. Acad. Sci. Paris*, Ser. A-304, **12** (1987) pp. 311–314

[Hi 87b] M. HINDRY, Points quadratiques sur les courbes, *C.R. Acad. Sci. Paris*, Ser. A-305 (1987) pp. 219–221

[Hi 88] M. HINDRY, Autour d'une conjecture de Serge Lang. *Invent. Math.* **94** (1988) pp. 575–603

[HiS 88] M. Hindry and J. Silverman, The canonical height and integral points on elliptic curves, *Invent. Math.* **93** (1988) pp. 419–450

[Hir 90a] N. Hirata-Kohno, Minorations effective de formes linéaires de logarithmes sur les groupes algébriques, to appear

[Hir 90b] N. Hirata-Kohno, Formes linéaires de logarithmes de points algébriques sur les groupes algébriques, to appear

[Ho 88] C. Hooley, On nonary cubic forms, *J. reine angew. Math.* **386** (1988) pp. 32–98

[Hr 85] P. Hriljac, Heights and Arakelov's intersection theory, *Amer. J. Math.* **107** No. 1 (1985) pp. 23–38

[Hu 1899] P. Humbert, Sur les fonctions abéliennes singulières I et II, *J. Math.* **5** (1899) p. 233; **6** (1900) p. 279

[Ig 78] J. Igusa, *Lectures on forms of higher degree*, Tata Institute for Fundamental Research, Springer-Verlag 1978

[Ii 76] S. Iitaka, Logarithmic forms of algebraic varieties, *J. Fac. Sci. Univ. Tokyo Sect. IA Math* **23** (1976) pp. 525–544

[Ii 77] S. Iitaka, On logarithmic Kodaira dimension of algebraic varieties, *Complex analysis and algebraic geometry*, Iwanami Shoten, Tokyo 1977, pp. 175–189

[Ii 82] S. Iitaka, Algebraic geometry, Grad. Texts Math. **76**, Springer-Verlag 1982

[Isk 71] V. A. Iskovskih, A counterexample to the Hasse principle for a system of two quadratic forms in 5 variables, *Mat. Zametki* **10** (1971) pp. 253–257; translation: *Math. Notes* **10** (1971) pp. 575–577

[Isk 72] V. A. Iskovskih, Birational properties of a surface of degree 4 in P_k^4, *Mat. Sbornik* **88** (1972) pp. 31–37; translation: *Math. USSR Sbornik* **17** (1972) pp. 30–36

[Jo 78] J. P. Jouanolou, Hypersurfaces solutions à une équation de Pfaff analytique, *Math. Ann.* **232** (1978) pp. 239–245

[Kam 90] S. Kamienny, Torsion points on elliptic curves and q-coefficients of modular forms, preprint

[Kan 87] D. Kanevsky, Application of the conjecture on the Manin obstruction to various diophantine problems, Journées Arithmétiques de Besançon 1985, *Astérisque* **147–148** (1987) pp. 307–314

[KatK 86] K. Kato and T. Kuyumaki, The dimension of fields and algebraic K-theory, *J. Number Theory* **24** (1986) pp. 229–244

[KaL 82] N. Katz and S. Lang, Finiteness theorems in geometric class field theory, *Enseign. Math.* (1982) pp. 285–314

[Kaw 80] Y. Kawamata, On Bloch's conjecture, *Invent. Math.* **57** (1980) pp. 97–100

[Kn 69] M. Kneser, *Lectures on Galois cohomology of classical groups*, Tata Institute, Bombay, 1969

[Kob 70] S. Kobayashi, *Hyperbolic manifolds and holomorphic mappings*, Marcel Dekker 1970

[Kob 75] S. Kobayashi, Negative vector bundles and complex Finsler structures, *Nagoya Math. J.* **57** (1975) pp. 153–166

[KoO 71] S. KOBAYASHI and T. OCHIAI, Mappings into compact complex manifolds with negative first Chern class, *J. Math. Soc. Japan* **23** (1971) pp. 136–148

[KoO 75] S. KOBAYASHI and T. OCHIAI, Meromorphic mappings into compact complex spaces of general type, *Invent. Math.* **31** (1975) pp. 7–16

[Koll 89] J. KOLLÁR, Minimal models of algebraic threefolds: Mori's program, *Séminaire Bourbaki* (1988–1989), *Astérisque* **177–178** (1989) pp. 303–326

[Koly 88] V. A. KOLYVAGIN, Finiteness of $E(\mathbf{Q})$ and $Ш(E/\mathbf{Q})$ for a class of Weil curves, *Izv. Akad. Nauk SSSR Ser. Mat.* **52** (1988) pp. 522–540; translation: *Math. USSR Izv.* **32** (1989) pp. 523–541

[Kol 90] V. A. KOLYVAGIN, *Euler systems* (1988), to appear in Birkhauser volume in honor of Grothendieck

[Ku 76] D. KUBERT, Universal bounds on the torsion of elliptic curves, *Proc. London Math. Soc.* **33** (1976) pp. 193–237

[Ku 79] D. KUBERT, Universal bounds on the torsion of elliptic curves, *Compositio Math.* **38** (1979) pp. 121–128

[KuL 75] D. KUBERT and S. LANG, Units in the modular function field I, Diophantine applications, *Math. Ann.* **218** (1975) pp. 67–96

[KuL 79] D. KUBERT and S. LANG, Modular units inside cyclotomic units, *Bull. Soc. Math. France* **107** (1979) pp. 161–178

[Lac 82] G. LACHAUD, Une présentation adélique de la série singulière et du problème de Waring, *Enseign. Math.* **28** (1982) pp. 139–169

[LandP 66] L. LANDER and T. PARKIN, Counterexamples to Euler's conjecture on sums of the powers, *Bull. Amer. Math. Soc.* **72** (1966) p. 1079

[La 51] S. LANG, On quasi algebraic closure, *Ann. of Math.* **55** (1951) pp. 373–390

[La 54] S. LANG, Some applications of the local uniformization theorem, *Amer. J. Math.* **76** (1954) pp. 362–374

[La 56a] S. LANG, Unramified class field theory over function fields in several variables, *Ann. of Math.* **64** No. 2 (1956) pp. 285–325

[La 56b] S. LANG, Sur les séries *L* d'une variété algébrique, *Bull. Soc. Math. France* **84** (1956) pp. 385–407

[La 56c] S. LANG, Algebraic groups over finite fields, *Amer. J. Math.* **78** No. 3 (1956) pp. 555–563

[La 59] S. LANG, *Abelian varieties*, Interscience, New York 1959; reprinted by Springer-Verlag, 1983

[La 60a] S. LANG, Integral points on curves, *Pub. Math. IHES*, 1960

[La 60b] S. LANG, Some theorems and conjectures in diophantine equations, *Bull. Amer. Math. Soc.* **66** (1960) pp. 240–249

[La 62] S. LANG, *Diophantine geometry*, Wiley–Interscience, 1962

[La 64] S. LANG, Diophantine approximations on toruses, *Amer. J. Math.* **86** (1964) pp. 521–533

[La 65a] S. LANG, Division points on curves *Ann. Mat. Pura Appl.* (IV) **LXX** (1965) pp. 229–234

[La 65b] S. LANG, Report on diophantine approximations, *Bull. Soc. Math. France* **93** (1965) pp. 177–192

[La 66] S. LANG, *Introduction to diophantine approximations*, Addison-Wesley, 1966

[La 70] S. LANG, *Algebraic Number Theory*, Addison-Wesley 1970, reprinted by Springer-Verlag, 1986

[La 71] S. LANG, Transcendental numbers and diophantine approximations, *Bull. Amer. Math. Soc.* **77** (1971) pp. 635–677

[La 74] S. LANG, Higher dimensional diophantine problems, *Bull. Amer. Math. Soc.* **80** No. 5 (1974) pp. 770–787

[La 75] S. LANG, Division points of elliptic curves and abelian functions over number fields, *Amer. J. Math.* **97** No. 1 (1975) pp. 124–132

[La 78] S. LANG, *Elliptic curves: Diophantine analysis*, Springer-Verlag 1978

[La 83a] S. LANG, *Fundamentals of diophantine geometry*, Springer-Verlag 1983

[La 83b] S. LANG, Conjectured diophantine estimates on elliptic curves, *Arithmetic and geometry*, papers dedicated to Shafarevich, edited by Artin and Tate, Birkhaüser 1983

[La 83c] S. LANG, *Complex multiplication*, Grūndlehren 255, Springer-Verlag 1983

[La 86] S. LANG, Hyperbolic and diophantine analysis, *Bull. Amer. Math. Soc.* **14** (1986) pp. 159–205

[La 87] S. LANG, *Introduction to complex hyperbolic spaces*, Springer-Verlag 1987

[La 88] S. LANG, *Introduction to Arakelov theory*, Springer-Verlag 1988

[La 90a] S. LANG, The error term in Nevanlinna theory II, *Bull. Amer. Math. Soc.* **22** (1990) pp. 115–125

[La 90b] S. LANG, Lectures on Nevanlinna theory, in: Topics on Nevanlinna theory, S. Lang, W. Cherry *Springer Lecture Notes* **1433** (1990) pp. 1–112

[La 90c] S. LANG, Old and new conjectures in diophantine inequalities, *Bull. Amer. Math. Soc.* **23** (1990) pp. 37–75

[LaN 59] S. LANG and A. NÉRON, Rational points of abelian varieties over function fields, *Amer. J. Math.* **81** No. 1 (1959) pp. 95–118

[LaT 58] S. LANG and J. TATE, Principal homogeneous spaces over abelian varieties, *Amer. J. Math.* **80** (1958) pp. 659–684

[Lange 87] H. LANGE, Abelian varieties with several principal polarizations, *Duke Math. J.* **55** (1987) pp. 617–628

[Lau 84] M. LAURENT, Equations diophantiènnes exponentielles, *Invent. Math.* **78** (1984) pp. 299–327

[Lau 86] M. LAURENT, Une nouvelle démonstration du théorème d'isogénie d'après D. V. et G. V. Choodnovsky, *Séminaire de théorie des nombres*, Paris 1985–1986

[Lew 52] D. J. LEWIS, Cubic homogeneous polynomials over *p*-adic number fields, *Ann. of Math.* **56** (1952) pp. 473–478

[Li 74] P. LIARDET, Sur une conjecture de Serge Lang, *C.R. Acad. Sci. Paris* **279** (1974) pp. 435–437

[Li 75] P. LIARDET, Sur une conjecture de Serge Lang, *Astérisque* **24–25**, (1975)

[LuY] S. LU and S. T. YAU, Holomorphic curves in surfaces of general type, *Proc. Nat. Acad. Sci. U.S.A.* **87** (1990) pp. 80–82

[Mae 83] K. MAEHARA, A finiteness property of varieties of general type, *Math. Ann.* **262** (1983) pp. 101–123

[Mah 33] K. MAHLER, Zur Approximation algebraischer Zähler I (Über den grössten Primteiler binärer Formen), *Math. Ann.* **107** (1933) pp. 691–730

[Man 58] J. MANIN, Algebraic curves, over fields with differentiation, *Izv. Akad. Nauk. SSSR Ser. Mat* **22** (1958) pp. 737–756; translation: Transl., II. Ser., *Amer. Math. Soc.* **37** (1964) pp. 59–78

[Man 63] J. MANIN, Rational points of algebraic curves over function fields, *Izv. Akad. Nauk SSSR Ser. Mat.* **27** (1963) pp. 1395–1440; translation: Transl., II. Ser., *Amer. Math. Soc.* **50** (1966) pp. 189–234

[Man 64] J. MANIN, The Tate height of points on an abelian variety, its variants and applications, *Izv. Akad. Nauk SSSR Ser. Mat.* **28** (1964), pp. 1363–1390; translation: Transl. II. Ser., *Amer. Math. Soc.* **59** (1966) pp. 82–110

[Man 69] J. MANIN, The p-torsion of elliptic curves is uniformly bounded, *Izv. Akad. Nauk SSSR Ser. Mat.* **33** (1969) pp. 459–465; translation: *Math. USSR, Izv.* **3** (1969) pp. 433–438

[Man 70] J. MANIN, The refined structure of the Néron–Tate height, *Mat. Sbornik* **83** (1970) pp. 331–348 translation: *Math. USSR, Sb.* **12** (1970) pp. 325–342

[Man 70b] J. MANIN, Le groupe de Brauer–Grothendieck en géométrie diophantienne, *Actes Congrès Intern. Math.* 1970 Tome 1 pp. 401–411

[Man 72] J. MANIN, Parabolic points and zeta functions of modular curves, *Izv. Akad. Nauk SSSR Ser. Mat.* **36** (1972) pp. 19–65; translation: *Math. USSR Izv.* **6** (1972) pp. 19–64

[Man 74] J. MANIN, *Cubic forms*, North-Holland, 1974; second edition (with appendix) 1986

[MaT 86] J. MANIN and J. A. TSFASMAN, Rational varieties, algebra, geometry and arithmetic, *Uspekhi Mat. Nauk* **41** (1986); translation: *Russian Math. Surveys* **41 : 2** (1986) pp. 51–116

[MaZ 72] J. MANIN and Y. ZARHIN, Heights on families of abelian varieties, *Mat. Sbornik* **89** (1972) pp. 171–181

[Mas 83] R. C. MASON, The hyperelliptic equation over function fields, *Math. Proc. Cambridge Philos. Soc.* **93** (1983) pp. 219–230

[Mas 84a] R. C. MASON, Equations over function fields, in: *Number theory*, Noordwijkerhout 1983, ed. H. Jager, *Springer Lecture Notes* **1068** (1984) pp. 149–157

[Mas 84b] R. C. MASON, Diophantine equations over function fields, *London Math. Soc. Lecture Notes Series* 96, Cambridge University Press 1984

[Mass 81] D. MASSER, Small values of the quadratic part of the Néron–Tate height, *Progr. Math.* **12**, Birkhäuser (1981) pp. 213–222

[Mass 84] D. MASSER, Small values of the quadratic part of the Néron–Tate height on an abelian variety, *Comps. Math.* **53** (1984) pp. 153–170

[Mass 85] D. MASSER, Small values of heights on families of abelian varieties, in: Diophantine approximation and transcendence theory, ed. G. Wüstholz *Springer Lecture Notes* **1290** (1985) pp. 109–148

[Mass 88] D. MASSER, Linear relations on algebraic groups, in: *New advances in transcendence theory*, ed. A. Baker, Cambridge University Press 1988, pp. 248–262.

[MaW 90a] D. MASSER and G. WÜSTHOLZ, Estimating isogenies on elliptic curves, *Invent. Math.* **100** (1990) pp. 1–24

[MaW 90b] D. MASSER and G. WÜSTHOLZ, Some effective estimates for elliptic curves, in: Arithmetic of complex manifolds, Erlangen 1988, eds. W. P. Barth, H. Lange *Springer Lecture Notes* **1399** (1990) pp. 103–109

[MaW 91] D. MASSER and G. WÜSTHOLZ, *Minimal period relations on abelian varieties and isogenies*, to appear

[May 64] W. L. MAY, Binary forms over number fields, *Ann. of Math.* **79** No. 3 (1964) pp. 597–615

[Maz 72] B. MAZUR, Rational points of abelian varieties with values in towers of number fields, *Invent. Math.* **18** (1972) pp. 183–266

[Maz 74] B. MAZUR, Arithmetic of Weil curves, *Invent. Math.* **25** (1974) pp. 1–61

[Maz 76] B. MAZUR, Rational points on modular curves, in: Modular functions of one variable, eds. J. P. Serre, D. B. Zagier *Springer Lecture Notes* **601** (1976) pp. 108–148

[Maz 77] B. MAZUR, Modular curves and the Eisenstein ideal, *Pub. Math. IHES* **47** (1977) pp. 33–186

[Maz 78] B. MAZUR, Rational isogenies of prime degree, *Invent. Math.* 44 (1978) pp. 129–162

[Maz 83] B. MAZUR, Modular curves and arithmetic, *Proc. International Congress Math. Warsaw* (1983) pp. 185–211

[Mes 86] J.-F. MESTRE, Formules explicites et minorations des conducteurs de variétés algébriques, *Comput. Math.* **58** (1986) pp. 209–232

[Mil 86a] J. S. MILNE, Abelian varieties, in: *Arithmetic geometry*, eds. G. Cornell, J. Silverman, Springer-Verlag 1986, pp. 103–150

[Mil 86b] J. S. MILNE, Jacobian varieties, in ditto, pp. 167–212

[MirP 89] R. MIRANDA and U. PERSSON, Torsion group of elliptic surfaces, *Compos. Math.* **72** (1989) pp. 249–267

[Miy 88] Y. MIYAOKA, On the Kodaira dimension of minimal threefolds, *Math. Ann.* **281** (1988) pp. 325–332

[MiyM 86] Y. MIYAOKA and S. MORI, A numerical criterion for uniruledness, *Ann. of Math.* **124** (1986) pp. 65–69

[Mord 22] L. J. MORDELL, On the rational solutions of the indeterminate equation of the third and fourth degrees, *Math. Proc. Cambridge Philos. Soc.* **21** (1922) pp. 179–192

[MorB 85] MORET-BAILLY, Pinceaux de variétés abéliennes, *Astérisque* **129** (1985)

[Mori 82] S. Mori, Threefolds whose canonical bundles are not numerically effective, *Ann. of Math.* (2) **116** (1982) pp. 133–176

[Mori 87] S. Mori, Classification of higher dimensional varieties, *Proceedings of Symposia in Pure Mathematics AMS* **46** (1987) pp. 269–288

[MoM 83] S. Mori and S. Mukai, The uniruledness of the moduli space of curves of genus 11, Appendix, in: Algebraic geometry, eds. M. Raynaūd, T. Shioda, *Springer Lecture Notes* **1016** (1983) pp. 334–352

[Mum 65] D. Mumford, A remark on Mordell's conjecture, *Amer. J. Math.* **87** No. 4 (1965) pp. 1007–1016

[Mum 66] D. Mumford, On the equations defining abelian varieties I, *Invent. Math.* **1** (1966) pp. 287–354

[Mum 74] D. Mumford, *Abelian varieties*, Oxford University Press 1974

[Mum 83] D. Mumford, *Tata lectures on Theta* I, Birkhäuser 1983

[NaN 81] M. Narasimhan and M. Nori, Polarizations on an abelian variety, *Proc. Indian Acad. Sci. Math. Sci* **90** (1981) pp. 125–128

[Ne 52] A. Néron, Problèmes arithmétiques et géometriques rattachés à la notion de rang d'une courbe algébrique dans un corps, *Bull. Soc. Math. France* **83** (1952) pp. 101–166

[Ne 55] A. Néron, Arithmétique et classes de diviseurs sur les variétés algébriques, *Proc. International Symposium on Algebraic Number Theory*, Tokyo-Nikko, Tokyo 1955, pp. 139–154

[Ne 64] A. Néron, Modèles minimaux des variétés abéliennes sur les corps locaux et globaux, *Pub. Math. IHES* **21** (1964) pp. 361–482

[Ne 65] A. Néron, Quasi-functions et hauteurs sur les variétés abéliennes, *Ann. of Math.* **82** No. 2 (1965) pp. 249–331

[Nev 25] R. Nevanlinna, Zur Theorie der meromorphen Funktionen, *Acta Math.* **46** (1925) pp. 1–99

[Nev 53] R. Nevanlinna, *Eindeutige analytische Funktionen*, 2nd ed. Springer-Verlag 1953

[Nev 70] R. Nevanlinna, *Analytic functions*, translation and revision of the above, Springer-Verlag 1970

[No 81a] J. Noguchi, Lemma on logarithmic derivative and holomorphic curves in algebraic varieties, *Nagoya Math. J.* **83** (1981) pp. 213–233

[No 81b] J. Noguchi, A higher dimensional analogue of Mordell's conjecture over function fields, *Math. Ann.* **258** (1981) pp. 207–212

[No 85] J. Noguchi, Hyperbolic fiber spaces and Mordell's conjecture over function fields, *Publ. Res. Inst. Math. Sci.* **21** No 1 (1985) pp. 27–46

[No 87] J. Noguchi, Moduli spaces of holomorphic mappings into hyperbolically imbedded complex spaces and locally symmetric spaces, *Invent. Math.* **93** (1988) pp. 15–34

[Och 77] T. Ochiai, On holomorphic curves in algebraic varieties with ample irregularity, *Invent. Math.* **43** (1977) pp. 83–96

[Ogg 67] A Ogg, Elliptic curves and wild ramification, *Amer. J. Math.* **89** (1967) pp. 1–21

[Os 81] C. F. Osgood, A number theoretic-differential equations approach to generalizing Nevanlinna theory, *Indian J. Math.* **23** (1981) pp. 1–15

[Os 85] C. F. Osgood, Sometimes effective Thue–Siegel Roth–Schmidt Nevanlinna bounds, or better, *J. Number Theory* **21** (1985) pp. 347–389

[Par 68] A. N. Parshin, Algebraic curves over function fields, *Izv. Akad. Nauk SSSR Ser. Mat.* **32** (1968); translation: *Math. USSSR Izv.* **2** (1968) pp. 1145–1170

[Par 73] A. N. Parshin, Modular correspondences, heights and isogenies of abelian varieties, *Trudy Mat. Inst. Steklova* **132** (1973); translation: *Proc. Steklov Inst. Math.* pp. 243–270

[Par 86] A. N. Parshin, Finiteness theorems and hyperbolic manifolds, Preprint IHES, 1986

[Par 89] A. N. Parshin, On the application of ramified coverings to the theory of Diophantine Geometry, *Math. Sbornik* **180** No. 2 (1989) pp. 244–259

[Pat 85] S. J. Patterson, The Hardy–Littlewood method and diophantine analysis in light of Igusa's work, *Math. Gottingensis, Schriften. Sonderforschungsbereichs Geom. Anal.* **11** (1985) pp. 1–45

[Pec 49] L. J. Peck, Diophantine equations in algebraic number fields, *Amer. J. Math.* **71** (1949) pp. 387–402

[Ra 64–65] M. Raynaud, Caractéristique d'Euler–Poincaré d'un faisceau et cohomologie des variétés abéliennes, *Sém. Bourbaki* 1964–65 #286

[Ra 74] M. Raynaud, Schémas en groupes de type $(p, p, \dots, p)$, *Bull. Sci. Math. France* **102** (1974) pp. 241–280

[Ra 83a] M. Raynaud, Courbes sur une variété abélienne et points de torsion, *Invent. Math.* **71** (1983) pp. 207–233

[Ra 83b] M. Raynaud, Sous-variétés d'une variété abélienne et points de torsion, *Arithmetic and geometry*, dedicated to I. Shafarevich, edited by Artin and Tate, Birkhäuser 1983

[Ra 83c] M. Raynaud, Around the Mordell conjecture for function fields and a conjecture of Serge Lang, in: *Algebraic geometry*, eds. M. Raynaũd, T. Shioda, *Springer Lecture Notes* **1016** (1983) pp. 1–19

[Ra 85] M. Raynaud, Hauteurs et isogénies, in Séminaire sur les pinceaux arithmétiques: La conjecture de Mordell, ed. Lucien Szpiro, *Astérisque* 127, 1985 pp. 199–234

[RDM 62] R. Richtmyer, M. Devaney and N. Metropolis, Continued fraction expansions of algebraic numbers, *Numer. Math.* **4** (1962) pp. 68–84

[Ri 90a] K. Ribet, On modular representations of Gal($\bar{\mathbf{Q}}/\mathbf{Q}$) arising from modular forms, *Invent. Math.* **100** (1990) pp. 431–476

[Ri 90b] K. Ribet, From the Taniyama–Shimura conjecture to Fermat's last theorem, Annales de la Fac. des Sci. Toulouse (1990), appearing with delays

[Rob 81a] P. Robba, Une propriété de spécialisation continue, *Groupe d'Analyse Ultramétrique*, Paris 1980/81, No. 26

[Rob 81b] P. ROBBA, Propriété d'approximation pour les éléments algébriques, *Compositio Math.* **63** (1987) pp. 3–14

[Rot 55] K. F. ROTH, Rational approximation to algebraic numbers, *Mathematika* **2** (1955) pp. 1–20

[RuW 90] M. RU and P.-M. WONG, Integral points of $\mathbf{P}^n - \{2n+1$ hyperplanes in general position$\}$, to appear

[Ru 87a] K. RUBIN, Tate–Shafarevich groups and L-functions of elliptic curves with complex multiplication, *Invent. Math.* **89** (1987) pp. 527–560

[Ru 87b] K. RUBIN, Global units and ideal class groups, *Invent. Math.* **89** (1987) pp. 511–526

[Sai 89] S. SAITO, Some observations on motivic cohomology of arithmetic schemes, *Invent. Math.* **98** (1989) pp. 371–404

[Sai$_T$ 88] T. SAITO, Conductor, discriminant, and the Noether formula for arithmetic surfaces, *Duke Math. J.* **57** (1988) pp. 151–173

[Sal 86] P. SALBERGER, On the arithmetic of conic bundle surfaces, *Séminaire de théorie des nombres* 1985–1986, Paris, Birkhauser 1986, pp. 175–197

[Sal 88] P. SALBERGER, Zero cycles on rational surfaces over number fields, *Invent. Math.* **91** (1988) pp. 505–524

[Sam 66a] P. SAMUEL, Lectures on old and new results on algebraic curves, Tata Institute of Fundamental Research, Bombay 1966

[Sam 66b] P. SAMUEL, Compléments à un article de Hans Grauert sur la conjecture de Mordell, *Pub. Math. IHES* **29** (1966) pp. 55–62

[San 81] J. J. SANSUC, Groupe de Brauer et arithmétique des groupes algébriques linéaires sur un corps de nombres, *J. reine angew. Math.* **327** (1981) pp. 12–80

[San 85] J. J. SANSUC, Principe de Hasse, surfaces cubiques et intersection de deux quadriques, Journées arithmétiques de Besançon 1985, *Astérisque* **147–148** (1987) pp. 183–207

[Sch 64] S. SCHANUEL, On heights in number fields, *Bull. Amer. Math. Soc.* **70** (1964) pp. 262–263

[Sch 79] S. SCHANUEL, Heights in number fields, *Bull. Soc. Math. France* **107** (1979) pp. 443–449

[Schm 60] W. SCHMIDT, A metrical theorem in diophantine approximation, *Canad. J. Math.* **22** (1960) pp. 619–631

[Schm 70] W. SCHMIDT, Simultaneous approximation to algebraic numbers by rationals, *Acta Math.* **125** (1970) pp. 189–201

[Schm 71] W. SCHMIDT, Approximation to algebraic numbers, *Enseign. Math.* **17** (1971) pp. 187–253

[Schm 80] W. SCHMIDT, Diophantine approximation, *Springer Lecture Notes* **785** (1980)

[Schm 84] W. SCHMIDT, *Analytische Methoden für Diophantische Gleichungen*, DMV Seminar, Band 5, Birkhäuser 1984

[Schm 85] W. SCHMIDT, The density of integer points on homogeneous varieties, *Acta Math.* (1985) pp. 244–296

[Schn 36] T. SCHNEIDER, Über die Approximation algebraischer Zahlen, *J. reine angew. Math.* **175** (1936) pp. 182–192

[Schapp 84] N. SCHAPPACHER, Tate's conjecture on the endomorphisms of abelian varieties, in Faltings–Wüstholz et al. *Rational points*, Vieweg, 1984

[See 67] R. SEELEY, Complex powers of an elliptic operator, in *Singular integrals, Proc. Sympos. Pure Math.* **10** (1967)

[Sel 51] E. SELMER, The diophantine equation $ax^3 + by^3 + cz^3 = 0$, *Acta Math.* **85** (1951) 203–362 and **92** (1954) 191–197

[Ser 68] J-P. SERRE, *Abelian l-adic representations and elliptic curves*, Benjamin 1968

[Ser 69–70] J-P. SERRE, Facteurs locaux des fonctions zeta des variétés algébriques, *Séminaire Delange–Pisot–Poitou* #**19**, 1969–1970

[Ser 72] J-P. SERRE, Propriétés galoisiennes des points d'ordre fini des courbes elliptiques, *Invent. Math.* **15** (1972) pp. 259–331

[Ser 87] J-P. SERRE, Sur les représentations modulaires de degré 2 de $\mathrm{Gal}(\bar{Q}/Q)$, *Duke Math. J.* **54** (1987) pp. 179–230

[Ser 89] J-P. SERRE, *Lectures on the Mordell–Weil theorem*, Vieweg 1989

[SeT 68] J-P. SERRE and J. TATE, Good reduction of abelian varieties, *Ann. of Math.* **88** (1968) pp. 492–517

[Sha 63] I. SHAFAREVICH, Algebraic number fields, *Proc. Int. Congr. Math.* 1962, Inst. Mittag-Leffler (1963) pp. 163–176. *Transl. II Ser. Am. Math Soc.* **31** (1963) pp. 25–39. *Collected Works*, Springer-Verlag, pp. 283–294

[ShT 67] I. SHAFAREVICH and J. TATE, The rank of elliptic curves, *Amer Math. Soc. Transl.* **8** (1967) pp. 917–920

[Shat 72] S. SHATZ, Profinite groups, arithmetic, and geometry, *Ann. of Math. Stud.* **67**, (1972)

[Shi 71a] G. SHIMURA, On elliptic curves with complex multiplication as factors of the Jacobian of modular function fields, *Nagoya Math. J.* **43** (1971) pp. 199–208

[Shi 71b] G. SHIMURA, *Introduction to the arithmetic theory of automorphic functions*, Princeton University Press, 1971

[Shi 73] G. SHIMURA, On modular forms of half-integral weight, *Ann. of Math.* **97** (1973) pp. 440–481

[Shi 89] G. SHIMURA, Yutaka Taniyama and his time, *Bull. London Math. Soc.* **21** (1989) pp. 186–196

[Shio 72] T. SHIODA, On elliptic modular surfaces, *J. Math. Soc. Japan* **24** (1972) pp. 20–59

[Shio 73] T. SHIODA, Algebraic cycles on certain K3 surfaces in characteristic *p*, *Manifolds-Tokyo*, Proc. Internat. Conf. Tokyo 1973, pp. 357–364; University of Tokyo Press, Tokyo (1975)

[Shio 89a] T. SHIODA, Mordell–Weil lattices and Galois representations I, II, III *Proc. Japan Acad.* **65** (1989) pp. 268–271, 296–299, 300–303

[Shio 89b] T. SHIODA, Mordell–Weil lattices and sphere packings, Preprint 1989

[Shio 89c] T. SHIODA, The Galois representation of type E_8 arising from certain Mordell–Weil groups, *Proc. Japan Acad.* **65**, Ser A, No. 6 (1989) pp. 195–197

[Shio 90a] T. SHIODA, Construction of elliptic curves over **Q**(t) with high rank: a preview, *Proc. Japan Acad.* **66**, Ser A, No. 2 (1990) pp. 57–60 (accompanied by a preprint: Construction of elliptic curves with high rank via the invariants of the Weyl groups, 1990)

[Shio 90b] T. SHIODA, Theory of Mordell–Weil lattices, ICM talk, Kyoto, 1990

[Shio 90c] T. SHIODA, On the Mordell–Weil lattices, preprint, 1990

[Sie 29] C. L. SIEGEL, Einige Anwendungen diophantischer Approximationen, *Abh. Preuss. Akad. Wiss. Phys. Math. Kl.* (1929) pp. 41–69

[Sie 69] C. L. SIEGEL, Abschätzung von Einheiten, *Nachr. Akad. Wiss. Göttingen* (1969) pp. 71–86

[Slbg 85] A. SILVERBERG, Mordell–Weil groups of generic abelian varieties, *Invent. Math.* **81** (1985) pp. 71–106

[Sil 81] J. SILVERMAN, Lower bound for the canonical height on elliptic curves, *Duke Math. J.* **48** (1981) pp. 633–648

[Sil 82] J. SILVERMAN, Integer points and the rank of Thue elliptic curves, *Invent. Math.* **66** (1982) pp. 395–404

[Sil 83a] J. SILVERMAN, Heights and the specialization map for families of abelian varieties, *J. reine angew. Math.* **342** (1983) pp. 197–211

[Sil 83b] J. SILVERMAN, Integer points on curves of genus 1, *J. London Math. Soc.* **28** (1983) pp. 1–7

[Sil 83c] J. SILVERMAN, Representations of integers by binary forms and the rank of the Mordell–Weil group, *Invent. Math.* **74** (1983) pp. 281–292

[Sil 84a] J. SILVERMAN, Lower bounds for height functions, *Duke Math. J.* **51** No. 2 (1984) pp. 395–403

[Sil 84b] J. SILVERMAN, The S-unit equation over function fields, *Math. Proc. Cambridge Philos. Soc.* **95** (1984) pp. 3–4

[Sil 85] J. SILVERMAN, Divisibility of the specialization map for families of elliptic curves, *Amer. J. Math.* **107** (1985) pp. 555–565

[Sil 86] J. SILVERMAN, *The arithmetic of elliptic curves*, Springer-Verlag 1986

[Sil 87] J. SILVERMAN, Integral points on abelian varieties are widely spaced, *Compositio Math.* **61** (1987) pp. 253–266

[Sil 87] J. SILVERMAN, A quantitative version of Siegel's theorem: integral points on elliptic curves and Catalan curves, *J. reine angew. Math.* **378** (1987) pp. 60–100

[Sil 87] J. SILVERMAN, Arithmetic distance functions and height functions in diophantine geometry, *Math. Ann.* **279** (1987) pp. 193–216

[Siu 87] Y. T. SIU, Defect relations for holomorphic maps between spaces of different dimensions, *Duke Math. J.* **55** No 1 (1987) pp. 213–251

[So 89] C. SOULÉ, Géométrie d'Arakelov des surfaces arithmétiques, *Séminaire Bourbaki*, 1989–1989

[St 89] G. STEVENS, Stickelberger elements and modular parametrizations of elliptic curves, *Invent. Math.* **98** (1989) pp. 75–106

[SWD 52] P. Swinnerton-Dyer, A solution of $A^5 + B^5 + C^5 = D^5 + E^5 + F^5$, *Math. Proc. Cambridge Philos. Soc.* **48** (1952) pp. 516–518

[Szp 81] L. Szpiro, Séminaire sur les pinceaux de courbes de genre au moins deux, *Astérisque* **86** (1981)

[Szp 84] L. Szpiro, Small points and torsion points, *Contemp. Math.* **58**, Lefschetz Centennial Conference, 1986, pp. 251–260

[Szp 85a] L. Szpiro, Présentation de la théorie d'Arakelov, *Contemp. Math.* **67**, Summer Conference at Arcata, 1985, pp. 279–293

[Szp 85b] L. Szpiro, Séminaire sur les pinceaux arithmétiques: La conjecture de Mordell, *Astérisque* **127** (1985)

[Ta 52] J. Tate, The higher dimensional cohomology groups of class field theory, *Ann. of Math.* **56** (1952) pp. 294–297

[Ta 58] J. Tate, WC groups over p-adic fields, *Séminaire Bourbaki*, 1958

[Ta 62] J. Tate, Duality theorems in Galois cohomology over number fields, *Proc. Internat. Cong. Math. Stockholm* (1962) pp. 288–295

[Ta 65] J. Tate, Algebraic cycles and poles of zeta functions, in *Arithmetical algebraic geometry*, Harper and Row 1963, pp. 93–111

[Ta 66a] J. Tate, Endomorphisms of abelian varieties over finite fields, *Invent. Math.* **2** (1966) pp. 134–144

[Ta 66b] J. Tate, On the conjectures of Birch and Swinnerton-Dyer and a geometric analogue, *Séminaire Bourbaki* 1965–1966, No. 306

[Ta 67] J. Tate, p-divisible groups, in *Proc. of Conference on Local fields (Driebergen)*, Springer-Verlag 1967

[Ta 68–79] J. Tate, letters to Serre, 21 June 1968 and 12 October 1979 (appendix to the first letter)

[Ta 74] J. Tate, The arithmetic of elliptic curves, *Invent. Math.* (1974) pp. 179–206

[Ta 83] J. Tate, Variation of the canonical height of a point depending on a parameter, *Amer. J. Math* **105** (1983) pp. 287–294

[Ter 66] G. Terjanian, Un contrexemple à une conjecture d'Artin, *C.R. Acad. Sci. Paris* **262** (1966) p. 612

[Ter 77] G. Terjanian, Sur la dimension diophantienne des corps p-adiques, *Acta Arith.* **34** No. 2 (1977) pp. 127–130

[Tu 83] J. Tunnell, A classical diophantine problem and modular forms, *Invent. Math.* **72** (1983) pp. 323–334

[Ue 73] K. Ueno, Classification of algebraic varieties I, *Comp. Math.* **27** No. 3 (1973) pp. 277–342

[Ul 90] D. Ulmer, On universal elliptic curves over Igusa curves, *Invent. Math.* **99** (1990) pp. 377–381

[vanD 81] L. van den Dries, A specialization theorem for p-adic power series converging on the closed unit disc, *J. Algebra* **73** (1981) pp. 613–623

[Vau 81] R. C. Vaughan, *The Hardy–Littlewood method*, Cambridge Tract in Mathematics **80**, Cambridge University Press 1981

[Vio 85] C. Viola, On Dyson's lemma, *Ann. Scuola Norm. Sup. Pisa* **12** (1985) pp. 105–135

[Vo 87] P. VOJTA, Diophantine approximations and value distribution theory, *Springer Lecture Notes* **1239**, (1987)

[Vo 88] P. VOJTA, Diophantine inequalities and Arakelov theory, in Lang, *Introduction to Arakelov theory*, 1988 pp. 155–178

[Vo 89a] P. VOJTA, Dyson's lemma for products of two curves of arbitrary genus, *Invent. Math.* **98** (1989) pp. 107–113

[Vo 89b] P. VOJTA, Mordell's conjecture over function fields, *Invent. Math.* **98** (1989) pp. 115–138

[Vo 89c] P. VOJTA, A refinement of Schmidt's subspace theorem, *Amer. J. Math.* **111** (1989) pp. 489–518

[Vo 90a] P. VOJTA, Siegel's theorem in the compact case, *Ann. of Math.*

[Vo 90b] P. VOJTA, Arithmetic discriminants and quadratic points on curves,

[Vo 90c] P. VOJTA, Algebraic points on curves,

[Vo 90d] P. VOJTA, A height inequality for algebraic points,

[Vol 85] J. VOLOCH, Diagonal equations over function fields, *Bol. Soc. Brasil. Mat.* **16** (1985) pp. 29–39

[Vol 90] J. VOLOCH, On the conjectures of Mordell and Lang in positive characteristic, to appear

[Wa 80] M. WALDSCHMIDT, A lower bound for linear forms in logarithms, *Acta Arithm.* **37** (1980) pp. 257–283

[Walp 81] J. L. WALDSPURGER, Sur les coefficients de Fourier des formes modulaires de poids demi entier, *J. Math. Pures Appl.* **60** (1981) pp. 375–484

[We 28] A. WEIL, L'arithmétique sur les courbes algébriques, *Acta Math.* **52** (1928) pp. 281–315]

[We 35] A. WEIL, Arithmétique et géométrie sur les variétés algébriques, *Act. Sc. et Ind.* No. 206, Hermann, Paris 1935

[We 51] A. WEIL, Arithmetic on algebraic varieties, *Ann. of Math.* **53** 3 (1951) pp. 412–444

[We 54] A. WEIL, Sur les critères d'équivalence en géometrie algébrique, *Math. Ann.* **128** (1954) pp. 95–127

[We 55] A. WEIL, On algebraic groups and homogeneous spaces, *Amer. J. Math.* **77** (1955) pp. 493–512

[We 57] A. WEIL, Zum Beweis des Torellisschen Satzes, *Gött. Nachr.* **2** (1957) pp. 33–53

[We 67] A. WEIL, Über die Bestimmung Dirichletscher Reihen durch Funktionalgleichungen, *Math. Ann.* **168** (1967) pp. 149–156

[Weng 91] L. WENG, The conjectural arithmetic Riemann–Roch–Hirzebruch–Grothendieck theorem, *preprint Max Planck Institut* 1991; submitted for publication in *C.R. Acad. Sci.* with the title: Le théorème conjectural arithmétique de R–R–H–G.

[Wi 87] K. WINGBERG, On the rational points of abelian varieties over $\mathbf{Z}_p$-extensions of number fields, *Math. Ann.* **279** (1987) pp. 9–24

[Wo 89] P. M. WONG, On the second main theorem of Nevanlinna theory, *Amer. J. Math.* **111** (1989) pp. 549–583

[Wu 84] G. WÜSTHOLZ, The finiteness theorems of Faltings, in *Rational points* by G. Faltings and G. Wüstholz, Aspects of Mathematics, Vieweg Verlag 1984, pp. 154–202

[Wu 85] G. WÜSTHOLZ, A new approach to Baker's theorem on linear forms in logarithms I and II, in Diophantine Approximation and Transcendence Theory *Springer Lecture Notes* **1290** (1985) pp. 189–211

[Wu 88] G. WÜSTHOLZ, A new approach to Baker's theorem on linear forms in logarithms III, in *New Advances in transcendence theory*, edited by A. Baker, Cambridge University Press 1988, pp. 399–410

[Xi 87] XIAO GANG, Fibered algebraic surfaces with low slope, *Math. Ann* **276** (1987) pp. 449–466

[ZaK 87] D. ZAGIER and G. KRAMARZ, Numerical investigations related to the *L*-series of certain elliptic curves, *J. Indian Math. Soc.* **52** (1987) pp. 51–69

[Zar 74a] J. ZARHIN, Isogenies of abelian varieties over fields of finite characteristic, *Mat. Sb.* **95** (1974) pp. 461–470; *Math. USSR Sb.* **24** (1974) pp. 451–461

[Zar 74b] J. ZARHIN, A finiteness theorem for isogenies of abelian varieties over function fields of finite characteristic, *Funct. Anal. Prilozh.* **8** (1974) pp. 31–34; translation: *Funct. Anal. Appl.* **8** (1974) pp. 301–303

[Zar 74c] J. ZARHIN, A remark on endomorphisms of abelian varieties over function fields of finite characteristic, *Izv. Akad. Nauk SSSR Ser. Mat.* **38** (1974); translation: *Math. USSR Izv.* **8** no. 3 (1974) pp. 477–480

[Zar 75] J. ZARHIN, Endomorphisms of abelian varieties over fields of finite characteristic, *Izv. Akad. Nauk SSSR Ser. Mat.* **39** (1975); translation: Math. *USSR Izv.* **9** no. 2 (1975) pp. 255–260

[Zar 76] J. ZARHIN, Abelian varieties in characteristic *p*, *Mat. Zametki* **19** 3 (1976) pp. 393–400; translation: *Math. Notes* **19** (1976) pp. 240–244

[Zar 77] J. ZARHIN, Endomorphisms of abelian varieties and points of finite order in characteristic *p*, *Mat. Zametki* **21** (1977) pp. 737–744; translation: *Math. Notes* **21** (1977) pp. 415–419

[Zar 85] J. ZARHIN, A finiteness theorem for unpolarized abelian varieties over number fields with prescribed places of bad reduction, *Invent. Math.* **79** (1985) pp. 309–321

[Zar 87] J. ZARHIN, Endomorphisms and torsion of abelian varieties, *Duke Math. J.* **54** (1987) pp. 131–145

[Zi 76] H. ZIMMER, On the difference of the Weil height and the Néron–Tate height, *Math. Z.* **147** (1976) pp. 35–51

Index

B

C

D

G

T